ADVANCES IN LACTOFERRIN RESEARCH

ADVANCES IN EXPERIMENTAL MEDICINE AND BIOLOGY

Recent Volumes in this Series

Volume 436
ASPARTIC PROTEINASES: Retroviral and Cellular Enzymes
Edited by Michael N. G. James

Volume 437
DRUGS OF ABUSE, IMMUNOMODULATION, AND AIDS
Edited by Herman Friedman, John J. Madden, and Thomas W. Klein

Volume 438
LACRIMAL GLAND, TEAR FILM, AND DRY EYE SYNDROMES 2: Basic Science and Clinical Relevance
Edited by David A. Sullivan, Darlene A. Dartt, and Michele A. Meneray

Volume 439
FLAVONOIDS IN THE LIVING SYSTEM
Edited by John A. Manthey and Béla S. Buslig

Volume 440
CORONAVIRUSES AND ARTERIVIRUSES
Edited by Luis Enjuanes, Stuart G. Siddell, and Willy Spaan

Volume 441
SKELETAL MUSCLE METABOLISM IN EXERCISE AND DIABETES
Edited by Erik A. Richter, Bente Kiens, Henrik Galbo, and Bengt Saltin

Volume 442
TAURINE 3: Cellular and Regulatory Mechanisms
Edited by Stephen Schaffer, John B. Lombardini, and Ryan J. Huxtable

Volume 443
ADVANCES IN LACTOFERRIN RESEARCH
Edited by Geneviève Spik, Dominique Legrand, Joël Mazurier, Annick Pierce, and Jean-Paul Perraudin

Volume 444
REPRODUCTIVE TOXICOLOGY: In Vitro Germ Cell Developmental Toxicology, from Science to Social and Industrial Demand
Edited by Jesús del Mazo

Volume 445
MATHEMATICAL MODELING IN EXPERIMENTAL NUTRITION
Edited by Andrew J. Clifford and Hans-Georg Müller

A Continuation Order Plan is available for this series. A continuation order will bring delivery of each new volume immediately upon publication. Volumes are billed only upon actual shipment. For further information please contact the publisher.

ADVANCES IN LACTOFERRIN RESEARCH

Edited by

Geneviève Spik
Dominique Legrand
Joël Mazurier
Annick Pierce

Université des Sciences et Technologies de Lille
Villeneuve d'Ascq, France

and

Jean-Paul Perraudin

Biopole
Brussels, Belgium

SPRINGER SCIENCE+BUSINESS MEDIA, LLC

Library of Congress Cataloging-in-Publication Data

Advances in lactoferrin research / edited by Geneviève Spik ... [et al.].
p. cm. -- (Advances in experimental medicine and biology ; v. 443)
"Proceedings of the Third International Congress on Lactoferrin, held May 5-9, 1997, at Le Touquet, France"--CIP t.p. verso.
Includes bibliographical references and index.

DOI 10.1007/978-1-4757-9068-9

1. Lactoferrins--Congresses. I. Spik, G. (Geneviève)
II. International Congress on Lactoferrin (3rd : 1997 : Le Touquet, France) III. Series.
QP552.L345.A38 1998S
572'.6--dc21 98-28776
CIP

Proceedings of the Third International Congress on Lactoferrin, held May 5–9, 1997, at Le Touquet, France

Originally published by Plenum Press, New York in 1998
MyCopy version of the original edition 1998

http://www.plenum.com

10 9 8 7 6 5 4 3 2 1

PREFACE

Following the two meetings on Lactoferrin Structure and Function that were held in Honolulu, Hawaii, in 1993 and 1995, the Third International Conference on Lactoferrin Structure and Function was held in Le Touquet, France, and has successfully reinforced and diversified the previously created bridges between biochemists, clinicians, and companies.

In fact, scientists, physicians, and people of industry from different domains have brought a wealth of recent information concerning biochemistry and technical advances in the identification of lactoferrin-derived compounds as well as cell biology, molecular biology, pathology, and medical applications of lactoferrin and lactoferrin-derived compounds.

We were so delighted with the rapid growth of knowledge concerning many biological and immunological functions of lactoferrins and the relationships between their structure and function, we wanted to share our pleasure with the readers interested in this field. The present book, which represents a review of some of the most exciting contributions, is intended to reflect the status of our knowledge and transmit our hopes for the future development of *in vivo* applications of natural and recombinant lactoferrins.

We would like to express our gratitude to the sponsors who contributed to the organization of the meeting in such a pleasant place and allowed the participation of several young researchers. We would also like to thank all the participants who have answered with enthusiasm our invitation and to every one of the Laboratoire de Chimie Biologique for the constant and efficient help.

One hopes that we will meet again in Sapporo and that the physiological proof of the beneficial applications of lactoferrins will allow a development of new approaches in the treatment of diverse pathologies such as bacterial and viral infections and cancer.

The Editors

ACKNOWLEDGMENTS

The organizers and participants would like to thank the following sponsors of the Third International Conference on Lactoferrin Structure and Function:

Morinaga Milk Industry CO
DMV International
Agennix Inc.
Armor Protéines
Biopole
Domo
FerroDynamics
Gist-Brocades
Nutricia/Milupa Research
Pharming B.V
Tatua Biologics
Tonipharm Laboratoires
Wyeth-Ayerst Research
Université des Sciences et Technologies de Lille
Centre National de la Recherche Scientifique

CONTENTS

1

THREE-DIMENSIONAL STRUCTURE OF LACTOFERRIN

Implications for Function, Including Comparisons with Transferrin

Edward N. Baker, Bryan F. Anderson, Heather M. Baker,
Ross T. A. MacGillivray, Stanley A. Moore, Neil A. Peterson,
Steven C. Shewry, and John W. Tweedie

Department of Biochemistry
Massey University
Palmerston North, New Zealand

1. INTRODUCTION

Lactoferrin has many demonstrated activities. Some of these undoubtedly correspond to important *in vivo* functions; others may only apply *in vitro*, but may nevertheless lead to possible uses for lactoferrin in medicine or in biotechnology. In either case, the key to understanding the molecular basis of these activities, and ultimately being able to manipulate them, resides in the three-dimensional structure of the protein.

The biological activities of lactoferrin clearly depend on its ability to bind other molecules or ions. The first such activity to be recognised was its ability to bind iron, very tightly but reversibly. Lactoferrin is now well established as a member of the transferrin family of proteins[1,2], with a similar size (80 kDa, ~690 residues), carbohydrate content (~6%), and amino acid sequence (~55% identity with serum transferrin) and an ability to specifically bind 2 Fe^{3+} and 2 CO_3^{2-} ions. In addition to its iron binding ability, however, lactoferrin has the ability to bind to a wide variety of cells[3] and to bind to various anionic molecules, including heparin and other glycosaminoglycans[4], lipopolysaccharide[5] and DNA[6]. Its ability to bind to these species implies that it is likely to possess specific binding motifs and/or receptor binding sites, dependent on its three-dimensional structure, as well as less specific binding regions which may be more dependent on surface charge. It also possesses a particular surface region which has been implicated as a bactericidal domain[7].

The close structural relationship of lactoferrin with transferrin is potentially misleading. The two proteins undoubtedly share some broad functional roles (for example their

Advances in Lactoferrin Research, edited by Spik *et al.*
Plenum Press, New York, 1998.

ability to control iron levels in body fluids and so protect against free radical formation[8]) but there are also distinct differences (for example lactoferrin's retention of iron at much lower pH than transferrin[9]) and many other properties are quite specific to lactoferrin. Thus it is necessary to understand the differences between lactoferrin and transferrin as well as the similarities. Likewise it will ultimately be necessary to understand the differences between lactoferrins of different species.

Here we outline the main features of the three-dimensional structure of lactoferrin, as a framework for understanding its various binding activities, and provide some specific comparisons between lactoferrin and transferrin and between human and bovine lactoferrins.

2. THREE-DIMENSIONAL STRUCTURE

2.1. Structural Data Available

The various forms of lactoferrin for which 3D structures are available to date are listed in Table 1. All have been determined by X-ray crystallography since neither the whole molecule (80 kDa) nor half-molecules (40 kDa) are small enough for analysis by NMR. However, it is important to note that for other proteins for which 3D structures have been determined both in solution (NMR) and in the crystal (X-ray crystallography) no significant differences have been found. The main effect of the crystal lattice (in which the molecules are loosely packed, separated by large regions of solvent) is to reduce the flexibility of the molecules somewhat. Thus there is little doubt that the structures seen by X-ray crystallography will be the same *in vivo*.

Table 1. Lactoferrin crystal structures[a]

	Reference	Code[b]
Whole molecule structures		
hLf / 2 Fe^{3+}	10	1LFG
bLf / 2 Fe^{3+}	14	—
hLf / apo, form 1	24	1LFH
hLf / apo, form 2	25	
hLf / 2 Cu^{2+}	19	1LFI
hLf / 2 Cu^{2+}, 1 oxalate	20	1LCF
hLf / 2 oxalate	21	1BKA
hLf / Ru(III) drug	22	—
rec.hLf / 2 Fe^{3+}	unpublished	
Half-molecule structures		
Lf_N/Fe^{3+}	15	1LCT
Lf_N/D60S / Fe^{3+}	29	1DSN
Lf_N/R121S / Fe^{3+}	30	1VFE
Lf_N/R121E / Fe^{3+}	30	1VFD
Lf_N/H253M / Fe^{3+}	32	1HSE
Lf_N/P251A / Fe^{3+}	unpublished	
Lf_N/N137S / Fe^{3+}	unpublished	
Lf_N/R210G / Fe^{3+}	unpublished	
Lf_N/R210K / Fe^{3+}	unpublished	
Lf_N/R210E / Fe^{3+}	unpublished	
Lf_N/R121D / apo	unpublished	

[a]hLf = human lactoferrin; bLf = bovine lactoferrin; rec. hLf = human lactoferrin from *A. awamori*; Lf_N = human lactoferrin N-lobe.
[b]Protein Data Bank entry codes. Other coordinates available from the authors.

2.2. Structural Organisation of Lactoferrin

The three-dimensional structure of human lactoferrin has been described in detail elsewhere[2,10,11]. The key feature of the organisation of the polypeptide chain, shown in Figure 1, is its division into four structural domains, two of which (N1 and N2) comprise the N-terminal half of the molecule (the N-lobe) and the other two (C1 and C2) the C-terminal half (the C-lobe). The two lobes are joined by a short connecting peptide, which in lactoferrin forms a 3-turn α-helix. In each lobe the two domains enclose a large central cleft, which houses the iron-binding site, and the conformational changes undergone by lactoferrin all seem to involve rigid-body movements of these domains, relative to each other (see below).

The basic folding of the lactoferrin molecule is the same in every structural or functional variant we have examined to date. There are no changes in secondary structure—with one minor exception. When different ligands are bound there is sufficient sidechain flexibility to adjust, without change to the polypeptide conformation. The protein structure is also not affected by differences in glycosylation and it is relatively unaffected by mutagenesis.

2.3. Recombinant Lactoferrin Expressed in *Aspergillus awamori*

Recombinant human lactoferrin is expressed at extremely high levels from *Aspergillus awamori*, following strain improvement by selection and mutagenesis[12]. It is glycosylated at the same sites as human milk Lf, but with carbohydrate of different composition. Nevertheless, its 3D structure is effectively identical to that of human milk Lf; crystals of rec-hLf, in its iron-bound form, are isomorphous with those of milk hLf, and determination of the structure by X-ray crystallography, to a resolution of 2.2 Å (R = 0.190), shows that both the iron sites and the whole polypeptide chain can be superimposed on those of milk hLf with a root-mean-square deviation of only 0.3 Å. The only apparent difference,

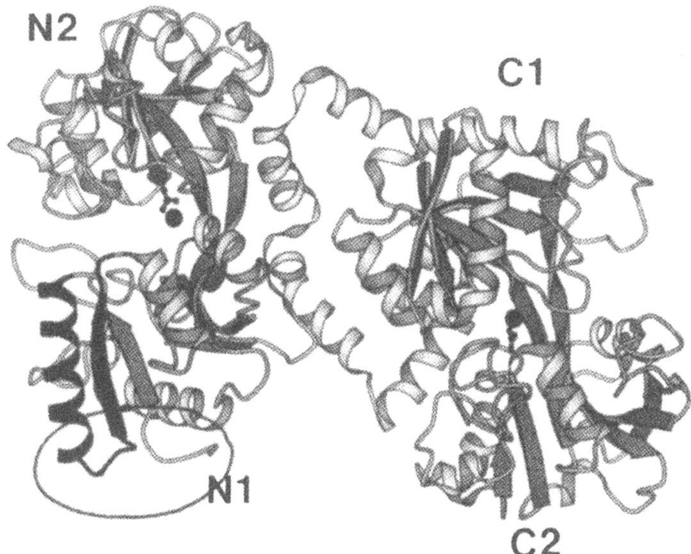

Figure 1. Organisation of the lactoferrin molecule. Diferric human lactoferrin is shown, with the four domains labelled. The postively-charged region comprising the N-terminus and the end of helix 1 is circled. The region containing the lactoferricin anti-bacterial peptide is shown in black (extreme left of the N1 domain).

in the N-terminal residues 1–4, is not a real difference because these four residues are highly flexible and a number of conformations can be modelled for them. At the glycosylation sites, Asn 137 and Asn 476, no clear carbohydrate structure is visible, either because of flexibility or because of heterogeneity. In these respects there is again no difference with milk hLf.

2.4. Bovine Lactoferrin

Bovine lactoferrin has approximately 69% amino acid sequence identify with hLf[13]. It is glycosylated at four sites, Asn 233, Asn 368, Asn 476 and Asn 545, only one of which (Asn 476) is shared by hLf, and it also has an extra disulphide bridge 160–183, not present in hLf. We have recently determined the 3D structure of bovine lactoferrin, in its diferric form, and refined it at 2.8Å resolution[14]. Despite the sequence differences the polypeptide chain conformation is the same as for hLf. The extra disulphide is accommodated without any disturbance, and the only differences are slight changes in the relative orientations of the 4 domains. The N-lobe of bLf is 5.3° more closed than that of hLf, and the C-lobe 3.7° more closed, while the two lobes have slightly different orientations; when the C-lobes of bLf and hLf are superimposed a further rotation of 11.3° is required to superimpose their N-lobes. These small differences which arise from sequence changes, and slight repacking, in the interfaces between the domains, do not appear to be of any functional significance. The iron binding sites appear identical but there is at least one specific amino acid substitution near the N-lobe iron site which seems to weaken iron binding. The carbohydrate structure of bovine lactoferrin is also much better defined than that of human lactoferrin.

2.5. Lactoferrin Half-Molecules

All of the site-specific mutants we have analysed to date are based on the N-lobe half-molecule of human lactoferrin, Lf_N. This species has the same structure as the N-lobe of the parent hLf molecule, with one exception; an α-helix (321–332) that is normally buried between the two lobes in intact hLf becomes unwound in Lf_N, possibly because of the loss of the stabilising interactions it would normally make with the C-lobe[15]. This illustrates the communication that exists between the two lobes, a phenomenon that has also been shown by mutagenesis and iron binding experiments[16].

2.6. Comparisons with Transferrin

Amino acid sequence identity between human lactoferrin and human transferrin is about 60%[2]. It is no surprise, therefore, that their 3D structures are also highly similar. We have recently determined the 3D structure of the N-terminal half-molecule of human transferrin, and refined it at 1.9 Å resolution (R = 0.190)[17]. Again, the folding closely matches that of the human lactoferrin N-lobe; allowing for a slight difference in domain closure the atoms of the polypeptide chain superimpose within ~0.4 Å. The iron ligands and their orientations are also the same, making the point that these are highly similar members of a homologous family of proteins.

The distinct functional differences that exist between lactoferrins and transferrins seem likely to derive from three main factors. First, the proteins are found in different bodily locations and this must have a major influence on their biological roles. Second, they have very different surfaces since this is where most of the amino acid changes occur; thus transferrin has no basic patch corresponding to the N-terminal and anti-bacterial region of human or bovine lactoferrin, and its potential receptor-binding surfaces are likely to be very different. Third, there are differences in specific interactions resulting from just a very few amino acid changes, that may have a profound affect on function. An example is the proposed iron-release "trigger"[18].

2.7. Specific and Non-Specific Binding

The very high iron affinity of lactoferrin, and other transferrins, implies that the two binding sites are highly specific. This is true in the sense that the ligands in each site (2 Tyr, 1 Asp, 1 His and the CO_3^{2-} ion) provide an environment that is ideal for binding iron, as Fe^{3+}. These ligands are almost totally invariant in the binding sites of transferrins[2]. Paradoxically, however, the binding sites of transferrins are also capable of considerable versatility.

In the absence of iron many other metals can be bound, including some that are of physiological importance such as manganese, copper, zinc and aluminium[2]. Given that lactoferrin usually has a very low saturation with iron (5–10%), the possibility exists that it may in some situations carry other metals. Likewise, although its anion affinity is greatest for carbonate, it is also capable of binding many other anionic species[2]. Our crystallographic studies have shown the reason for this versatility; immediately adjacent to the iron binding site, and within the interdomain cleft, is a large solvent-filled cavity. This allows considerable internal flexibility to exist, even within the stable, closed form of the protein, and adjustments to occur to accommodate a different metal ion or anion[2,19,21].

These observations imply that lactoferrin could be used to carry other species. We have recently shown that lactoferrin will bind certain Ru(III) anti-tumour drugs[22], and as the receptor-binding properties of lactoferrin become better understood the possibility exists that it could be used therapeutically as a drug delivery agent. There is in fact one remarkable natural analogue: a newly-discovered member of the transferrin family called saxiphilin, with 44% sequence identity to human lactoferrin, which instead of binding iron, binds a marine toxin called saxitoxin[23].

Finally the existence of non-specific binding sites on the surface of the lactoferrin molecule should not be forgotten. These result from the fortuitous proximity of residues such as His, Met, Asp or Glu in pockets on the protein surface. Such sites occur on most proteins, and will differ from one species of lactoferrin to another, as sequences vary. They are likely to be of low affinity, but still could be functionally important by providing sites through which, for example, metal ions such as zinc or calcium could bridge between lactoferrin and other molecules. These sites can be identified by crystallography, and the binding studies with Ru(III) compounds[22] did, in fact, identify a number of non specific sites on the human lactoferrin surface.

2.8. Importance of Dynamics

Given that the iron atom bound by each lobe is buried deep within the interdomain cleft, how does this iron get out? The answer lies in two antiparallel polypeptide strands that run behind the iron site in each lobe. A hinge in these strands allows the two domains to move apart in a rigid-body motion, opening up the cleft and releasing the iron[24]. This neatly explains the conformational changes known to be associated with iron binding; the protein moves between open and closed states. In the N-lobe the movement is large (a 55° rotation of one domain relative to the other) whereas in the C-lobe it is smaller (a 20° rotation)[25]. A similar movement is seen for the isolated lactoferrin N-lobe half-molecule[26] (a 49° rotation,

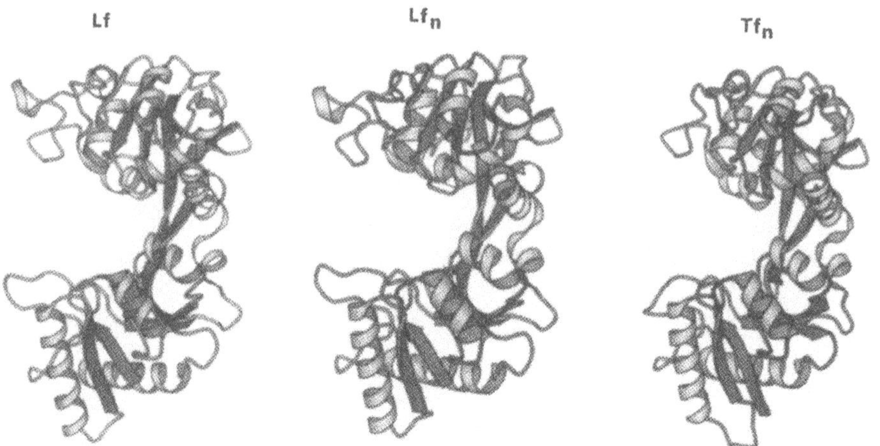

Figure 2. Open forms of transferrins. From left to right are shown the open N-lobe of human apo-lactoferrin[24]; the iron-free R121D mutant N-lobe half-molecule of human lactoferrin[26]; and the iron-free half-molecule of human transferrin[27].

comparing $FeLf_N$ with apo-R121D/Lf_N), and for the human transferrin N-lobe half-molecule[27] (a 60° rotation). These structures are shown in Figure 2, and make the point that this conformational change seems likely to be highly similar in nature in all transferrins.

The two apo-lactoferrin structures we have determined offer another very important insight into lactoferrin dynamics, however[25]. In one of them (Form 2) both lobes are open, as expected. In the other (Form 1), however, the N-lobe is open but the C-lobe is closed. This observation indicates that for the iron-free protein there is very little energy difference between open and closed forms. A similar conclusion has been reached for the related bacterial periplasmic binding proteins[28]. Our conclusion is that in solution, iron-free lactoferrin can explore a range of different structures, as shown in Figure 3. The predomi-

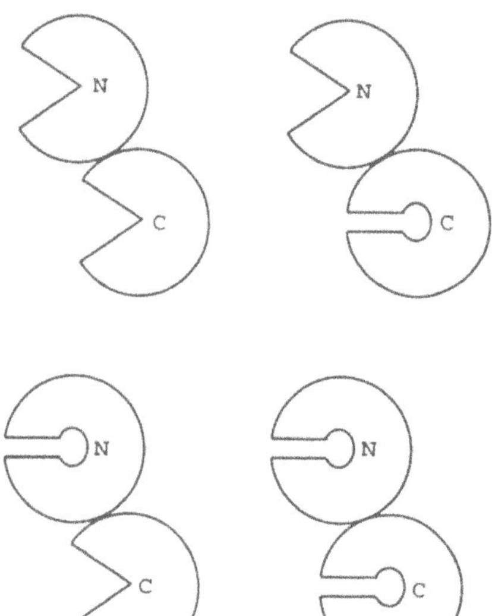

Figure 3. Accessible structures for apo-lactoferrin. The major form is assumed to be that with both lobes open (top, left) but the form with the N-lobe open and the C-lobe closed (top, right) has also been seen crystallographically[24]. The other two forms could also exist transiently in solution.

nant form is probably the "both open" one, but others can exist such as the "N-lobe open, C-lobe closed" structure seen in the Form 1 crystals.

There are important implications for the binding properties of lactoferrin. If there is little difference between the energies of these different structures as our results suggest, a molecule binding to lactoferrin could bind preferentially to one or other of them. Thus a molecule binding in the open cleft could prevent it from closing and so prevent stable iron binding. We have already shown that anions and anionic species such as Cl^- and EDTA have a strong tendency to bind in the open cleft (G.B. Jameson and E.N. Baker, unpublished). Alternatively a molecule such as DNA might prefer to bind elsewhere and might prefer to bind to one of the other forms shown in Figure 3, thereby stabilising it. The conclusion is that binding to apo-Lf could result in the preferential stabilisation of any of these structural forms, although binding to Fe_2Lf could only be to the fully-closed structure.

2.9. Surface Structure

Two important surface features are apparent on the lactoferrin molecule. First, two high concentrations of positive charge are apparent. The larger of these is on the surface of the N1 domain, involving the polypeptide N-terminus with its four Arg sidechains and the C-terminal end of helix 1, with Arg 25, Arg 28, Lys 29 and Arg 31 (see Figure 1). This region is already known to be the site of glycosaminoglycan binding and is a likely binding site for DNA and other anionic molecules[4,6]. It is far removed from the iron-binding cleft of the N-lobe and thus seems unlikely to influence (or be influenced by) iron binding. The second, smaller, basic patch is between the two lobes, associated with Arg residues at the C-terminal end of the inter-lobe connecting helix. Binding here could affect iron binding by perturbing the ease of conformational change.

The second important surface feature is the "antibacterial domain" first recognised by Bellamy *et al.*[7] Comprising most of helix 1 and the following β-strand (residues 14–39) this structural element includes a disulphide bridge (19–36) and a number of projecting basic sidechains which could allow it to bind to bacterial cell surfaces. The separated peptide, known as lactoferricin (residues 17–41 of bovine lactoferrin, for example), is strongly antibacterial and this property has been linked to its helical nature. We stress, however, that (i) the peptide will not necessarily have the same 3D structure on its own as it does in the context of the whole molecule, and (ii) the antibacterial action of the peptide may not involve the same mechanism as in the intact lactoferrin molecule.

2.10. Site-Specific Mutants

We have now determined crystal structures of no fewer than 10 site-specific mutants of lactoferrin (Table 1). All were based on the N-terminal half-molecule of lactoferrin, and all have been expressed using the baby hamster kidney (BHK) system, with the expression vector pNUT[29,30]. Mutations have been designed to address the determinants of iron binding (D60S and H253M), anion binding (R121S and R121E), the molecular hinge in the N-lobe (P251A), the N-lobe glycosylation site (N137S) and a residue (Arg 210) involved in the modulation of iron binding and release.

The effects of mutations of Arg 210 are discussed below. A general observation, however, is that in none of the 10 mutants is the folding significantly changed. This makes the point that, as in other proteins, the structure is remarkably tolerant of amino acid substitutions, which cause only very minor local effects. The only conformational differences are slight changes in the extent of domain closure in the mutants; the largest is in the D60S mutant which is 7.5° more closed than wild-type as a result of the loss of a specific interdomain interaction [26], but in other mutants the changes in domain closure are at most 2–3° and in no case is there any change to the secondary structure or the overall fold.

3. IRON RELEASE

3.1. Comparisons with Transferrin

One of the most striking features that differentiates lactoferrins from transferrins is the difference in their pH dependence of iron release[9]. For human lactoferrin, iron release from both lobes occurs practically together, over the pH range 4.0 to 2.5. For serum transferrin however, iron release begins at pH 6.0. Moreover there is a distinct difference between its two sites, in that as the pH is lowered the N-lobe releases iron first, over the pH range 6.0 to 5.0, followed by the C-lobe over the pH range 5.0 to 4.0. Thus lactoferrin retains iron to much lower pH and does not show the early release that is shown particularly by the transferrin N-lobe. These differences are demonstrated by pH/release profiles such as shown in Figure 4. It should be noted, however, that such profiles are obtained by equilibrating the protein at each pH over a period of several days; as such they are sensitive to the precise conditions of the experiment (buffers, time, whether a chelator is used etc), they are thermodynamic rather than kinetic analyses, and they are complicated by the solution chemistry of Fe^{3+}.

The origins of this characteristic difference have recently been attributed to a pair of lysine residues, one from each domain, that lie behind the N-lobe iron site in the interdomain space[18]. In both human transferrin and chicken ovotransferrin these lysines (Lys 206 and Lys 296 in transferrin, Lys 210 and Lys 301 in ovotransferrin) are hydrogen bonded, implying that at least one of the pair is not charged. It has thus been suggested that they

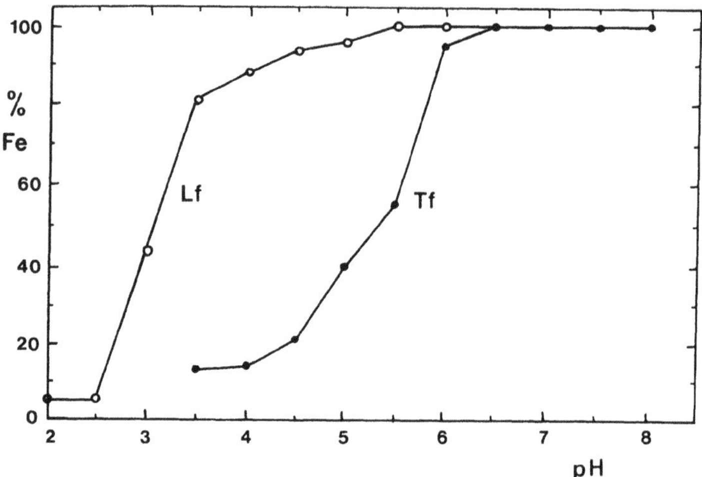

Figure 4. pH dependence of iron release from lactoferrin and transferrin. At each pH the iron saturation is determined from the absorbance at the visible absorption maximum after incubation of the (initially iron-loaded) protein in a suitable buffer for 1–2 days.

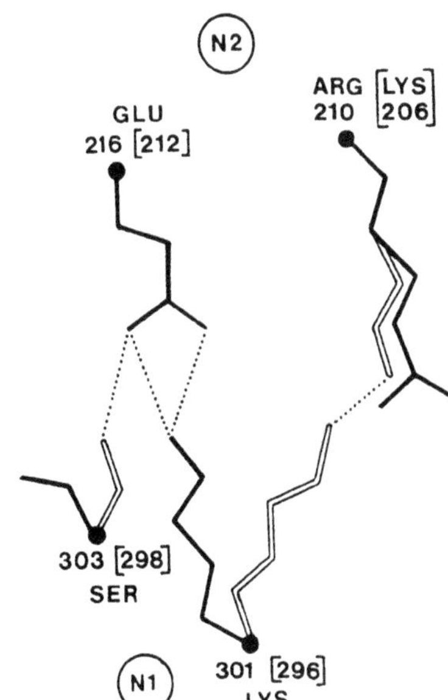

Figure 5. Basic sidechains behind the iron sites of lactoferrin and transferrin. The different orientations of Arg 210 and Lys 301 in human lactoferrin[10] (solid bonds) compared with Lys 206 and Lys 296 in human transferrin[17] (open bonds) are shown.

form a "dilysine trigger"; protonation at low pH should result in a pair of positively charged residues which should then repel and trigger the moving apart of the domains[18]. Such a pair does not exist in the transferrin C-lobe or in lactoferrin. In human lactoferrin the equivalent residues are Arg 210 and Lys 301. Both are basic, but they do not interact as in transferrin. Instead they are differently oriented, so that their positive charges are more than 5 Å apart (cf 2.8 Å in transferrin); Arg 210 makes a complex set of interactions with the iron ligands Tyr 92 and Tyr 192 and with Tyr 82 in the opposing domain, and Lys 301 forms an interdomain salt bridge with Glu 216 (see Figure 5).

We have tested the role of Arg 210 and the significance of the "dilysine trigger" in two ways, (i) by site-specific mutagenesis of Arg 210 and (ii) by X-ray crystallographic analysis of bovine lactoferrin, in which residues 210 and 301 are both Lys[14]. To correlate the structural results with iron release properties we have also used a spectrofluorometric method to measure the kinetics of iron release. This method monitors the increase in fluorescence (due to the loss of quenching of intrinsic Trp fluorescence) as iron is released, and has the advantage that only very small quantities of protein are needed.

3.2. Spectrofluorometric Measurement of Iron Release Kinetics

The method is adapted from that of Egan *et al.*[31], used originally to measure iron release from transferrin. The protein solution (80 mg of iron-saturated lactoferrin in 10 ml of 0.05 M Tris-HCl pH 7.9) is mixed with 3 ml of buffer (typically 0.05 M acetate, plus 0.1 M NaCl, at the desired pH) containing 0.01 M sodium citrate, and the increase in fluorescence is monitored. The citrate acts as an "iron sink" to prevent precipitation or re-binding of the iron. Similar measurements are made at a series of different pH values to determine how the rate of release varies with pH. Typical iron release curves, which follow simple first order kinetics, are shown in Figure 6a. In order to compare different lactoferrins and their relative stabilities to pH the rate constants measured at different pH values can be used, as in Figure 6b.

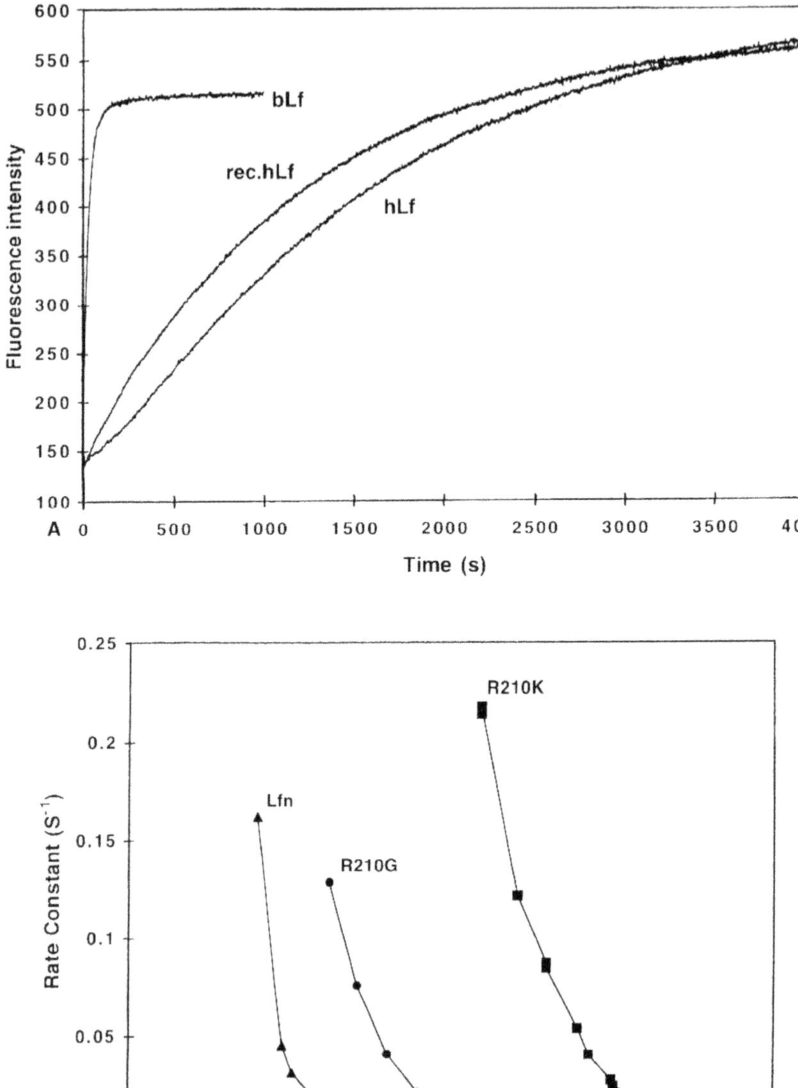

Figure 6. Kinetics of iron release. In (a) typical rate curves are shown, comparing iron release from human milk lactoferrin (hLf), recombinant hLf from *A. awamori* (rec. hLf) and bovine lactoferrin (bLf), all at pH 2.5. The native and recombinant hLf have essentially identical rates of release, but bLf is much faster. In (b) the pH dependence of the rate constants for iron release from the wild-type N-lobe half-molecule of human lactoferrin (Lfn) and its R210G and R210K mutants is shown.

3.3. Effect of Mutations of Arg 210

Site-specific mutants were prepared, based on the N-terminal half-molecule of human lactoferrin Lf_N. Oligonucleotide-directed mutagenesis was used to introduce single-site mutations to the cDNA, which encoded residues 1–333 of the native molecule; this DNA was then incorporated into the expression vector pNUT and the recombinant proteins expressed in baby hamster kidney (BHK) cells, as before[29,30]. Seven different mutations were made at position 210, with the Arg being changed to Lys (basic, as in transferrin), Glu (acidic), Ala and Gly (small, neutral), Gln (polar, uncharged), Leu and Met (hydrophobic). Expression levels were in the range 10–20 mg/l and the final yields of purified proteins were 61 mg (R210K), 156 mg (R210E), 121 mg (R210G), 98 mg (R210A), 41 mg (R210Q), 75 mg (R210L) and 45 mg (R210M).

The mutants have been characterised in two ways. Their rates of iron release, as a function of pH, were determined. They have also been crystallized and 3D structures determined for the R210K, R210G and R210E mutants. The crystals were grown by a novel batch technique in which the concentrated protein solution (50 mg/ml) was added drop-wise to a low ionic strength buffer solution (10 mM Tris-HCl, pH 8.0). Crystals formed where the protein solution settled at the bottom of the tube. The crystal structures were determined by molecular replacement and refined at high resolution, 2.0Å for the R210K mutant (final R factor 0.20), 1.95Å for R210G (R factor 0.23) and 2.2Å for R210E (R factor 0.23) [Peterson, N.A., Tweedie, J.W. and Baker, E.N., unpublished].

The effects of the mutations at this site are striking. Removal of the Arg 210 sidechain, as in the Gly mutant, R210G, does not change the structure (the folding is the same and two water molecules fill the position formerly occupied by the Arg sidechain) but iron binding is substantially destabilised. When the pH dependence of the rate constant for iron release from the wild-type half-molecule is compared with that for R210G (Figure 6b) it is clear that R210G is destabilised by approximately 1 pH unit as a result of removal of the Arg sidechain. This demonstrates the stabilising role of Arg 210, which is conserved in the N-lobes of most lactoferrins. The addition of a lysine sidechain, as in transferrin, further destabilises iron binding, as release from the R210K mutant is shifted a further 2 pH units higher, compared with R210G. The new Lys 210 sidechain does not hydrogen bond to Lys 301 as in transferrin, however (they are still 4.6 Å apart); we conclude that the "dilysine trigger" concept is an over-simplication, but that the presence of two Lys residues at this position, behind the iron site, clearly does stimulate iron release at higher pH.

3.4. Bovine Lactoferrin

Iron release from lactoferrin depends on two factors. Specific residues around the iron sites play a part, as shown by the site-specific mutants described above. The stabilisation of the N-lobe by the C-lobe, in full-length lactoferrins, also contributes significantly, however, as shown both by structure/function studies of the recombinant lactoferrin N-lobe[15] and by mutagenesis of full-length human lactoferrin[16]. This combination of factors is seen in the properties of bovine lactoferrin. In bLf residue 210 is Lys, just as in transferrin and in the R210K mutant of the lactoferrin half-molecule. The presence of the two Lys residues (Lys 210 and Lys 301) behind the iron site in bLf does destabilise iron binding compared with hLf, as shown in Figure 6a. The pH of iron release in bLf is raised by nearly 1.0 pH units ($t_{1/2}$ for hLf is 1200s at pH 2.6, whereas for bLf $t_{1/2}$ is 1200s at pH 3.5). bLf is much more stable towards acid than the R210K mutant, however, and much more stable than transferrin, despite the fact that all three proteins have the same pair of Lys residues. The lower stability of the R210K mutant compared with bLf, even though the structures around the iron sites are identical, is attributed to the loss of the stabilising effect of the C-lobe. The lower stability of transferrin relative to bLf is less easily explained and there are clearly other factors at work here, most probably electrostatic effects from other sequence changes.

4. CARBOHYDRATE STRUCTURE

The role of the carbohydrate chains on lactoferrin remains an enigma. From the 3D structural point of view it is clear that the carbohydrate does not affect the protein structure. This is shown by the comparisons of (i) human milk lactoferrin and the recombinant human lactoferrin from *A. awamori* (identical polypeptide structure despite different glycan chains), (ii) the glycosylated N-lobe of human milk lactoferrin and the deglycosylated N-lobe half-molecule Lf_N (same structure despite removal of the carbohydrate) and (iii) human and bovine lactoferrins (same polypeptide folding despite different numbers of glycan chains with different sites of attachment). The only difference we have found to result from removal of the carbohydrate is that the polypeptide chain in the vicinity of the glycosylation site becomes somewhat more flexible; the average B factor of residues 134–144 increases from ~35 Å^2 to ~50 Å^2 on removal of the carbohydrate at Asn 137. Otherwise, iron binding and release, heat stability and stability to proteolysis of human lactoferrin are hardly altered by deglycosylation.

The two carbohydrate chains on human lactoferrin show little evidence of a strongly defined three-dimensional structure, from our crystallographic studies at least. This may derive partly from flexibility of the chains and partly from chemical heterogeneity which would have the effect of "smearing out" their electron density. In any event, the carbohydrate structure of hLf is poorly defined in the crystal structure, and it does not have any significant interaction with the protein structure.

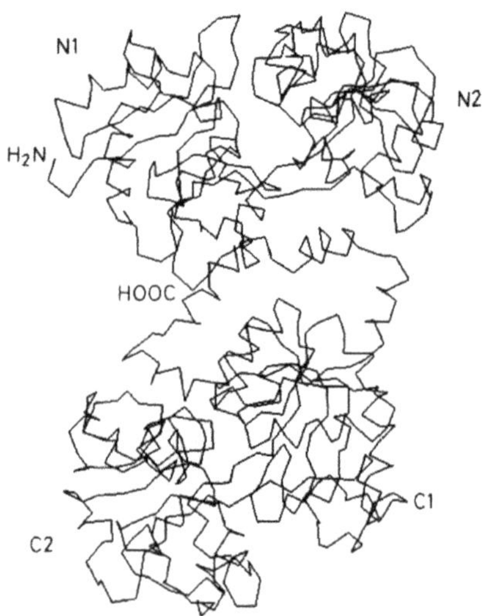

Figure 7. Ribbon diagram of bovine lactoferrin. Carbohydrate visible in the crystal structure is shown in stick representation, with the Asn 545 glycan chain on the right.

This is not true of bovine lactoferrin, however. A feature of the bLf structure is the presence of one very well defined glycan chain attached to Asn 545. This has a biantennary structure with the sequence Asn–(GlcNAc)$_2$–Man–(Man–GlcNAc) (Man) and is characterised by both internal hydrogen bonds and hydrogen-bonded and stacking interactions with the protein. Remarkably, its conformation proves to be extremely similar to that of the glycan attached to Asn 395 in a completely unrelated protein, glucoamylase from *Aspergillus awamori*[14]. This suggests that carbohydrate chains can indeed encode quite specific structural information. This Asn 545 glycan is interesting for two other reasons; it lies between the two domains of the C-lobe (see Figure 7), behind the iron site, and thus may have a role in stability and in modulating iron binding and release by bLf. Its location in a fairly deep cleft also explains the difficulty experienced in fully deglycosylating bLf. Four other sugar residues can be defined on bLf, one at Asn 361 and three at Asn 476, giving the impression that the carbohydrate may play a greater role in the stability of bLf than is the case for hLf.

5. CONCLUDING COMMENTS

Our overall conclusion is that the lactoferrin structure is highly conserved. The secondary structure does not alter significantly in any of the forms examined and functional differences between different lactoferrins and between lactoferrins and transferrins seem likely to result from quite subtle effects of individual amino acid sequence changes. An important challenge for the future is to define the ways in which lactoferrin interacts with other molecules, but it seems unlikely that significant structural change will occur when it does so. The most likely effects may be changes to the relative domain orientations or to its dynamics. The three-dimensional structure thus provides a framework through which the functional properties of lactoferrin should be able to be understood, and it should provide a focus for hypotheses concerning its mode(s) of action.

ACKNOWLEDGMENTS

We gratefully acknowledge the substantial contributions made by colleagues at Massey University towards the development of this work over many years, notably Dr Sylvia Rumball who stimulated it in the first place. We also acknowledge research support from the US National Institute of Child Health and Human Development, the Wellcome Trust, the Health Research Council of New Zealand, the NZ Dairy Research Institute and Agennix Inc. ENB also receives research support as an International Research Scholar of the Howard Hughes Medical Institute.

REFERENCES

1. Brock, J.H. (1985). In Metalloproteins (Harrison, P.M., ed.), Vol. 2, pp. 183–262, MacMillan London.
2. Baker, E.N. (1994). Adv. Inorg. Chem. 41, 389–463.
3. Birgens, H.S., Hansen, N.E., Karle, H. and Ostergaard Kristensen, L. (1983). Brit. J. Haematol 54, 383–391.
4. Mann, D.M., Romm, E and Migliorini, M. (1994). J. Biol. Chem 269, 23661–23667.
5. Elass-Rochard, E., Roseanu, A., Legrand, D., Trif, M., Salmon, V., Motas, C., Montreuil, J. and Spik, G. (1995). Biochem. J. 312, 839–845.

6. Hutchens, T.W., Maguson, Y.S. and Yip, T-T. (1989). Pediatr. Res. 26, 618–622.
7. Bellamy, W., Takase, M., Yamauchi, K., Wakabayashi, H., Kawase, K. and Tomita, M. (1992). Biochim. Biophys. Acta 1121, 130–136.
8. Baldwin, D.A., Jenny, R.R. and Aisen, P. (1984). J. Biol. Chem. 259, 13391–13394.
9. Mazurier, J. and Spik, G. (1980). Biochim. Biophys. Acta 629, 399–408.
10. Anderson, B.F., Baker, H.M., Norris, G.E., Rice, D.W. and Baker, E.N. (1989). J. Mol. Biol. 209, 711–734.
11. Haridas, M., Anderson, B.F. and Baker, E.N. (1995). Acta Cryst D51, 629–646.
12. Ward, P.P., Piddington, C.S., Cunningham, G.A., Zhou, Z., Wyatt, R.D. and Conneely, O.M. (1995). Bio/Technology 13, 489–503.
13. Pierce, A., Colavizza, D., Benaisser, M., Maes, P., Tartar, A., Montreuil, J. and Spik, G. (1991). Eur. J. Biochem. 196, 177–184.
14. Moore, S.A., Anderson, B.F., Groom, C.R., Haridas, M. and Baker, E.N. (1997). Submitted for publication.
15. Day, C.L., Anderson, B.F., Tweedie, J.W. and Baker, E.N. (1993). J. Mol. Biol. 232, 1084–1100.
16. Ward, P.P., Zhou, X. and Conneely, O.M. (1996). J. Biol. Chem. 271, 12790–12794.
17. MacGillivray, R.T.A., Anderson, B.F., Brayer, G.D., Mason, A.B., Moore, S.A., Woodworth, R.C. & Baker, E.N. (1997). Manuscript in preparation.
18. Dewan, J.C., Mikami, B., Hirose, M. and Sacchettini, J.C. (1993). Biochemistry 32, 11963–11968.
19. Smith, C.A., Anderson, B.F., Baker, H.M. and Baker, E.N. (1992). Biochemistry 31, 4527–4533.
20. Smith, C.A., Anderson, B.F., Baker, H.M. and Baker, E.N. (1994). Acta Cryst. D50, 302–316.
21. Baker, H.M., Anderson, B.F., Brodie, A.M., Shongwe, M.S., Smith, C.A., and Baker, E.N. (1996). Biochemistry 35, 9007–9013.
22. Smith, C.A., Sutherland-Smith, A.J., Keppler, B.K., Kratz, F. and Baker, E.N. (1996). J. Biol. Inorg. Chem 1, 424–431.
23. Morabito, M.A. and Moczydlowski, E. (1994). Proc. Natl. Acad. Sci. USA 91 2478–2482.
24. Anderson, B.F., Baker, H.M., Norris, G.E., Rumball, S.V. and Baker, E.N. (1990). Nature 344, 784–787.
25. Baker, E.N., Anderson, B.F., Baker, H.M., Faber, H.R., Smith, C.A. and Sutherland-Smith, A.J. (1997). In Lactoferrin: Interactions and Biological Functions (Hutchens, T.W. and Lonnerdal, B., eds), pp. 177–191, Humana Press, New Jersey.
26. Jameson, G.B., Breyer, W.E., Anderson, B.F., Day, C.L., Kingston, R.L. & Baker, E.N. (1997). Manuscript in preparation.
27. Jeffrey, P.J., Bewley, M.C., MacGillivray, R.T.A., Mason, A.B., Woodworth, R.C. & Baker, E.N. (1997). Manuscript in preparation.
28. Flocco, M.M., and Mowbray, S.L. (1994). J. Biol Chem 269, 8931–8936.
29. Faber, H.R., Bland, T., Day, C.L., Norris, G.E., Tweedie, J.W. and Baker, E.N. (1996). J. Mol. Biol. 256, 352–363.
30. Faber, H.R., Baker, C.J., Day, C.L., Tweedie, J.W. and Baker, E.N. (1996). Biochemistry 35, 14473–14479.
31. Egan, T.J., Zak, O. and Aisen, P. (1993). Biochemistry 32, 8162–8167.
32. Nicholson, H., Anderson, B.F., Bland, T., Shewry, S.C., Tweedie, J.W. and Baker, E.N. (1997) Biochemistry 36, 341–346.

STRUCTURES OF BUFFALO AND MARE LACTOFERRINS

Similarities, Differences, and Flexibility

A. K. Sharma, S. Karthikeyan, S. Sharma, S. Yadav, A. Srinivasan, and T. P. Singh

Department of Biophysics
All India Institute of Medical Sciences
New Delhi 110 029, India

1. INTRODUCTION

Lactoferrin (Lf), an iron binding glycoprotein found in the external secretions and neutrophilic leucocytes of mammals, is thought to be responsible for primary defence against microbial infection, mainly as a result of lactoferrin sequestration of iron required for microbial growth (Weinberg, 1978). Many other functions have been attributed to lactoferrin, including immunomodulation and cell growth regulation (Lönnerdal & Iyer, 1995). The lactoferrin has molecular mass of 80 kDa. The protein folds into two globular lobes, the N-lobe comprising the N-terminal half of the polypeptide chain and C-lobe comprising the C-terminal half of the polypeptide chain. The lobes are connected by a 3_{10}-helical 10–12 residue peptide. Each lobe is further subdivided into two domains. Each domain contains a single iron binding site, located in the interdomain cleft.

The detailed three-dimensional structures of human lactoferrin at 2.2 Å resolution (Haridas et al., 1996), bovine lactoferrin at 2.5 Å resolution (Haridas et al., 1994), equine lactoferrin at 2.65 Å resolution (Sharma et al., 1997) and buffalo lactoferrin at 3.3 Å resolution (Karthikeyan et al., 1997) are available. The structural reports of N-lobe (Day et al., 1993) and C-lobe (Sharma et al., 1997) are also available. The overall structural organizations in lactoferrins are similar. The C-lobe displays a remarkable structural similarity in various lactoferrins while the N-lobe differs significantly especially in the loop regions. A detailed comparison of the structures of human, buffalo and equine lactoferrins is presented here.

Advances in Lactoferrin Research, edited by Spik *et al.*
Plenum Press, New York, 1998.

Table 1. Comparison of N-terminal sequences in various lactoferrins

		1				5					10					15					20
Mare lactoferrin		A	P	R	K	S	V	R	W	C	T	I	S	P	A	E	A	A	K	C	A
Buffalo lactoferrin		A	R	A	G	K	S	G	W	C	T	I	A	A	P	E	G	T	A	C	A
Bovine lactoferrin		A	P	R	K	N	V	R	W	C	T	I	S	Q	P	E	W	F	K	C	R
Human lactoferrin	G	R	R	R	R	S	V	Q	W	C	A	V	S	Q	P	E	A	T	K	C	F

2. SEQUENCE HOMOLOGY IN LACTOFERRINS

The full sequence data are available on human and bovine lactoferrins while partial sequence data have been reported on equine and buffalo lactoferrins. The human lactoferrin contains significantly more arginine residues than rest of the three species. A comparison of 20 N-terminal residues is presented in Table 1. An arginine rich cluster found in human lactoferrin is no longer present in other lactoferrins. It might have important functional implications.

3. STRUCTURE OF BUFFALO LACTOFERRIN

Lactoferrin was isolated from the colostrum/milk of Murrah Buffaloes. The purified samples of the protein were crystallized using microdialysis method. The protein was dialysed against low ionic strength buffer solution. The crystal was stable in the X-ray beam and diffracted upto 3.3 Å resolution. The crystal belongs the to monoclinic system with cell dimensions a = 56.79 Å, b = 101.44 Å, c = 76.27 Å, space group $P2_1$, with one protein molecule per asymmetric unit and a solvent content of 54%. The structure of protein has been determined using molecular replacement method with human lactoferrin as a model (poly ALA). The refinement of structure is in progress and current R-factor is 0.25 for

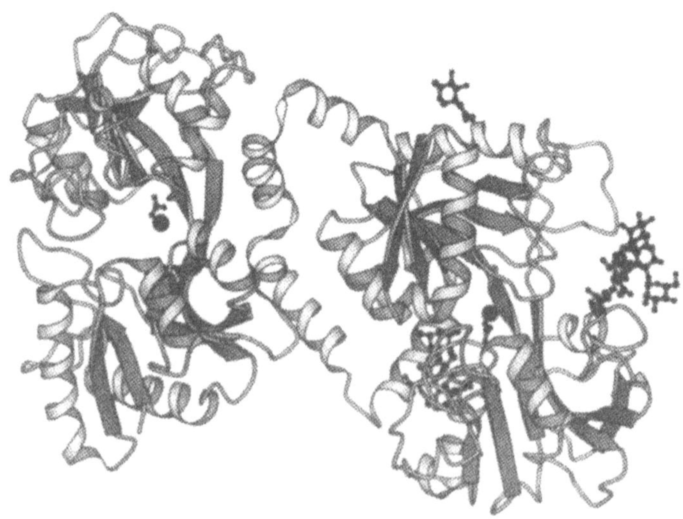

11,273 reflections in the resolution range 12.0 to 3.4 Å. Further refinement are progress. The backbone tracing is shown in Figure 1.

4. STRUCTURE OF MARE LACTOFERRIN

Mare lactoferrin was isolated from colostrum of mixed breed mares. The protein was purified using ion exchange and gel filtration chromatographies. The crystals of diffraction quality were obtained by dialysing protein against low ionic strength buffer containing 10% ethanol at pH 8 and pH 8.5. Two forms of crystals were obtained. Form I: obtained at pH 8.0, diffracted up to 4.0 Å resolution and Form II: obtained at pH 8.5, diffracted up to 2.65 Å resolution. Both crystallised in orthorhombic form with cell dimensions, form I: a = 79.8 Å, b = 103.5 Å, c = 112.0 Å and form II: a = 85.5 Å, b = 99.5 Å, c = 103.1 Å, space group $P2_12_12_1$ with one molecule in asymmetric unit. The structure of both forms have been solved by molecular replacement method using AMORE package with diferric human lactoferrin (polyALA) as a model. The refinement and model building of both forms is in progress. The current R-factor for form I is 24% for 7974 reflections and form II is 27% for 22820 reflections. Further refinement is in progress. The backbone tracing is shown in Figure 2.

5. STRUCTURAL COMPARISONS

The structures of human, bovine, buffalo and mare lactoferrins have been determined. As seen from Figures 1 and 2, the overall structures of lactoferrins are very similar. The polypeptide chain in lactoferrin folds into two globular lobes, the N-lobe and the C-lobe representing the N-terminal and C-terminal halves of the molecule, and each lobe is further subdivided into two domains (N1 and N2, C1 and C2) with specific iron binding

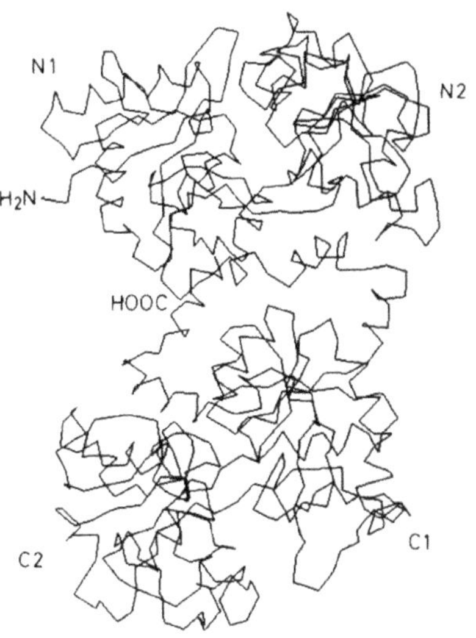

Figure 2. Cα tracing of diferric mare lactoferrin.

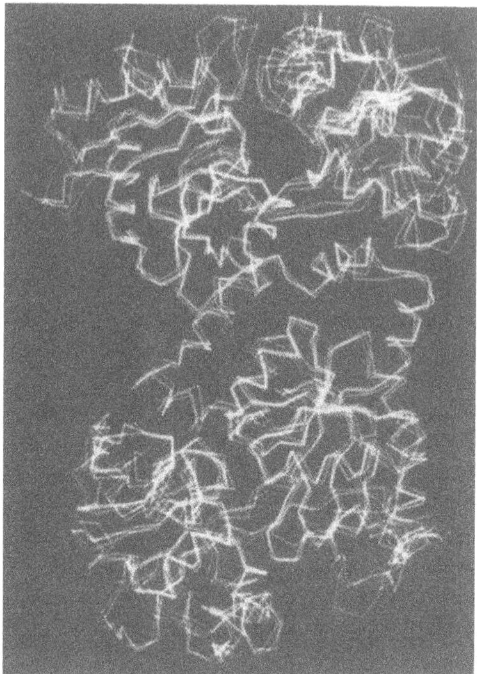

Figure 3. Superposition of Cα tracings of human, buffalo, and mare diferric lactoferrin.

sites in the interdomain cleft of each lobe (Figures 1 and 2). The connecting peptide between the lobes is helical in all the structures.

As seen from Figure 3, the superposition by least-squares fit of the backbone of human lactoferrin, buffalo lactoferrin and mare lactoferrin indicates that the secondary structure contents of these proteins are similar. However, the significant differences are observed in the loop regions. The mLf particularly shows large deviations. The most striking differences are observed in the N-lobe.

6. INTERLOBE INTERACTIONS

The interactions between the N-terminal and C-terminal lobes of lactoferrins show a distinct lack of specificity apart from two salt bridges: Arg 313–Asp379 and Asp 315–Lys 386, and a hydrogen bond. There are several hydrophobic interactions and water mediated contacts. Because of these non specific interactions, different lobe orientations are observed in various lactoferrins.

7. COMPARISON OF THE TWO LOBES

The two lobe, four domain structure of lactoferrin allows considerable scope for flexibility, which could include relative movement of the N- and C-lobes, or of two domains which comprise each lobe. The two lobes match extremely closely, as expected from their high level of sequence identity (40%). They are related by an approximate 2-fold screw axis namely a rotation of 180.4° followed by translation of 25.2Å for diferric lactoferrin. A comparison of human, bovine, mare and buffalo diferric lactoferrins shows a significant difference in the relative orientations of two lobes, which is of the order of 10–12°. These relative movements could be possible because there is a cushion of mainly hydrophobic side chains between two lobes and because a single covalent connection between them lies on the outside of the contact area. The superposition of N- and C-lobes shows that the corresponding individual domains superimpose better than do the whole lobes (r.m.s deviations 1.06Å for domain C1 on domains N1, 1.06 Å for C2 on N2 and 1.32 Å for entire C-lobe on N-lobe). This results from the fact that the C-lobe is more closed than, N-lobe, the difference being a relative domain rotation of 6°.

The other conformational differences between two lobes are (i) a difference in orientation of small loop 311 to 314 (654 to 657) joining a β–strand (304 to 310) to a helix (315 to 321). (ii) a difference in orientation of final helix (334 to 344 and 680 to 691) are not structurally comparable. They have no sequence similarity and fill different structural roles. Helix 334 to 344 projects out from the N-lobe as connecting helix to C-lobe, whereas the C-terminal helix 680 to 691 folds back across the surface of the C-lobe in the interface between two lobes. It is fastened down by two additional disulfide bridges not present in N-lobe (677 to 483 and 686 to 405) and provide link between two domains of C-lobe, which reduces domain movement significantly.

8. COMPARISON OF INTERDOMAIN INTERACTIONS

The binding cleft of each lobe in all lactoferrins is predominantly polar in nature and is formed by inner surfaces of its two domains. The most striking feature of the binding cleft is the lack of direct protein–protein interactions. The structural studies of binding cleft of different lactoferrins show the distinct differences in interdomain interactions which can have important implications on iron binding.

In the N-lobe of hLf, the two salt bridges cross between the two walls of the cleft, Lys 296...Asp 217 and Lys 301...Glu 216. The former is absent in bovine, mare and buffalo lactoferrins. The residue 217 is changed to Asn in bovine, 297 is changed to Ser in mare and 217 is changed to Asn in buffalo. In mLf, there is an additional salt bridge at the mouth of the cleft, Arg296...Asp 187, which is absent in all other lactoferrins. The loops of two domains come very close to each other at the mouth of cleft Also, a hydrophobic contact between Pro 42 and Phe 183 at the mouth of the cleft is found only in hLf. In bLf and mLf Phe 183 is changed to Cys which forms disulfide bridge with 161 Cys of same domain. The function of the disulfide bridge is not very clear. The structures of apo mLf and bLf can throw light on functional aspect of S-S bond. Other direct interdomain interaction in N-lobe involves Arg 210 behind iron site in mare and human, which in bovine, residue is changed to Lys which does not have direct interaction.

In C lobe, there are no interdomain salt bridges. There exist only three hydrogen bonds at mouth of the cleft. Glu 216 and Asp 217 are replaced by amides, Lys 301 is replaced by Asn 644. Arg 210 is replaced by Lys 216. Most of the interdomain interactions are through water bridges.

9. METAL AND ANION SITE

The iron binding sites are extremely similar in all the lactoferrins. The metal is coordinated by four protein side chains (i.e.) 2 Tyr, 1 Asp and 1 His (Tyr 92 and 192, Asp 60 and His 253 in N-lobe and corresponding Tyr 435 and 528, Asp 395 and His 597 in the C-lobe). The two oxygens from bidentate carbonate ion complete a distorted octahedral geometry at the metal ion. It also provides two ligands to the iron atom. The carbonate ion occupies a pocket formed by two positively charged groups of Arg side chain (Arg 121 in N-lobe and Arg 465 in C-lobe) and N-terminus of helix 121–136, making good hydrogen bonds with both. There are few difference in anion binding site between two lobes. The C-lobe appears to provide a slightly tighter anion site, the hydrogen bonds being on average approximately 0.1 Å shorter. The carbonate coordination is less symmetric in N-lobe site. In the N-lobe, Arg121 hydrogen bonds through NH1 to the hydroxyl group of Ser191, but this interaction is not possible in the C-lobe where equivalent residue is Glycine. In the C-lobe on the other hand, NH1 of equivalent Arg (Arg465) is instead hydrogen bonded to the phenolic O atom of Tyr526 (Phe 190 in the N-lobe). The different positions of these hydrogen bonding partners for the arginine may be responsible for small variations in the two anion binding sites.

The comparison of the diferric and dicupric structures of lactoferrin shows the effect of substituting Fe^{3+} by Cu^{2+}. The protein structure is essentially undisturbed (r.m.s deviation between c^{a} positions for both the structures being 0.25 Å). The only significant difference is in the positions of the metal ions. In the N-lobe Cu atom is displaced 0.7 Å from the iron position in a direction away from Tyr 92. This also causes the anion to become monodentate and the Cu geometry becomes square pyramidal. In the C-lobe, the Cu atom is displaced by 0.2 Å from the iron position and remains six coordination with slightly different geometry.

Thus different metal ion stereochemistries and variations in anion coordinations can be accommodated without any change in the protein structure and in particular without change in closure of two domains over the metal. This may be not true for ions as large as lanthanides, which have radii of 0.3 Å larger than Fe^{3+} and this presumably explains their weaker binding.

The adjustments of the structure to accept anions larger than carbonate is illustrated by Cu_2Ox Lf complex, which has carbonate ion in N-lobe site and oxalate in C-lobe. The oxalate ion binds in 1,2-bidentate fashion with one carboxylate group hydrogen bonded to, like carbonate, to helix N-terminus. The extra bulk of oxalate ion requires that the Arg 465 side chain move some 2 Å away from metal site. This, in turn, displaces the side chain of adjacent Tyr 398, but these movements can be accommodated within large solvent filled cavity, in interdomain cleft, without any change in protein structure.

10. COMPARISON OF DIFERRIC AND APO LACTOFERRIN

The comparison of structures of diferric and apo lactoferrin demonstrates the importance of large scale domain movements for metal binding and release and suggests that in solution an equilibrium exists between open and closed forms, with the open form being the active binding species.

The apo-Lf structure shows two striking features when compared with Fe_2Lf. The large scale conformational change in N-lobe and closed C-lobe even without metal. There are small differences (~8°) between the orientations of the N1 domain relative to the C-lobe domains, presumably as a result of some readjustments in inter lobe contact area, but the most striking change is the 54° rotation of the N2 domain relative to N1 domain. This causes the N-lobe binding cleft to be opened wide. That this rigid body movement is shown by the very close correspondence when individual domains of apo-Lf and Fe_2-Lf are superimposed the r.m.s deviation is only 0.45 Å for N1 and 0.51 Å for N2, for all C^{α} atoms.

The nature of conformational change in the N-lobe has elements both of the simple hinge and of helix mediated domain closures. The two extended backbone strands do show a quite abrupt hinge at about residue His 91 and Val 250. This is achieved with generally small changes in main chain torsion angles, the major difference being 45° in ψ (90), 26° in ϕ (91), 20° in ϕ (250) and 24° in ψ (250). No Gly residues are involved and no change in hydrogen bonding pattern occurs; it is a simple flexing of an antiparallel β ribbon. The second feature of domain movement involves two helices. Helices 121–136 and 321–332, which lie across each other at an angle of ~75°. Helix 321–332 which provides a third cross over between the domains does not move significantly during conformational change, remaining helix 121–136 which is centrally involved in the organisation of domain N2 and which has the anion binding site at its N-terminus pivots on helix 321–332 and carries rest of domain N2 with it; the rotation axis lies approximately along helix 321–332. In diferric lactoferrin, hydrogen bonds link the side chains of Thr 122 and Asn 126 near the N-terminus of the helix 321–332. In apo-Lf, these hydrogen bonds are broken but new salt bridge links Arg 133, at C-terminus of helix 121–136, to Glu 335, just following C-terminal of helix 321–332. These hydrogen bonds may help to stabilize two structures.

The open N-lobe cleft suggests a sequence of steps in the binding process. A consequence of the α/β folding pattern is that a number of helix in N-termini line the walls of the binding cleft. These together, with several basic side chains (Arg 121, Arg 210, Lys 296, Lys 308) should provide a positive potential which attracts the CO_3^{2-} ion into the open cleft, to bind to the specific pocket provided by Arg 121 and N-terminus of helix 121–136. With CO_3^{2-} ion bound, four of six iron ligands would then be in place of domain N2 (i.e.) two CO_3^{2-} ion oxygens together with Tyr 92 and Tyr 192, which in open structure remain with domain N2. The metal binding initially occurs at domain N2, which then rotates to enable the metal to complete its coordination, to Asp 60 and His 253. Support for this sequence of events also comes from the spectroscopic detection of quarternary protein-anion-metal-chelate complex when metal ion is added as chelate complex.

The C-lobe binding cleft is closed even though no metal is bound. In C-lobe a disulfide bridge 483–677 link C-terminal end of helices 465–481 and 664–677. These helices are equivalent to 121–136 and 321–332 of N-lobe. This is consistent with the known tighter iron binding and shown iron release.

REFERENCES

Day, C.L., B.F. Anderson, J.W. Tweedie, and E.N. Baker. J. Mol. Biol. 232:1084 (1993).
Haridas, M. B.F. Anderson, H.M. Baker, G.E. Norris, and E.N. Baker. Adv. Exp. Med. Biol. 357:235 (1994).
Haridas, M., B.F. Anderson, and E.N. Baker. Acta. Crystallogr. D51:629 (1995).
Karthikeayn, S., S. Yadav, A. Srinivasan, and T.P. Singh. (Manuscript under preparation) (1997).
Lonnerdal, B., and S. Iyer. Annu. Rev. Nutr. 93 (1995).
Sharma, A.K., A. Srinivasan, and T.P. Singh. (Manuscript under preparation) (1997).
Sharma, S., T.P. Singh, and K.L. Bhatia. Acta. Crystallogr. D53:116 (1997)
Weinberg, E.D. Microbial. Rev. 42:232 (1978).

3

DIRECT DETECTION AND QUANTITATIVE DETERMINATION OF BOVINE LACTOFERRICIN AND LACTOFERRIN FRAGMENTS IN HUMAN GASTRIC CONTENTS BY AFFINITY MASS SPECTROMETRY

Hidefumi Kuwata,[1,2] Tai-Tung Yip,[3,4] Christine L. Yip,[3,4] Mamoru Tomita,[2] and T. William Hutchens[1,4]

[1]Department of Food Science and Technology
University of California, Davis
Davis, California
[2]Nutritional Science Laboratory
Morinaga Milk Industry Co., Ltd
Kanagawa, Japan
[3]Molecular Analytical Systems Inc.
Davis, California
[4]Current address: Ciphergen Biosystems, Inc.
490 San Antonio Rd.
Palo Alto, California 94306

1. ABSTRACT

Lactoferricin (Lfcin®) is a bioactive fragment of lactoferrin derived from the bactericidal and putative lymphocyte receptor binding domain(s) located within the N-lobe of lactoferrin. Although known to be liberated from at least three species of lactoferrin, conditions leading to Lfcin generation *in vivo* and factors affecting its distribution are still not known. Recently, we have developed a method of surface-enhanced laser desorption/ionization (SELDI®) affinity mass spectrometry using n-butyl terminal groups for surface-enhanced affinity capture (SEAC) to quantify not only Lfcin generated in vivo but also other lactoferrin fragments. Unlike previous efforts to detect lactoferrin and Lfcin with specific antibodies, the SELDI affinity assay distinguished lactoferrin, lactoferrin fragments, Lfcin and unrelated peptides without their interference with each other. To evaluate Lfcin generation *in vivo,* the experimental design involved feeding 200 mL of 10 mg/mL (1.22×10^{-4} mol/L)

Advances in Lactoferrin Research, edited by Spik *et al.*
Plenum Press, New York, 1998.

bovine lactoferrin to an adult. Gastric contents were recovered 10 min after ingestion. Lfcin produced *in vivo* was directly captured by the SEAC device. The amount of Lfcin in the gastric contents was 16.91 ± 2.65 μg/mL (5.350 ± 0.838 × 10^{-6} mol/L). However, a large proportion of the ingested lactoferrin was not completely digested. Lactoferrin fragments containing the Lfcin region were analyzed by *in situ* hydrolysis with pepsin after being captured by the SEAC device. As much as 5.740 ± 0.702 × 10^{-5} mol/L of the partially degraded lactoferrin fragments were found to contain the Lfcin region, including peptide domains 17–43, 17–44, 12–44, 9–58, and 16–76 of bovine lactoferrin. These results show that bovine Lfcin can be produced in the human stomach after ingestion of an infant formula supplemented with bovine lactoferrin. It is now important to determine whether Lfcin is generated in the intestinal tract of formula-fed and breast-fed infants, and geriatric patients consuming foods enriched with lactoferrin.

2. INTRODUCTION

Lactoferrin, an 80 kDa major protein component of mammalian colostral whey, is also present in several other biological fluids and in granules of neutrophils[1–3]. It has been associated with a wide variety of biologically important processes, including host defense, regulation of cell function, cell growth, and cell differentiation[4].

Though it has been thought that its ability to bind and sequester iron prevent microbial growth, a bactericidal domain has been identified and named lactoferricin (Lfcin®)[5]. The antimicrobial activity of Lfcin is much stronger than that of an equimolar amount of intact lactoferrin. bLfcin (bovine lactoferricin) consists of 24 or 25 amino acid residues including a loop of 18 amino acid residues derived from the N-terminal region of the lactoferrin molecule. Like various other antimicrobial peptides it contains a high proportion of basic amino acid residues. It has been reported that lymphocyte receptor, glycosaminoglycan and lipopolysaccharide (LPS) binding site are located in the Lfcin region[6–8].

Lfcin displays a broad spectrum activity against enteric bacteria, while some species of *Bifidobacterium* spices, which comprise the major part of the flora of breast-fed infant, are highly resistant to bLfcin. The intestinal flora of bottle-fed babies is quite different from that of breast-fed infants, because the *Bifidobacterium* is rarely predominant. Acting in combination with other factors such as lysozyme, lactoperoxidase and sIgA, Lfcin produced by hydrolysis of lactoferrin in the stomach could influence the formation of the flora in upper intestine, and possibly in the large intestine during the early stage after birth.

The effects on the intestinal ecosystem of a milk supplemented with different amounts of bovine lactoferrin have been studied[9–11]. However, a limited amount of information exists about the actual generation of Lfcin in the stomach and the general process of lactoferrin degradation in the gastrointestinal tract. It is well known that lactoferrin is unusually resistant to proteolytic attack by digestive proteases *in vivo* as well as *in vitro*[11–13]. It seems that part of the lactoferrin ingested by infants could survive passage through the gastrointestinal tract as partially degraded forms and perform biological functions relating to formation of the intestinal flora, absorption of iron, cell growth and cell differentiation.

In spite of the fact that ELISA (enzyme linked immunosorbent assay), Western Blotting, immunodiffusion and direct purification methods have been most commonly used for detection and quantitative determination of lactoferrin and its fragments, presence of various lactoferrin forms, including intact lactoferrin, partially degraded lactoferrin and its oligopeptides, has complicated precise characterization of the peptides in physiological fluids. In addition, the interaction between lactoferrin and polysaccharide such as heparin affects the quantitative assay of antibody binding.

Quantitative measurement by matrix assisted laser desorption/ionization (MALDI) mass spectrometry using an internal standard has been reported[14–16]. This offers a solution to the problems associated with cross reactivity because of the ability to measure precise molecular mass by mass spectrometry. When analyzing a highly complex biological fluid by MALDI mass spectrometry, in order to overcome signal suppression and low reproducibility, affinity mass spectrometry which encompasses a new desorption strategy: Surface Enhanced Affinity Capture (SEAC) is introduced. SEAC is one of the Surface Enhanced Laser Desorption (SELDI[8]) method, which consists of affinity capture the analyte and ordered presentation of the molecules for subsequent analysis by laser desorption/ionization mass spectrometry[17,18].

Recently, we have developed an affinity mass spectrometry method for detection, characterization and quantitative determination of Lfcins using n-butyl terminal groups as the affinity capture molecule. The n-butyl terminal group captures only the Lfcin region from lactoferrin hydrolysate with a sensitivity of 200 pg/mL of Lfcin as measured directly in serum[19]. Precise mass analysis differentiates Lfcin from partially degraded lactoferrin and its oligopeptides fragments. This report describes the actual quantitative determination of bLfcin and lactoferrin fragments in gastric contents from a human subject fed bovine lactoferrin using affinity mass spectrometry. This provides direct evidence of the generation of a functional peptide domain *in vivo* and allows characterization of the process of digestion of lactoferrin in the stomach.

3. EXPERIMENTAL

3.1. Materials

bLfcin was purified as described previously[5]. Since bovine lactoferrin from commercial sources shows batch-to-batch variation in the degree of iron saturation, and since the enzymatic digestibility is influenced by the degree of iron saturation, a single lot (product number 8X31) of lactoferrin from Morinaga Milk Industry, having an iron saturation of 15.2% as estimated by direct quantitative determination of iron, and 17.1% as estimated by measuring the absorption at 405 nm, was used for all experiments. The 21-residue HRG (histidine rich glycoprotein) 4mer $(GHHPH)_4G$ was prepared by procedures described previously[20]. The molecular weight of the peptide is 2337.4.

3.2. Instrumentation

Both MALDI and SELDI were performed by means of a modified Hewlett Packard model 1700XP laser desorption time-of-flight mass spectrometer (LDTOF MS) operated with a 3-stage ion optic assembly at high voltage (27 kV).

3.3. Feeding and Sampling

A healthy man starved for 12 h was fed 200 mL of 10 mg/ml bovine lactoferrin. After 10 min, the mouth was washed with water and gastric contents were recovered by vomiting. A total 40 mL of gastric contents was collected. The pH of the contents estimated by pH strip was 3.5–4.0. For the prevention of further proteolysis, pepstatin A was added to a final concentration of 0.01% and the pH was adjusted to neutral. The gastric contents were frozen at –20°C before use.

3.4. SELDI Affinity Assay for Lfcin and Lactoferrin

The SELDI affinity assay for Lfcin and lactoferrin has been described previously[18]. Briefly, 10 μL of Butyl Toyopearl 650 M chromatography gel (Toso Hass) was mixed with the gastric contents, standard lactoferrin solution, or standard Lfcin solution. For quantitative determination of Lfcin and for identification of lactoferrin fragments generated *in vivo*, a portion of the n-butyl gel was directly transferred to the probe surface and analyzed by SELDI after being washed with water. The proteins bound to the gel were hydrolyzed by porcine pepsin and analyzed in the same manner so as to quantify intact lactoferrin and lactoferrin fragments containing the Lfcin region. Alpha-cyano-4-hydroxycinnamic acid (SIGMA) solution in acetonitrile/water/trifluoroacetic acid (60/40/0.05) was used as the matrix solution. HRG 4 mer peptide at a final concentration of 200 fmol/mL was added to the matrix solution as an internal standard. Each spectrum was normalized with respect to the peak intensity of the internal standard.

4. RESULTS AND DISCUSSION

4.1. Administration of Lactoferrin

Bovine lactoferrin dissolved in saline at 10 mg/mL was fed to a man to mimic the ingestion of human milk or infant formula supplemented with bovine lactoferrin. 40 mL of gastric contents were recovered 10 min after ingestion of 200 mL of 10 mg/mL bovine lactoferrin.

Figure 1 shows the SDS-PAGE and immunoblot analysis of the gastric contents. A complex pattern of protein bands was observed on the SDS-PAGE due to the high com-

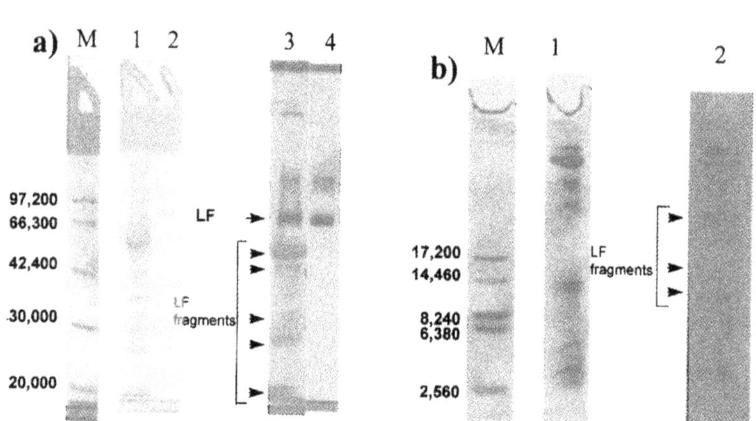

Figure 1. SDS-PAGE (a, lanes 1 and 2), urea SDS-PAGE (b, lane 1), and immunoblot analysis (a, lanes 3 and 4; b, lane 2). Gastric contents were boiled in SDS sample buffer with 2-mercaptoethanol. Proteins and peptides were revealed by staining with Coomassie Brilliant Blue. Polyclonal antibodies against synthetic bLfcin were used as specific antibodies for lactoferrin fragments as well as Lfcin. **a.** lane M, molecular weight markers; lane 1, gastric contents (Coomassie Brilliant Blue stained); lane 2, purified lactoferrin (Coomassie Brilliant Blue stained); lane 3, gastric contents (immunostained);lane 4, purified lactoferrin (immunostained). **b.** lane M, molecular weight markers; lane 1, gastric contents (Coomassie Brilliant Blue stained); lane 2, gastric contents (immunostained).

plexity of the gastric contents (Figure 1a, lane 1; Figure 1b, lane 1). Intact lactoferrin (estimated molecular weight 80 kDa) and its fragments (estimated molecular weight 60, 52, 36, 29, 24 kDa) that would contain Lfcin region in their sequence were revealed by immunoblot analysis using antibodies against synthetic bLfcin (Figure 1a, lane 3 indicated by arrows). Peptides with molecular weight in the range of 2,000–20,000 in the gastric contents were also separated by urea SDS-PAGE and immunoblotted (Figure 1b, lanes 1 and 2). Although some weakly immunostained bands were observed on the immunoblot (Figure 1b, lane 2), the resolution was too low to detect bLfcin and peptides containing the bLfcin region. The great variety of the fragments derived from orally administered lactoferrin prevented precise characterization of the process of hydrolysis of lactoferrin by the western blotting technique. In addition, the abundance of contaminants such as mucin and other proteins in the gastric contents made separation by SDS-PAGE very poor.

4.2. SELDI Affinity Assay for Lfcin and Lactoferrin

In order to overcome the problems associated with the sample complexity of the gastric contents, we employed the SELDI affinity assay for Lfcin and lactoferrin. Figure 2 shows the outline of the method employed for affinity mass spectrometry to detect and quantify bLfcin, intact lactoferrin and lactoferrin fragments containing Lfcin in their sequence. The n-butyl terminal group was used because there is a region of lactoferrin with affinity for binding the n-butyl group (i.e., FKCRRWQWR), even in 6 M urea, which is sufficient to denature almost all proteins. This sequence FKCRRWQWR includes the whole antimicrobial active center (i. e., RRWQWR) of bLfcin. After washing to remove unbound and nonspecifically bound peptides, affinity captured bLfcin, intact lactoferrin and peptides containing the FKCRRWQWR amino acid sequence were desorbed and ionized from SEAC device by laser pulses in the presence of matrix molecules (Figure 2). Since all of the captured proteins and peptides have different masses, multiple peaks were detected. To detect and quantitate all proteins and peptides containing the bLfcin sequence at once, they were converted to bLfcin by hydrolysis with porcine pepsin (Figure 2). The concentrations of bLfcin and lactoferrin fragments containing the Lfcin region were calculated from the normalized signal intensity of the molecular ion peak of bLfcin with and without hydrolysis, respectively.

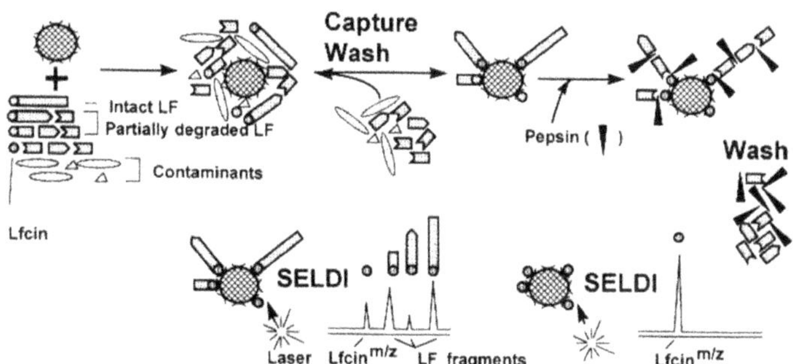

Figure 2. Schematic outline of the method used for quantitative determination of bLfcin and lactoferrin fragments in human gastric contents.

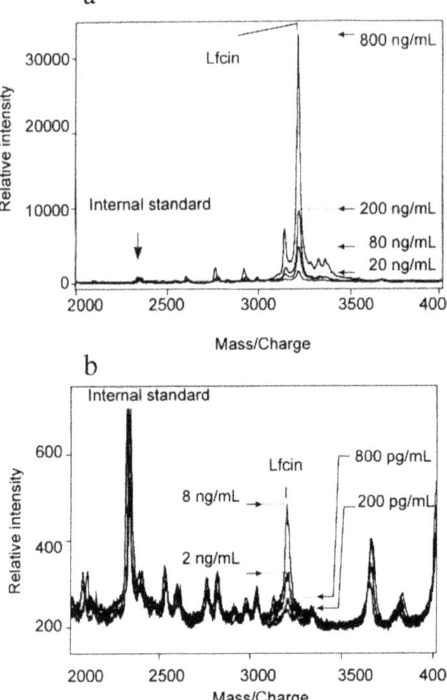

Figure 3. Composite mass spectrum obtained from 20–800 ng/mL bLfcin (a) and 200 pg-8 ng/mL bLfcin (b) affinity captured from 10 mL of serum as a model mixture. The signal intensity of each spectrum was normalized with respect to the internal standard.

Figure 3 shows the selectivity of the n-butyl SEAC device for Lfcin peptides in the model mixture solution (whole serum) containing Lfcin. Mainly the Lfcin peptides were specifically captured from the mixture solution. Normalizing the spectrum intensity with respect to the signal of the internal standard peptide added in the same amount in each instance, the signal intensity of Lfcin was quantitative from 200 pg/mL to 800 ng/mL. The standard curve prepared by plotting normalized Lfcin peak area versus concentration of Lfcin was linear over 3 orders of magnitude, and the regression coefficient was 0.9727 (data not shown). The concentration of Lfcin in the gastric contents was calculated from the obtained standard curve (described later).

Figure 4 shows the mass spectrum of peptides or proteins captured from gastric contents on the SEAC device. Using molecular weight matching of the peptides, 8 of 12 molecular ion peaks of peptides captured by the SEAC device were identified as peptides containing the n-butyl binding region (i.e., FKCRRWQWR). Molecular ion peaks corresponding to bLfcin produced *in vivo* were observed (m/z; 3122.9, 3195.0). The following are the identified peptides: 3308.4 Da corresponding to sequence 17–43 of lactoferrin (expected mass 3309.0 Da), 3441.8 Da corresponding to sequence 17–44 of lactoferrin (expected mass 3438.2 Da), 4061.9 Da corresponding sequence 12–44 of lactoferrin (expected mass 4065.8 Da), 5853.9 Da corresponding to sequence 9–58 of lactoferrin (expected mass 5852.0 Da) and 7341.4 Da corresponding to sequence 16–79 of lactoferrin (expected mass 7338.7 Da). The molecular ion peak which gave the strongest peak intensity in the spectrum was predicted to be the 13–36 region of bovine lactoferrin (expected mass 3035.6 Da). This peptide was not produced by porcine pepsin hydrolysis of bovine lactoferrin *in vitro*. There is the possibility that *in vivo* other proteolytic enzymes were involved, such as carboxypeptidases in fluid that had flowed backward from the duodenum.

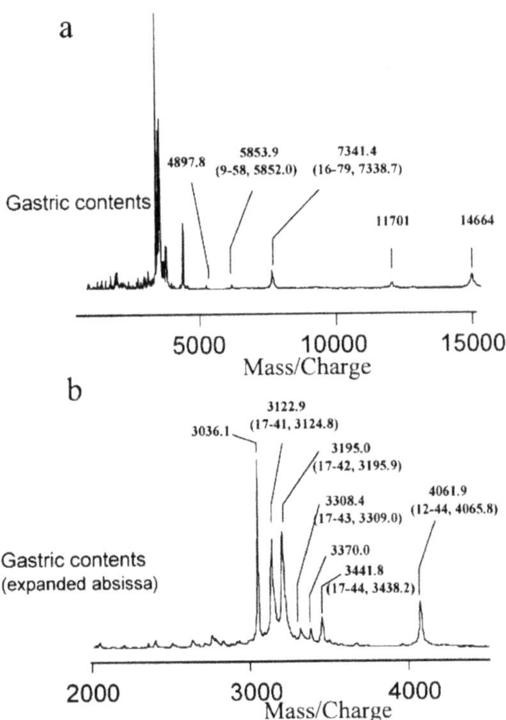

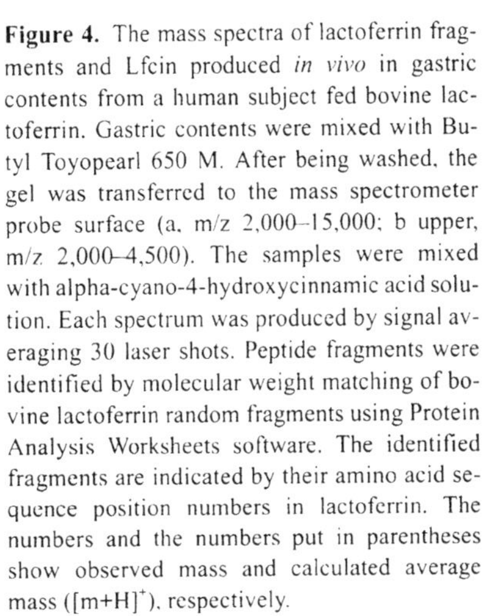

Figure 4. The mass spectra of lactoferrin fragments and Lfcin produced *in vivo* in gastric contents from a human subject fed bovine lactoferrin. Gastric contents were mixed with Butyl Toyopearl 650 M. After being washed, the gel was transferred to the mass spectrometer probe surface (a, m/z 2,000–15,000; b upper, m/z 2,000–4,500). The samples were mixed with alpha-cyano-4-hydroxycinnamic acid solution. Each spectrum was produced by signal averaging 30 laser shots. Peptide fragments were identified by molecular weight matching of bovine lactoferrin random fragments using Protein Analysis Worksheets software. The identified fragments are indicated by their amino acid sequence position numbers in lactoferrin. The numbers and the numbers put in parentheses show observed mass and calculated average mass ($[m+H]^+$), respectively.

Pepsin shows a broad specificity of digestion sites as compared to other digestive proteases like trypsin. In addition, pepsins purified from different sources vary in specificity. The finding that bLfcin is produced from ingested bovine lactoferrin by human pepsin digestion *in vivo* suggests that in a similar manner hLfcin (human lactoferricin) could be generated in the stomach of infants from lactoferrin in breast milk, or even in the stomach of human adults from lactoferrin secreted in the saliva.

The composite mass spectra in Figure 5 resulted from peptides affinity captured from human gastric contents, both diluted 10 fold with (thick line)/without (thin line) *in situ* porcine pepsin hydrolysis after capture on the SEAC device. These two spectra were normalized in signal intensity with respect to the internal standard (Figure 4, arrow). Clearly the signal intensity of the peaks of bLfcin (sequence 17–41 and 17–42 of lactoferrin) in the gastric contents after *in situ* pepsin hydrolysis increased to a level 10 times greater than that without pepsin hydrolysis. This observation indicates that 90% (mol) of the lactoferrin existed as partially degraded forms that could be potentially processed further to Lfcin in the stomach. The intensity of a molecular ion peak which could be the peptide with the sequence of residues 13–36 of lactoferrin was not affected by porcine pepsin hydrolysis (Figure 4). Since there did not seem to be a pepsin digestion site between W16 and F17, this peptide could not be the fragment 13–35 of ingested lactoferrin, and it seems that this peptide is of human origin.

The concentrations of Lfcin and lactoferrin fragments in the gastric contents were calculated from the normalized peak intensity of Lfcin using the described standard curve. The gastric contents were diluted 10 fold in 6 M urea and analyzed by affinity mass spectrometry. The concentration of bLfcin in gastric contents was 16.9 ± 2.65

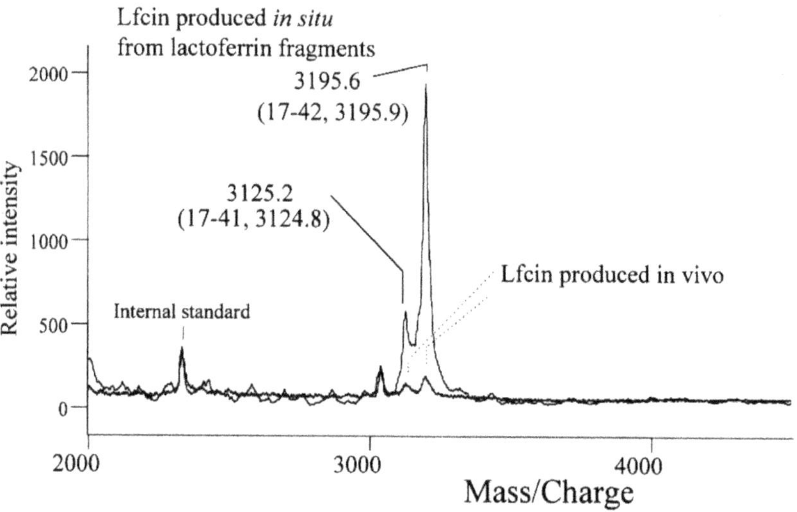

Figure 5. Characterization of lactoferrin hydrolysis products in human gastric contents before and after hydrolysis with exogenous pepsin. Gastric contents were diluted 10-fold with 6 M urea. 10 μL of the diluted sample was mixed with 10 μL of Butyl Toyopearl 650 M as the affinity capture device. After being washed, a portion of the sample-loaded device was transferred to the mass spectrometer probe surface and analyzed by SELDI (thin line). The rest of the sample-loaded device was mixed with porcine pepsin in 10 mM HCl and allowed to react 37°C for 1 h. The peptides were recaptured by the SEAC device after the pH of the mixture was neutralized. The peptides recaptured were analyzed as described previously (thick line). In each case, the same amount of the gel (0.5 μL of 67% gel suspension) was mixed with 1 μL of alpha-cyano-4-hydroxycinnamic acid solution containing 200 fmol/mL HRG 4 mer as an internal standard. The spectra were normalized with respect to the signal intensity of the internal standard.

μg/mL ($5.450 \pm 0.838 \times 10^{-6}$ mol/L) , which was calculated from the sum of the normalized area of the peptides 17–41 and 17–42 of lactoferrin. Recovery of standard Lfcin added to the gastric contents determined from the signal increment was 93.3 ± 14.1%. Gastric contents were diluted 30 fold for quantitative determination of lactoferrin and lactoferrin fragments containing the Lfcin region. The concentration of intact lactoferrin and lactoferrin fragments containing the Lfcin region was 181.4 ± 22.2 μg/mL as Lfcin or 4543 ± 556 μg/mL as lactoferrin ($5.740 \pm 0.702 \times 10^{-5}$ mol/L). Recovery of standard added to the gastric contents was 100.8 ± 16.2% under such condition. Nearly 100% of the Lfcin derived from added lactoferrin was recovered. This shows that all added lactoferrin was captured by the SEAC device and hydrolyzed by pepsin *in situ* on the surface of the capture device. However, recovery of the Lfcin signal obtained from added lactoferrin using *in situ* pepsin hydrolysis was 73.7 ± 12.0% when the gastric contents were diluted 10 fold. Complex biological samples such as the gastric contents may require substantial dilution to overcome signal repression.

Ten minutes after ingestion of lactoferrin, the concentration of Lfcin had already reached a sufficiently high concentration to exert antimicrobial activity against some sensitive bacteria[21] With further digestion in the stomach, the concentration of Lfcin should increase[21], depending on the amount of lactoferrin fragments present that potentially could be further digested to produce Lfcin, pepsin activity, gastric fluid dilution and the speed of gastric emptying.

Studies in animals and *in vitro* experiments suggest that lactoferrin and its peptides could be responsible for some of the differences observed in the fecal flora comparing breast-fed and bottle-fed infants[22]. Lfcin produced in the stomach could influence the formation of the intestinal flora in the upper intestine and possibly in the large intestine, in mammals which have adequate amounts of lactoferrin secreted into milk. With regard to human infants, though the concentration of lactoferrin in breast milk is relatively high (i.e., 1–8 mg/mL), low pepsin activity in the stomach and the existence of proteases cleave within the Lfcin sequence may result in only moderate amounts of Lfcin generation in the gastrointestinal tract. Many important questions remain unanswered about the production of Lfcin in the stomach, its metabolic fate after leaving the stomach and its possible effect on the host defense in early stage of birth. The technology outlined in this paper will be facilitate further extensive studies in human infants fed bovine/human lactoferrin or breast milk to answer the general questions about the process of digestion of ingested lactoferrin.

5. CONCLUSIONS

1. Lfcin was produced from ingested bovine lactoferrin in the human stomach.
2. At 10 min post-administration, the concentration of Lfcin had reached 16.91 ± 2.65 μg/mL, which is higher than the minimum inhibitory concentration for some bacteria.
3. More than 90% (mol) of the ingested lactoferrin still remained as partially degraded forms.
4. SELDI affinity mass spectrometry may be a useful tool to evaluate the fate of lactoferrin beyond the stomach.

6. ABBREVIATIONS

The abbreviations used are: bLfcin, bovine lactoferricin; ELISA, enzyme linked immunosorbent assay; hLfcin, human lactoferricin; HRG, histidine rich glycoprotein; LDTOF MS, laser desorption time of flight mass spectrometer; LPS, lipopolysaccharide; MALDI, matrix assisted laser desorption /ionization.

REFERENCES

1. Masson, P. L., Heremans, J. F., and Dive, J. H. (1966) An iron-binding protein common to many external secretions. Clin.Chim. Acta 14, 735–739.
2. Masson, P. L., Heremans, J. F., and Schonne, E. (1969) Lactoferrin, an iron-binding protein in neutrophilic leukocytes. J. Exp. Med. 130, 643–658.
3. Bennett, R. M., Eddie-Quartey, A. C., and Holt, P. J. L. (1973) Lactoferrin—an iron binding protein in synovial fluid. Arthritis and Rheumatism 16, 186–190.
4. Sanchez, L., Calvo, M., and Brock, J. H. (1992) Biological role of lactoferrin. Arch. Dis. Child 67, 657–661.
5. Bellamy, W., Takase, M., Yamauchi, K., Wakabayashi, H., Kawase, K., and Tomita, M. (1992) Identification of the bactericidal domain of lactoferrin. Biochim. Biophys. Acta 1121, 130–136.
6. Legrand, D., Mazurier, J., Elass, A., Rochard, E., Vergoten, G., Maes, P., Montreuil, J., and Spik, G. (1992) Molecular interactions between human lactotransferrin and the phytohemagglutinin-activated human lymphocyte lactotransferrin receptor lie in two loop-containing regions of the N-terminal domain of human lactotransferrin. Biochemistry 31, 9243–9251.

7. Mann, D. M., Romm, E., and Migliorini, M. (1994) Delineation of the glycosaminoglycan-binding site in the human inflammatory response protein lactoferrin. J. Biol. Chem. 269, 23661–23667.
8. Elass-Rochard, E., Roseanu, A., Legrand, D., Trif, M., Salmon, V., Motas, C., Montreuil, J., and Spik, G. (1995) Lactoferrin-lipopolysaccharide interaction: involvement of the 28–34 loop region of human lactoferrin in the high-affinity binding to Escherichia coli O55B5 lipopolysaccharide. Biochem. J. 312, 839–845.
9. Roberts, A. K., Chierici, R., Sawatzki, G., Hill, M. J., Volpato, S., and Vigi, V. (1992) Supplementation of an adapted formula with bovine lactoferrin: 1. Effect on the infant faecal flora. Acta Paediatr. 81, 119–124.
10. Teraguchi, S., Ozawa, K., Yasuda, S., Shin, K., Fukuwatari, Y., and Shimamura, S. (1994) The bacteriostatic effects of orally administered bovine lactoferrin on intestinal Enterobacteriaceae of SPF mice fed bovine milk. Biosci., Biotechnol., Biochem. 58, 482–487.
11. Hentges, D. J., Marsh, W. W., Petschow, B. W., Thal, W. R., and Carter, M. K. (1992) Influence of infant diets on the ecology of the intestinal tract of human flora-associated mice. J. Pediatr. Gastroenterol Nutr. 14, 146–152.
12. Spik, G., Brunet, B., Mazurier-Dehaine, C., Fontaine, G., and Montreuil, J. (1982) Characterization and properties of the human and bovine lactotransferrins extracted from the faeces of newborn infants. Acta Paediatr. Scand. 71, 979–985.
13. Hutchens, T. W., Henry, J. F., Yip, T. T., Hachey, D. L., Schanler, R. J., Motil, K. J., and Garza, C. (1991) Origin of intact lactoferrin and its DNA-binding fragments found in the urine of human milk-fed preterm infants. Evaluation by stable isotopic enrichment. Pediatr. Res. 29, 243–250.
14. Duncan, M. W., Matanovic, G., and Cerpa-Poljak, A. (1993) Quantitative analysis of low molecular weight compounds of biological interest by matrix-assisted laser desorption ionization. Rapid Commun. Mass Spectrom. 7, 1090–1094.
15. Gusev, A. I., Wilkinson, W. R., Proctor, A., and Hercules, D. M. (1993) Quantitative analysis of peptides by matrix-assisted laser desorption ionization time-of-flight mass spectrometry. Appl. Spectrosc. 47, 1091–1092.
16. Nelson, R. W., Mclean, M. A., and Hutchens, T. W. (1994) Quantitative determination of proteins by matrix-assisted laser desorption ionization time-of-flight mass spectrometry. Anal. Chem. 66, 1408–1415.
17. Hutchens, T. W. and Yip, T. T. (1993) New desorption strategies for the mass spectrometric analysis of macromolecules. Rapid Commun. Mass Spectrom. 7, 576–580.
18. Yip, T. T., Van deWater, J., Gershwin, M. E., Coppel, R. L., and Hutchens, T. W. (1996) Cryptic antigenic determinants on the extracellular pyruvate dehydrogenase complex/mimeotope found in primary biliary cirrhosis. A probe by affinity mass spectrometry. J. Biol. Chem. 271, 32825–32833.
19. Kuwata, H., Yip, T. T., Yip, C. L., and Hutchens, T. W. Bactericidal domain of lactoferrin. Detection, quantitation, and characterization of lactoferricin in serum by SELDI affinity mass spectrometry. Manuscript in preparation
20. Hutchens, T. W., Nelson, R. W., Allen, M. H., Li, C. M., and Yip, T. T. (1992) Peptide-metal ion interactions in solution:detection by laser desorption time-of-flight mass spectrometry and electrospray ionization mass spectrometry. Biol. Mass Spectrom. 21, 151–159.
21. Bellamy, W., Takase, M., Wakabayashi, H., Kawase, K., and Tomita, M. (1992) Antibacterial spectrum of lactoferricin B, a potent bactericidal peptide derived from the N-terminal region of bovine lactoferrin. J. Appl. Bacteriol. 73, 472–479.
22. Yoshioka, H., Iseki, K., and Fujita, K. (1983) Development and differences of intestinal flora in the neonatal period in breast-fed and bottle-fed infants. Pediatrics 27, 317–320.

4

ANALYSIS OF BOVINE LACTOFERRIN IN WHEY USING CAPILLARY ELECTROPHORESIS (CE) AND MICELLAR ELECTROKINETIC CHROMATOGRAPHY (MEKC)

Peter Riechel,[1] Torsten Weiß,[1] Roland Ulber,[1] Heinrich Buchholz,[2] and Thomas Scheper[1]

[1]Institut für Technische Chemie
Universität Hannover
Callinstraße 3, 30 167 Hannover, Germany
[2]BIOLAC GmbH
Am Bahnhof 1, 31 097 Harbansen, Germany

1. INTRODUCTION

Lactoferrin (Lf) is found in milk and external secretions in high concentrations[1]. In human milk Lf is one of the major protein compounds ranging from 1000 μg/ml in mature milk to 7000 μg/ml in colostrum[2] instead of only 20–200 μg/ml in bovine[3]. Due to its high isoelectric point the protein is almost positively charged and can interact with other proteins in milk whey[4].

Analytical techniques for the determination of Lf are reported using high performance liquid chromatography[5] or fast protein liquid chromatography in combination with ion-exchange chromatography[6]. The use of affinity interactions for isolation of Lf have also been reported[7–10].

Capillary electrophoresis (CE) combines the quantification and handling benefits of liquid chromatography with the separation power of traditional gel electrophoresis. With its simple instrumental setup and the potential of short analysis times and highly automated analysis CE is an obvious choice for very fast monitoring of biologically interesting processes during the production or isolation of proteins. Analysis of proteins in complex media such as blood[11,12] or milk have be shown applicable for CE, up to now no report has been published for determination of minor whey proteins with CE especially of bovine lactoferrin (bLf). Heegaard & Brinnes reported 1997 the monitoring of the affinity interac-

Advances in Lactoferrin Research, edited by Spik *et al.*
Plenum Press, New York, 1998.

tion between heparin and human Lf[13]. For analysis of complex mixtures the resolution of the target protein is often hampered by comigration with other compounds in the mixture. With the use of micellar buffers in micellar electrokinetic chromatography (MEKC) the resolution could be increased. Through the use of mixed micells highest selectivity and resolution could be achieved[14]. In this communication we describe our development of a new method for the determination of b-Lf based on MEKC separation techniques.

2. MATERIALS AND METHODS

2.1. Materials

Chemicals: bovine Lactoferrin, Brij 35, sodium hydroxide, SDS, sodium sulfate, 2-propanole, Na_2HPO_4 (12 H_2O), NaH_2PO_4 (12 H_2O) and boric acid were purchased from Fluka (Deisenhofen, Germany) or Sigma/Aldrich. Whey samples were a gift of the BioLac GmbH (Harbansen, Germany).

2.2. Capillary Electrophoresis

All separations were carried out on a Beckman P/ACE 2100 instrument (Beckman, Palo Alto, CA, USA) using UV detection at 200 nm equipped with a mercury lamp (190–480 nm emission). System Gold software (Beckman, Palo Alto, CA; USA) and an IBM PS/2 personal computer for data collection, data analysis and system controlling. Fused silica capillaries were purchased from Polymicro Technologies (Phoenix, AZ, USA) with an inner diameter of 50 μm. The total capillary length was 57 cm otherwise cited, i.e. 50 cm from the capillary inlet. Before using the capillaries first time they were etched with 1 M NaOH for 15 minutes followed by 5 minutes rinsing with deionized water and 5 min electrophoresis buffer. After every run the capillaries were rinsed 5 minutes with 0.05 M NaOH and buffer. Sample injection was performed by applying 50 mbar pressure (0.5 psi) for 5 seconds to the sample vial placed at the grounded end of the capillary. The injected volume was estimated to be approximately 15 nl. The controlled temperature was 25.0 ± 0.1°C during all experiments. The pH of the electrophoresis buffer (see titles of electrophorerogramms) were adjusted with 1 M NaOH to the needed pH. Tested buffers for the determination of bLF with CE in chronological order.

1. 50 mM bistris pH 2.0;
2. 200 mM L-lysine pH 10.2 or pH 11.2
3. 50 mM 2-morpholinoethane suflonic acid monohydrate (MES) pH 8.2;
4. 200 mM carbonate pH 9.0;
5. 100 mM borate, 30 mM Na_2SO_4 or 60 mM Na_2SO_4, pH 8.3; 100 mM borate, 30 mM
6. 100 mM tris-HCl pH 7.4;
7. 100 mM phosphate pH 3.0 or pH 7.4;
8. 40 mM borate pH 9.0 or 10.6;
9. 40 mM borate/phosphate pH 9.0;
10. 40 mM borate/phosphate, 20 mM brij 35 pH 9.0;
11. 40 mM borate/phosphate, 50 mM SDS pH 9.0;
12. 40 mM phosphate, 50 mM SDS, 20 mM Brij 35, 30 mM Na_2SO_4 or 60 mM Na_2SO_4, 4 % 2-propanole pH 9.0.

3. RESULTS AND DISCUSSION

To prevent adsorption effects of milk proteins on fused silica capillary tubings several techniques have been reported. De Jong et al.[15] used a hydrophilically coated capillary yielding in a separation of serum proteins and caseins including some of the genetic casein variants. Paterson et al.[17] prevent the adsorption with hydronamic coating of the capillaries and separated β-lactoglobulin variants while Recio et al.[18] added sodium sulfate to the CE buffer to minimize adsorption effects on the capillary wall. To circumvent long analysis times by preparing coated capillaries we used basic buffers above the isoelectric point of Lf yielding in a repelling effect between the both negatively charged capillary walls and the proteins[19]. Figure 1 shows the reproducibility of the analysis of the pure commercial available bLF using an borate buffer with pH of 10.6. As it could be clearly seen the reproducibility from run to run is very good over 90 determinations.

The calculated standard deviations of the peak heights was 2.98 % and 6.58 % for the peak areas. The bLf peak is detected after 2.2 minutes with an applied voltage of 10 kV with the potential for shorter analysis times by using higher voltages up to 30 kV. However, by variation of the concentration the resulting calibration fit with bLf concentrations between 1 and 120 µg/ml achieved a sufficient regression coefficient of R=0.993 (see Figure 2 inset for calibration fit).

The limit of detection for the bLf Peak was obtained between 1 and 10 µg/ml as it is shown in Figure 2 and the calibration fit in the inset. If the concentration is as low as 10 µg/ml only a very small peak of the bLf could be detected in the resulting electropherogramm. The peak area and heights of 1 µg/ml is only slighty more then the noise of the UV-detector. The determination of bLf in whey samples under these conditions yielded only a very broad and non resolved peak with high intensity of the whey proteines at the same migration time then the bLF (data not shown). This behaviour is caused by comigration of allmost all whey proteines and the separation power of the used buffer is insuffi-

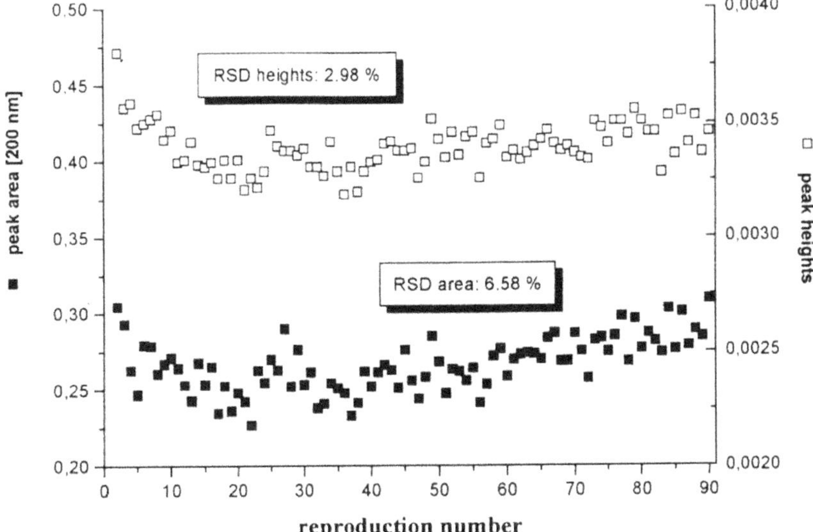

Figure 1. Reproducibility of 150 µg/ml b-Lf protein standard solution; CE buffer: 40 mM borate buffer pH 10.6, 10 kV voltage applied; capillary dimensions: 50 µm inner diameter 27 cm length, fused silica.

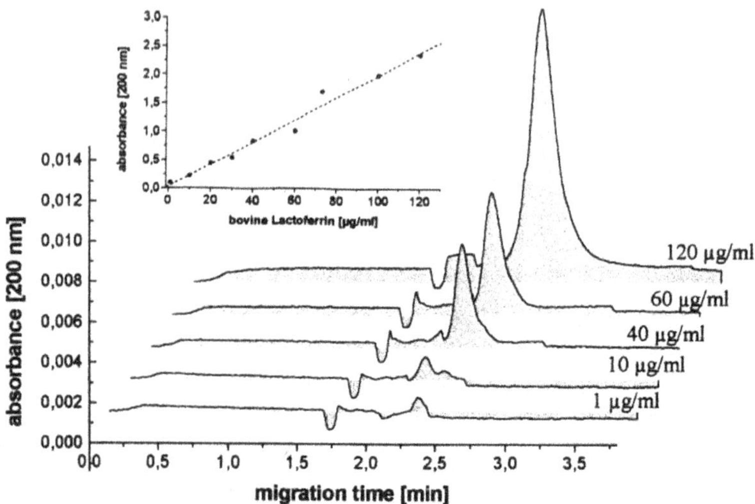

Figure 2. Calibration with b-Lf protein standards using borate buffer pH 10.6. Concentration bLf: 1–120 µg/ml; CE buffer: 40 mM borate buffer pH 10.6, 10 kV voltage applied; capillary dimensions: 50 µm × 27 cm fused silica. Inset: Calibration curve with b-Lf protein standards using the same conditions.

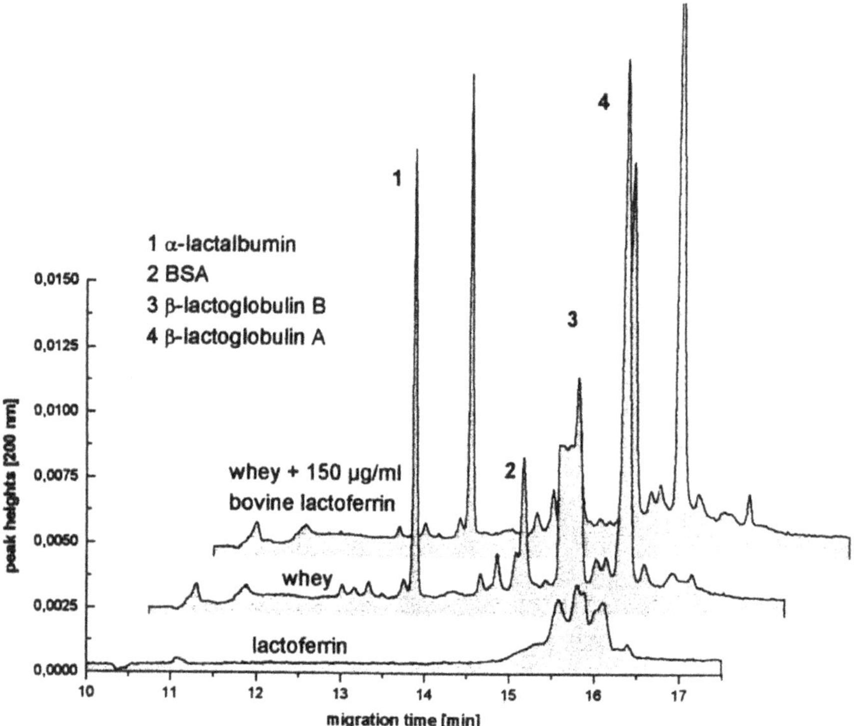

Figure 3. Comparison of bLf, whey and a mixture of bLf and whey. Concentration bLf: 150 µg/ml; CE buffer: 100 mM borate buffer pH 8.2 + 30 mM Na_2SO_4, 10 kV voltage applied; capillary dimensions: 50 µm × 57 cm fused silica.

cient for the determination of bLf in whey samples. By changing the separation conditions as published by Recio et al.[16] a differentiation of three peaks for a bLF standard is obtained as it is shown in Figure 3.

The three peaks may be due to the different electrophoretic migration behaviour of three lactoferrin glycoforms with a different carbohydrates content at the asparagin-linked residues. The determination of the lactoferrin glycoforms are now under examination and the following results will be reported separately. The separation of the whey compounds α-lactalbumin, BSA and β-lactoglobulin A and B is enabled using the conditions of [16] while the lactoferrin now comigrates with β-lactoglobulin A as indicated in the electropherogramms of Figure 3. Although spiking with bLf (150 μg/ml) yielded no significantly increasing of the peaks with the bLf migration time in the upper electropherogramm. The peak pattern between 15.2 and 16.5 minutes are nearly the same in the normal and the bLf spiked whey samples. To enable the separation we tested several different buffers both with acidic and basic pH. However, non of the 14 tested buffer systems (for details see experimental section) achieved a separation of the minor protein from the other whey compounds. After the addition of SDS as a negative micellar compound to the buffer no increased separation in micellar electrokinetic chromatography mode of capillary electrophoresis was detected (data not shown). Figure 4 shows the effect of using mixed micellar buffers for the determination of bLf in whey samples.

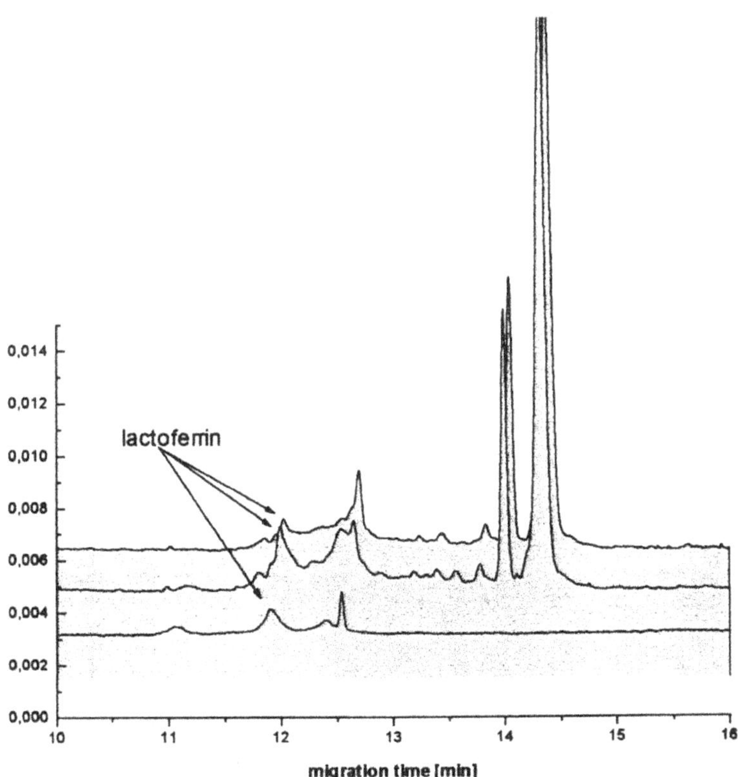

Figure 4. Comparison of bLf, whey and a mixture of bLf and whey. Concentration bLf: 150 μg/ml; CE buffer: 40 mM borate/phosphate pH 9.0 + 50 mM SDS + 20 mM Brij 35 + 4 % Isopropanol + 30 mM Na_2SO_4, 15 kV voltage applied; capillary dimensions: 50 μm × 57 cm fused silica.

The lower electropherogramm shows two peaks which are also present in the whey sample in the upper electropherogramm. In the center electropherogramm the whey was spiked with 150 µg/ml bLf. Due to the higher bLf concentration it could be clearly seen, that the bLf peak is increased in comparison to the non spiked whey of approximately 2fold. Despite of this the bLf is not correctly separated from the other whey compounds which is indicated by the additional peak sholder in the upper electropherogramm. All separations shown to this point were carried out with non-concentrated whey. However, nearly all whey samples in the cheese industry are concentrated 5 fold with a resulting total protein content of about 80 g/l as determined by the BCA-Protein test. Due to this high protein content the dilution factor used for measurements of concentrated whey samples was 50 to lower the UV absorbance of the major whey proteins. After the dilution the bLf concentration of 100–1000 µg/ml in the concentrated whey is decreased to the level of 2–20 µg/ml. Because of the detection limit of 10 µg/ml and the not well-resolved and comigrating whey compounds it is still not possible to determine bLf in whey using capillary zone electrophoresis. On the other hand Figure 5 shows that there are slight differences in the peak pattern of different sweet cheese wheys collected with our mixed micellar buffer systems using the MEKC mode of capillary electrophoresis.

This observations was confirmed by our FPLC-measurements and SDS-PAGE (data not shown). But to enhance the resolution a more suitable method must be developed using a preconcentrating step as it is possible with capillary isoelectric focussing or capillary isotachophoresis. Another promissing and much more sensitive method is the use of laser-induced fluorescence detection and FITC-conjugated antibodies against bLf to detect the formed immunocomplex of bLf and the FITC-conjugated antibody which are now under examination[20].

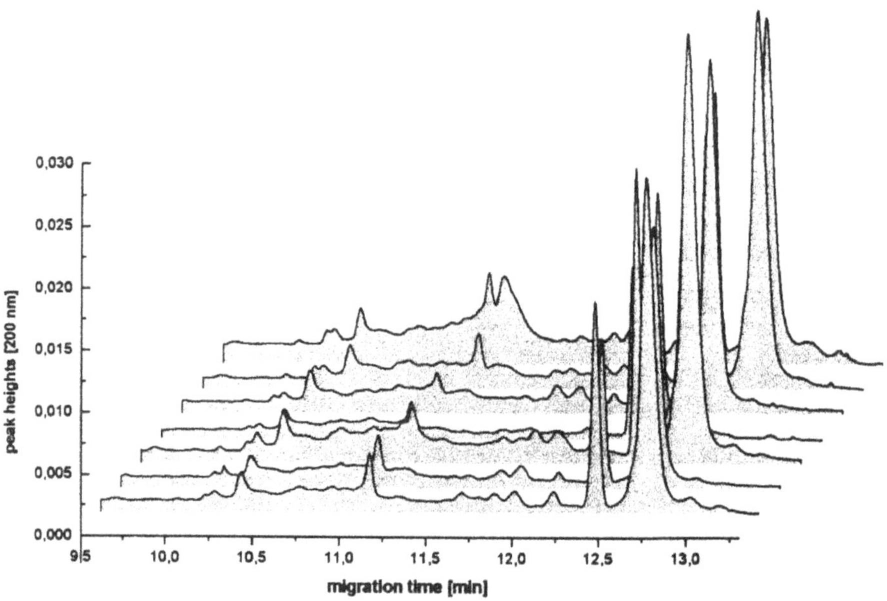

Figure 5. Comparison of whey concentrates from different cheese manufacturers. CE buffer: 40 mM borate/phosphate pH 9.0 + 50 mM SDS + 20 mM Brij 35 + 4% Isopropanol + 30 mM Na_2SO_4, 15 kV voltage applied; capillary dimensions: 50 µm × 57 cm fused silica.

REFERENCES

1. J. Carlsson, J. Porath, B. Lonnerdal: Isolation of lactoferrin from human milk by metal-chelate affinity chromatography, FEBS Letters 75 (1977) 89–92.
2. P.L. Masson, J.F. Heremans: Lactoferrin in milk from different species, Comp. Biochem. Physiol 39B (1971) 119–129.
3. B.A. Law, B. Reiter: The isolation and bacteriostatic paroperties of lactoferrin from bovine milk whey, J. Dairy Res. 44 (1977) 595–599.
4. F. Lampreave, A. Piñeiro, J.H. Brock, H. Castillo, L. Sánchez, M. Calvo: Interaction of bovine lactoferrin with other proteins of milk whey, Int. J. Biol. Macromol. 12 (1990) 2–5.
5. Y. Makino, S. Nishimura: High-performance liquid chromatography separation of human apolactoferrin and monoferric and diferric lactoferrins. J. Chromatogr. 579 (1992) 346–349.
6. B. Ekstrand, L. Björk: Fast protein liquid chromatography of antibacterical components in milk, J. Chromatogr. 358 (1986) 429–433.
7. K.-I. Shimazaki, N. Nishio: Interaction properties of bovine lactoferrin with immobilized Cibacron Blue F3GA in column chromatography, J. Dairy Sc. 74 (1991) 404–408.
8. Y.S. Kim, S.M. Cramer: Experimental studies in metal affinity displacement chromatography of proteins, J. Chromatogr. A 686 (1994) 193–203.
9. P. Aranda, L. Sánchez, M.D. Pérez, J.M. Ena, P. Puyol, R. Oria, M. Calvo: Rapid immunoenzymatic method for detecting adulteration in ewe's milk, Food Control 4 (1993) 101–104.
10. T. Hutchens, J.S. Magunson, T.T. Yip: Rapid purification of porcine colostral whey lactoferrin by affinity chromatography on single-stranded DNA-agarose. Characterization, amino acid composition and N-terminal amino acid sequence, Biochim. Biophys. Acta 999 (1989) 323–329.
11. M.A. Jenkins, M.D. Guerin: Capillary electrophoresis as a clinical tool, J Chromatogr. B 682 (1996) 23–34.
12. F.T.A. Chen, J.C. Sternberg: Characterization of proteins by capillary electrophoresis in fused-silica columns: Review on serum protein analysis and application to immunoassays, Electrophoresis 15 (1994) 13–21.
13. N.H.H. Heegaard & J. Brimnes: Comparison of heparin-binding to lactoferrin from human milk and from human granulocytes by means of affinity capillary electrophoresis, Electrophoresis 17 (1996) 1916–1920.
14. H. Engelhardt, W. Beck, T. Schmitt: Capillary electrophoresis, methods and potentials, Vieweg Verlag, Braunschweig, 1996.
15. M.A. Strege, A.L. Lagu: Micellar electrokinetic chromatography of proteins, Anal. Biochem. 210 (1993) 402.410.
16. N. de Jong, S. Visser, C. Olieman: Determination of milk proteins by capillary electrophoresis, J. Chromatogr. A 652 (1993) 207–213.
17. G.R. Paterson, J.P. Hill, D.E. Otter: Separation of beta-lactoglobulin A, B and C variants of bovine whey using capillary electrophoresis, J. Chromatogr. A 700 (1995) 105–110.
18. I. Recio, E. Molina, M. Ramos, M. de Frutos: Quantitative analysis of major whey proteins by capillary electrophoresis using uncoated capillaries, Electrophoresis 16 (1995) 654–658.
19. H.H. Lauer & D. McManigall: Capillary zone electrophoresis of proteins in untreated fused silica tubing, Anal. Chem. 58 (1986) 166–170.
20. P. Riechel, T. Weiß, R. Ulber, H. Buchholz, T. Scheper: Affinity Interactions for the Detemination of Bovine Lactoferrin in Whey, submitted to Electrophoresis (1997).

5

STRUCTURAL AND IMMUNOCHEMICAL STUDIES ON BOVINE LACTOFERRIN FRAGMENTS

Kei-ichi Shimazaki,[1] Makoto Kamio,[1] Myong Soo Nam ,[2] Shinji Harakawa,[3] Tetsuya Tanaka,[3] Yoshitaka Omata,[3] Atsushi Saito,[3] Haruto Kumura,[1] Katsuhiko Mikawa,[1] Ikuo Igarashi,[4] and Naoyoshi Suzuki[4]

[1]Dairy Science Laboratory
Faculty of Agriculture
Hokkaido University
Sapporo, 060 Japan
[2]Korea Research Institute of Bioscience and Biotechnology
KIST, Taejon 305-600, Korea
[3]Department of Veterinary Physiology
Obihiro University of Agriculture and Veterinary Medicine
Obihiro, 080 Japan
[4]The Research Center for Protozoan Molecular Immunology
Obihiro University of Agriculture and Veterinary Medicine
Obihiro, 080 Japan

1. INTRODUCTION

Lactoferrin (Lf) is a metal-binding protein found in milk and other secretory fluids and also in blood. It shows multifunctional properties but the mechanism of developing its function in living systems has not been resolved yet. It is known to exert bacteriostatic effects due to its ability to bind environmental iron. Moreover, apo-lactoferrin has been shown to bind to microbial membranes and causes the direct destruction of microorganisms. Other biological functions attributed to lactoferrin include roles in modulation of the inflammatory response, activation of the immune system, and control of myelopoiesis or cell growth. This molecule is constructed with N- and C-lobes, each of which is composed of 3 domains[1]. The function of each lobe has been studying and there are certain differences. The biologically significant function has been found mainly in N-lobe. For the aids of resolving their functional analysis, the authors prepared the monoclonal antibodies (mAb) against N-lobe and C-lobe of bovine lactoferrin. To prepare the mAb specific to N-lobe, we used lactoferricin®

Advances in Lactoferrin Research, edited by Spik *et al.*
Plenum Press, New York, 1998.

B (bLfcin , an anti-microbial peptide isolated from N-lobe[2]) as an antigen[3]. In this paper, the characterization of the mAb against lactoferrin fragments has been examined and the structure of the mAb-recognition site on lactoferrin molecule was identified.

2. EXPERIMENTAL

2.1. Materials

Bovine lactoferrin (bLf) and bLfcin were kindly supplied by the Nutritional Science Laboratory, Morinaga Milk Industries Inc. C-lobe was prepared by mild tryptic digestion and isolated by ion-exchange chromatography as reported previously[4] or by reverse phase (RP) chromatography.

2.2. Preparation of Monoclonal Antibodies

Monoclonal antibodies against bLfcin or C-lobe were prepared according to the method of Oi et al.[5] bLfcin or C-lobe was injected intravenously, the spleen was aseptically removed and processed for screening positive hybridomas. bLfcin -keyhole limpet hemocyanin complex was used for immunization.

2.3. Reverse Phase HPLC

RP-HPLC was carried out using ODS column. For elution, a mixture of the eluents A (0.1% TFA in water) and B (0.1% TFA in acetonitrile) was employed, using a linear or a convex gradient of A and B.

2.4. Protein Concentration Determination

Concentrations of bovine lactoferrin, C-lobe and bLfcin were determined by UV absorption at 280 nm using the extinction coefficient (1 mg/ml) of 1.27[6], 1.23 and 3.02, respectively.

2.5. Chemical Modification and Cleavage Reactions

Disulfide bonds of peptides were reduced and then pyridylethylated (Pe) with 4-vinylpyridine or acetylated with monoiodoacetamide. Lys, Arg and Trp residues were modified using succinic anhydride, 1,2-cyclohexanedione and N-bromosuccinimide, respectively. CNBr cleavage of the peptide was carried out in 70% formic acid solution. The peptide bond cleavage between aspartic acid and proline was done with the acetic acid treatment in 10% acetic acid containing 7 M guanidium hydrochloride. The chemically treated peptide were separated from unreacted peptide by RP-HPLC.

2.6. Enzymatic Digestion

Completely denatured Pe-C-lobe by 8 M urea was digested by trypsin at 37°C for 9 hour in the presence of 2 M urea. Carbohydrate moiety of peptide was removed by endoglycosidase H. The removal of sugar chain was detected by the staining method using periodic acid-schiff reagent after SDS-PAGE.

2.7. Amino Acid Sequence Analysis and Peptide Synthesis by SPOTs™

N-terminal amino acid sequences were determined using an Applied Biosystems Model 492A protein sequencer or Tosoh sequencing system. For the epitope determination, peptides were synthesized from F-moc amino acid active esters on a pre-activated cellulose membrane using SPOTs™ (Genosys Biotechnologies, Inc.).

2.8. Mass Spectrometry

Matrix-Assisted Laser Desorption Time-of-Flight Mass Spectrometry (MALDI-TOF MS Voyager™ RP, PerSeptive Biosystems/Vectec Products) was used to measure the molecular mass (m/z) of peptides. Sinapinic acid was used as the matrix and angiotensin I was used as the molecular mass standard.

2.9. Antibacterial Activity Measurements

E. coli O111 was used to measure the antibacterial activities of lactoferrin fragment with the modified method of Tomita et al.[7] After incubation, turbidity was measured.

3. RESULTS

3.1. Monoclonal Antibody Binding Site of bLfcin

After screening and cloning, 4 colonies were chosen and the mAb (IgG1) produced by 5F12.1.2 cells was used mainly for further experiments. All 4 mAb's showed reactivity against both native and chemically synthesized bLfcin, of which the sulfhydryl groups are acetamidomethylated, by ELISA. None of the 4 antibodies showed reactivity against human lactoferrin or hLfcin.

The reactivity of the mAb against bLfcin derivatives of chemically modified Lys, Trp or Arg residues, as compared with intact bLfcin, was 96.5, 30.1 and 27.9%, respectively. The ratio of Trp residue modified was estimated to be 70% from the decrease in absorbance at 280 nm. This observation suggests that the Trp and Arg residues of bLfcin are mainly involved in the epitopic region recognized by the mAb. By ELISA, the mAb did not show any reactivity against CNBr-cleaved fragments. Fifty kinds of peptides corresponding to the region around the Trp and Met residues of bLfcin were synthesized and the reactivity of anti-bLfcin mAb against each of these peptides was estimated. The common sequence found in each of the peptides recognized by the mAb is "QWR" as shown in Figure 1.

3.2. Monoclonal Antibody Binding Site of C-lobe

The colony displayed higher absorbance in the ELISA against C-lobe was chosen and this clone showed high specificity to anti-mouse IgG1 subclass antibody. The mAb showed reactivity against both C-lobe and intact lactoferrin by ELISA or Western-blotting. Human lactoferrin and transferrin, bovine transferrin and ovotransferrin did not react with this mAb. The partial N-terminal amino acid sequence of C-lobe used in this experiment (Figure 2A) was YTRVVWXAVX and this fragment begins from 342Tyr of bovine lactoferrin[8]. To determine the mAb-binding site of C-lobe, denatured C-lobe was digested into the smaller fragment by trypsin. Each fragment was fractionated by RP-HPLC and as-

No.	Peptide	Peptide	No.
	KCRRWQWRMKKLGAPSIT		
1	**WQWR**		
2	QWRM		
3	WRMK		
4	RMKK		
5	MKKL		
6	KKLG		
9	**RWQWR**	KLGA	7
10	**WQWRM**	LGAP	8
11	**QWRMK**		
12	WRMKK		
13	RMKKL		
14	MKKLG		
15	KKLGA		
18	**RRWQWR**	KLGAP	16
19	RWQWRM	LGAPS	17
20	WQWRMK		
21	QWRMKK		
22	WRMKKL		
23	RMKKLG		
24	MKKLGA		
25	KKLGAP		
28	**CRRWQWR**	KLGAPS	26
29	RRWQWRM	LGAPSI	27
30	RWQWRMK		
31	**WQWRMKK**		
32	**QWRMKKL**		
33	WRMKKLG		
34	RMKKLGA		
35	MKKLGAP		
36	KKLGAPS		
37	KLGAPSI		
39	**KCRRWQWR**	LGAPSIT	38
40	CRRWQWRM		
41	**RRWQWRMK**		
42	**RWQWRMKK**		
43	WQWRMKKL		
44	QWRMKKLG		
45	WRMKKLGA		
46	RMKKLGAP		
47	MKKLGAPS		
48	KKLGAPSI		
50	KLGAPSIT		

Figure 1. SPOTs™ analysis of bLfcin peptides. Peptides reacted with mAb are expressed in bold.

sayed with the reactivity against anti-C-lobe mAb by ELISA. The fraction eluted at 24 min. in Figure 2B showed reactivity against anti-C-lobe mAb and named CLT. The partial N-terminal amino acid sequences of both CLT and acetic acid-treated CLT were determined to be TAGWNIPMGLI. This fragment remained the reactivity against mAb after deglycosylation. The molecular mass of deglycosylated CLT was 4069.4 as determined by MALDI-TOF mass spectral analysis (Figure 3). No mAb-binding ability was changed on the chemically modified CLT with its Lys or Arg residues, although, the further digestion of CLT by pepsin, α-chymotrypsin or endoproteinase Glu-C made CLT very weak or not reactive against mAb. None of the peptides separated by RP-HPLC of CNBr-treated CLT showed any reactivities against anti-C-lobe mAb. Acetic acid-treated CLT lost its reactivity. Then, 38 kinds of peptides corresponding to the region from 464Thr to 508Asp of bovine lactoferrin were synthesized on the membrane and the reactivity of anti-C-lobe mAb was assayed (Figure 4). The common sequence found in the peptides recognized by the mAb is "WNIPMGL".

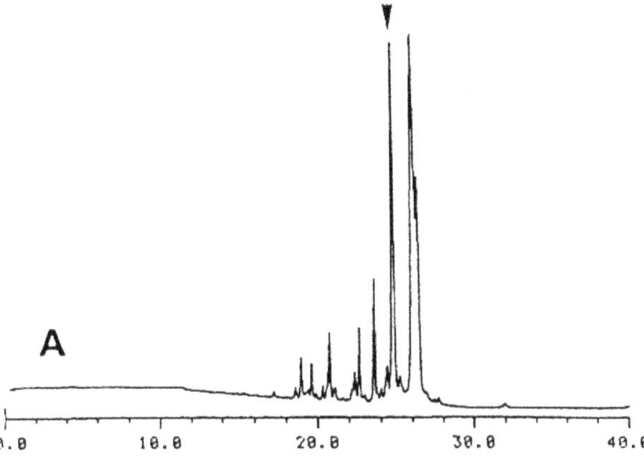

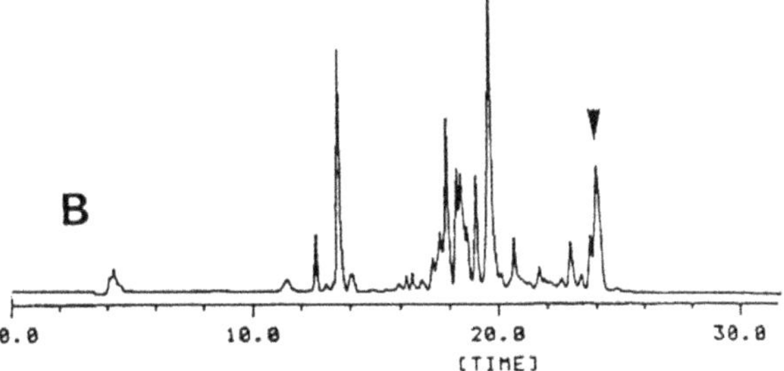

Figure 2. Separation of C-lobe (A) and CLT (B) by RP-HPLC using Capcell Pak C18 SG300 column (4.6 mm ID × 25 cm) at 40°C with a flow rate of 1 ml/min. C-Lobe and CLT are shown by arrows.

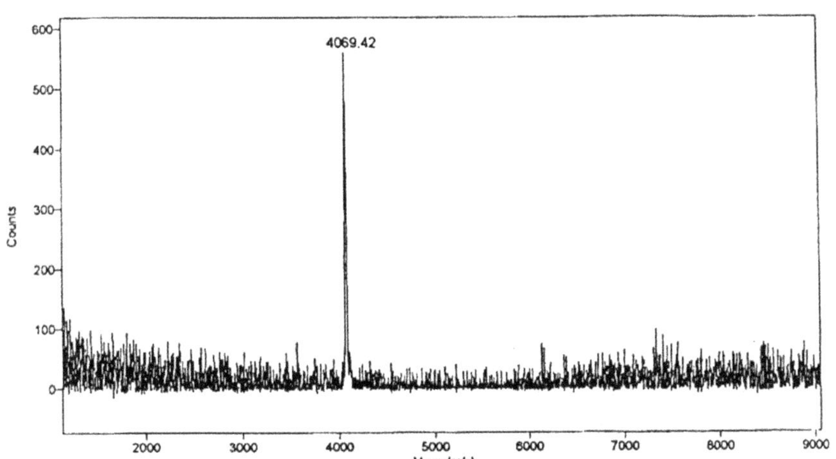

Figure 3. MALDI-TOF mass spectrum of pyridylethylated and deglycosylated CLT.

1 TAGWNIPM	9 GLIVNQTG	17 SCAFDEFF	25 SQSCAPGR	33 DPKSRLCA
2 AGWNIPMG	10 LIVNQTGS	18 CAFDEFFS	26 QSCAPGRD	34 PKSRLCAL
3 **GWNIPMGL**	11 IVNQTGSC	19 AFDEFFSQ	27 SCAPGRDP	35 KSRLCALC
4 **WNIPMGLI**	12 VNQTGSCA	20 FDEFFSQS	28 CAPGRDPK	36 SRLCALCA
5 NIPMGLIV	13 NQTGSCAF	21 DEFFSQSC	29 APGRDPKS	37 RLCALCAG
6 IPMGLIVN	14 QTGSCAFD	22 EFFSQSCA	30 PGRDPKSR	38 LCALCAGD
7 PMGLIVNQ	15 TGSCAFDE	23 FFSQSCAP	31 GRDPKSRL	
8 MGLIVNQT	16 GSCAFDEF	24 FSQSCAPG	32 RDPKSRLC	

Figure 4. Synthesized peptides for epitope determination by SPOTs™ analysis. The bold type peptides are reactive peptides.

3.3. Antimicrobial Activity of C-lobe

Antimicrobial activity of C-lobe was compared with apo- and holo-lactoferrin. C-lobe showed about 80 % turbidity of control and intact lactoferrin of apo- and holo-types showed 20–30% of control at the concentration of 5 mg/ml.

4. DISCUSSION

To investigate the relation between the biological function and structure of lactoferrin, mAb would be a useful tool. In this study, we prepared anti- bLfcin mAb and anti-C-lobe mAb. From the experiments described above, we concluded that the sequence "QWR" is the binding site with anti- bLfcin mAb. CNBr-cleaved fragments showed no reactivity against this mAb and this should be explained as follows: As the mAb binding site locates neighboring to the N-terminal side of Met, the peptidyl homoserine lactone formed as a result of CNBr treatment[9] hindered the antibody binding. This "QWR" sequence could not been found in human lactoferrin, human transferrin, melanotransferrin or ovotransferrin, all of which are a members of the transferrin family of proteins.

bLfcin has been shown to have an affinity for certain substances concerning living cells, membranes and others. It binds directly to lipopolysaccharide to disrupt the outer membrane of Gram-negative bacteria[10]. It is reported that the sequence "RRWQWR" is the subregion essential for antimicrobial activity of bLfcin[11]. The mAb binding site determined in this experiment is included in this subregion. Moreover, we have determined the heparin-binding site of bovine lactoferrin (in preparation). A fraction from pepsin hydrolysate of lactoferrin that binds to immobilized heparin column was separated. By sequence analysis and MALDI-TOF mass spectrometry, this heparin-binding peptide has the sequence of FKCRRWQWRMKKLGAPSITCVRRAFA, and this corresponds to the peptide of 17Phe to 42Ala, the same as bLfcin. We suggest that the heparin-binding site of bovine lactoferrin is at KCRR(18–21), RMKK(25–28) and RR(38–39) as shown in Figure 5. Heparin-binding site of human lactoferrin is reported to be at 1Gly-2–5Arg[12] or 5Arg, 25Arg-XX-28Arg-29Lys and 31Arg (BXXBBXB[13]). Here, B means basic residues. The consensus sequences of XBBXBX or XBBBXXBX has been reported for heparin-binding sites[14]. In bovine lactoferrin, the sequence of BXBB is found only at 2 locations of N-lobe and one in C-lobe.

bLfcin displayed no α-helix but rich in β-structure (ca. 50%) in the aqueous solution and circular dichroic spectra of bLfcin changed reversibly when heparin was mixed with bLfcin. In the presence of 50 % trifluoroethanol, a helix forming solvent, the peptide displayed only 6.5% of α-helical conformation and disulfide bond did show little restriction to prevent the formation of α-helix.

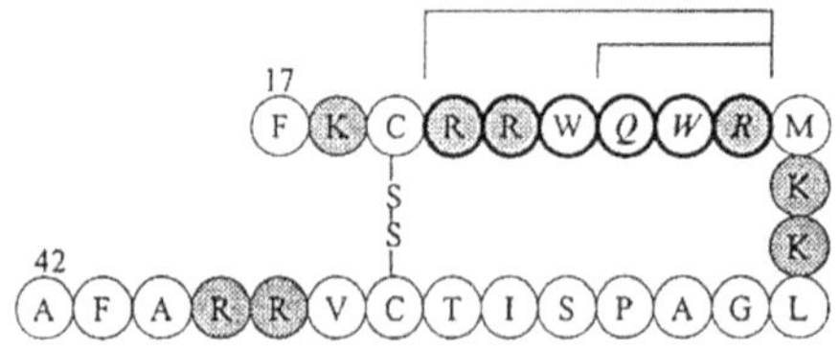

Figure 5. Binding sites of anti-bLfcin mAb (thick circle and *italic*) and heparin (thick circle), and antimicrobial activity subregion (gray dotted circle) of bLfcin.

As CLT fragment isolated by Figure 2B reminded the reactivity against mAb after removal of sugar moiety, it is excluded the possibility that carbohydrate chain is recognized by mAb. Partial N-terminal amino acid sequence of CLT was determined to be TAGWNIPMGLI and the molecular mass of deglycosylated CLT was 4069.4 as determined by MALDI-TOF mass spectrometry. Due to the enzymatic specificity of the endoglycosidase H, one residual N-acetylglucosamine (GluNAc, MW 221.2) should be remained at 476Asn[15]. Therefore, it is concluded CLT is composed of the residues from 464Thr to 498Lys of lactoferrin, i.e., TAGWNIPMGLIVNQTGSDAFDEFFSQSCAPGRDPK which attach one residual GluNAc and 2 sulfhydryl groups of Cys pyridylethylated (Figure 6). This 464Thr should be adjacent to the 463Arg which is interacting with the carbonate ion for iron ion binding[16].

By SPOTs analysis, the common sequence found in these peptides recognized by the mAb is "WNIPMGL" (467 to 473 of bovine lactoferrin) as in Figure 4. This sequence contains no Lys nor Arg. Therefore, we concluded that this sequence is the antigenic determinant or epitopic site of the C-lobe. This "WNIPMGL" sequence is found in transferrin of human, horse, rabbit, rat and pig and in lactoferrin of pig, mouse, goat (Saanen) and Korean native goat[17]. Melanotransferrin or ovotransferrin have sequences that one amino acid residue is replaced. All of the proteins described above are members of the transferrin family proteins. By ELISA, the mAb against bovine C-lobe did not bind to some of these proteins, that are commercially available. This may mean that the location of the mAb binding site of these proteins are buried from the molecular surface or not accessible by antibody molecule. There observed some disagreement among the results of peptide bond cleavage or enzymatic treatment. These may be explained by the steric hindrance occurring among the bulky side chain groups apart to disturb the binding of mAb and epitopic site. The structural confirmation and biological significance of these epitopic site has not been resolved yet.

Additionally, antimicrobial activity of C-lobe was measured to see one of the biological functions of C-lobe, and it showed lower activity than those of intact apo- and holo-lactoferrin. As C-lobe purified by RP-HPLC was used for the assay, it remains the possibility that C-lobe was not in the native conformation even after the removal of acetonitrile followed by dialysis against phosphate buffer for 2 days at refrigerator. This is confirmed by the measurement of CD spectra at the range of 250 to 350 nm.

5. CONCLUSION

Monoclonal antibodies (mAb) against bovine lactoferrin C-lobe and N-lobe were prepared. To prepare the mAb specific to N-lobe, lactoferricin® (bLfcin) coupled with KLH was used as an antigen. The anti- bLfcin mAb showed reactivity against both natural

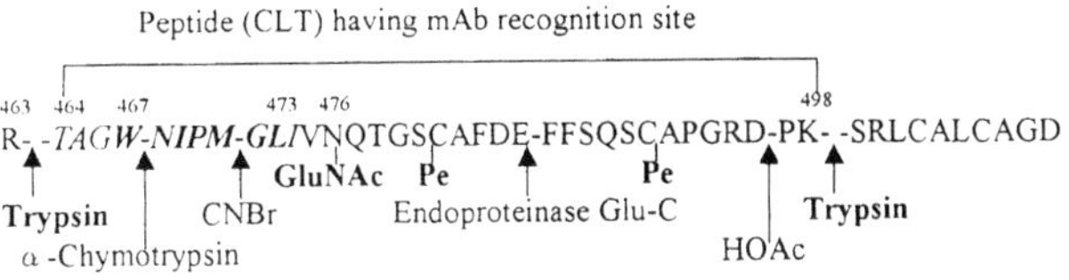

Figure 6. Amino acid sequence of CLT and mAb recognizing site (bold). The italic residues were determined by sequence analysis.

and chemically synthesized bLfcin, but not reacted with human lactoferrin nor hLfcin. By analyses with the synthetic peptides by SPOTs™ and the reactivity of chemically modified bLfcin, the mAb recognition site was identified to be the sequence of "QWR" of this peptide. C-lobe was cleaved into the smaller fragment by trypsin digestion and a small peptide that showed the binding activity against anti-C-lobe mAb was obtained. It is composed of the residues from 464 to 498 of lactoferrin judged from N-terminal amino acid sequence and molecular mass determined by MALDI-TOF mass spectrometry. By SPOTs™ analysis, the binding site of C-lobe was determined to be in the sequence of "WNIPMGL".

ACKNOWLEDGMENTS

We are very grateful to Dr. Y.C. Yoo, Mr.Y. Maki, Mr.Y. Hoshi, Miss T. Tazume and Mr. Y. Okouchi for their useful discussions and technical support.

REFERENCES

1. Baker, E.N., et al. Inter. J. Biol. Macromol. 13, 122–129 (1991).
2. Bellamy, W., et al. Biochim. Biophys. Acta 1121, 130–136 (1992).
3. Shimazaki, K., et al. J. Vet. Med. Sci. 58, 1227–1229 (1996).
4. Shimazaki, K., et al. J. Dairy Sci. 76, 946–955 (1993).
5. Oi, V.T. & Herzenberg, L.A. in Selected Methods in Cellular Immunology (eds. Mishell, B.B. & Shiigi, S.M.) 351–372 (W.H.Freeman and Co., New York, 1980).
6. Aisen, P. & Leibman, A. Biochim. Biophys. Acta 257, 314–323 (1972).
7. Tomita, M., et al. J. Dairy Sci. 74, 4137–4142 (1991).
8. Pierce, A., et al. Eur. J. Biochem. 196, 177–184 (1991).
9. Gross, E. Methods in Enzymology 11, 238–255 (1967).
10. Yamauchi, K., et al. Infect. Immun. 61, 719–728 (1993).
11. Tomita, M., et al. Acta Paediat. Japon. 36, 585–591 (1994).
12. Mann, D.M., et al. J. Biol. Chem. 269, 23661–23667 (1994).
13. Wu, H.F., et al. Arch. Biochem. Biophys. 317, 85–92 (1995).
14. Cardin, A.D. & Weinstraub, H.J.R. Arteriosclerosis 9, 21–32 (1989).
15. Spik, G., et al. Primary and three-dimensional structure of lactotransferrin (lactoferrin) glycans, 21–32 (Plenum Press, Hawaii, 1994).
16. Baker, E.N., et al. ibid, 1–12.
17. Yu, D.Y., et al. Animal Genetics, in press (1997).

6

ROLE OF THE FIRST N-TERMINAL BASIC CLUSTER OF HUMAN LACTOFERRIN ($R^2R^3R^4R^5$) IN THE INTERACTIONS WITH THE JURKAT HUMAN LYMPHOBLASTIC T-CELLS

D. Legrand,[1] P. H. C. van Berkel,[2] V. Salmon,[1] H. A. van Veen,[2]
M. C. Slomianny,[1] J. H. Nuijens,[2] and G. Spik[1]

[1]Laboratoire de Chimie Biologique and U.M.R. du CNRS n°111
Université des Sciences et Technologies de Lille
59655 Villeneuve d'Ascq, Cedex, France
[2]Pharming BV
Niels Bohrweg 11-13
2333 CA, Leiden, The Netherlands

1. SUMMARY

We previously characterized a receptor of Mr 105,000 for human lactoferrin (hLf) on Jurkat human lymphoblastic T-cells. To delineate the role of $R^2R^3R^4R^5$ of hLf in the interaction with cells, we studied the binding of hLf variants obtained either by tryptic proteolysis (hLf^{-2N}, hLf^{-3N} and hLf^{-4N}) or by mutagenesis ($rhLf^{-5N}$). Consecutive removal of N-terminal arginine residues from hLf progressively increased the binding affinity but decreased the number of binding sites on the cells. The binding parameters of bovine Lf and native hLf did not differ, whereas the binding parameters of murine Lf resembled those of $rhLf^{-5N}$. Culture of Jurkat cells in the presence of chlorate, which inhibits sulfation, reduced the number of binding sites for both native hLf and hLf^{-3N} but not for $rhLf^{-5N}$ indicating that the hLf binding sites include sulfated molecules. The results suggest that the interaction of hLf with about 80,000 binding sites per Jurkat cell, mainly sulfated molecules, is dependent on $R^2R^3R^4$, but not on R^5. Interaction with about 20,000 binding sites per cell, presumably the hLf receptor, does not require the first N-terminal basic cluster of hLf. We conclude that the deletion of R^2-R^5 from hLf may serve to modulate the nature of its binding to cells and thereby its effects on cellular physiology.

Advances in Lactoferrin Research, edited by Spik *et al.*
Plenum Press, New York, 1998.

2. INTRODUCTION

Some biological activities of human lactoferrin (hLf) are linked to its ability to strongly chelate iron, whereas others relate to the interactions of hLf with host cells[1–7] or its binding to bacterial lipopolysaccharides[8–9], proteoglycans[10,11], DNA[12] and human lysozyme[13,15]. The highly positively charged N-terminus of hLf may be involved in these interactions[4,8–12,14,15].

The N-terminal portion of hLf contains a unique cluster of four consecutive arginine residues ($R^2R^3R^4R^5$) as well as a second basic cluster ($R^{28}K^{29}V^{30}R^{31}$), that is also present in Lf from other species[16–19]. Mann *et al.*[10] suggested that the heparin-binding site in hLf represents a 'cationic cradle' formed by juxtaposition of the first and second basic clusters. We have previously reported a specific hLf receptor of 105 kDa on activated lymphocytes[6] and the Jurkat T-cell line that binds hLf via the loops involving residues 28–34 and 39–42[20]. The binding of hLf to lymphocytes promotes their differentiation[21].

The interaction of hLf with cells may involve multiple classes of binding sites[22]. It was postulated that the rapid hepatic clearance of hLf from the rat circulation involves at least two classes of hLf binding sites, proteoglycans and the chylomicron remnant receptor and/or the LDL-receptor-related protein (LRP)[4,22,23]. The RK-rich sequence ($R^{25}N^{26}M^{27}R^{28}K^{29}V^{30}R^{31}$) in hLf, which resembles the receptor recognition structure of apolipoprotein-E2, presumably mediates binding and internalization into the hepatocytes by the chylomicron remnant receptor and/or LRP, whereas the $R^2R^3R^4R^5$ stretch may play an important role in the massive low-affinity interaction of hLf with the large number of cell-associated chondroitin sulfate-type proteoglycans[4,22,23]. The contribution of each R residue of the first basic cluster to binding of hLf to hepatocytes or other cells has not been elucidated yet.

In the present paper, cell binding experiments were performed with N-terminally deleted hLf species lacking two to five N-terminal residues by tryptic proteolysis or recombinant DNA technology, in presence or in absence of sodium chlorate which inhibits sulfation[24]. The results allow us to discern the role of the first basic cluster in binding to cell-associated proteoglycans and the hLf lymphocyte receptor.

3. MATERIAL AND METHODS

3.1. Lactoferrins

Native hLf and mLf were purified from milk as previously described[25,26]. Bovine Lf was kindly provided by Biopole (Brussels, Belgium). Lactoferrins variants lacking the two (hLf^{-2N}), three (hLf^{-3N}) and four (hLf^{-4N}) N-terminal residues of hLf were produced by limited tryptic hydrolysis of the protein[26]. A recombinant hLf lacking the five N-terminal residues (hLf^{-5N}) was obtained by mutagenesis of the hLf cDNA and expression in Sf-9 insect cells infected by Baculovirus[26]. Non-modified recombinant hLf (rhLf) was prepared as described in [27]. SDS-PAGE of the purified protein preparations showed no other protein bands than those characteristic for each of the lactoferrins. For the cell binding experiments, proteins were labeled with ^{125}I using Iodo-beads as a catalyst[26].

3.2. Cell Culture

Jurkat cells were routinely grown as previously described[28]. Cells were kept in the logarithmic growth phase and diluted to a cell density of 4×10^5/ml in the absence or presence of 30 mM sodium chlorate. After 24 h, cell viability was checked using the Trypan Blue stain. Cells were then washed twice in ice-cold serum-free RPMI 1640 and harvested by centrifugation at 200 g for 10 min at 4°C.

3.3. Cell Binding Experiments

Equilibrium binding experiments were performed in serum-free RPMI 1640 containing 0.4% (w/v) human serum transferrin to prevent aspecific binding. Aliquots (100 μl) containing 5×10^5 cells were incubated with serial dilutions of ^{125}I-labeled protein (concentrations ranging from 0 to 80 nM). Incubation of cells with proteins was performed for 1 h at 4°C in the presence of 0.01% (w/v) sodium azide to prevent ligand internalization. Cells were washed three times with 1 ml RPMI, resuspended in 0.5 ml PBS and bound radioactivity was measured. Non-specific binding measured in the presence of a 100-fold molar excess of unlabeled hLf was typically around 25% of the total binding and was substracted from total binding to obtain the specific binding. All binding experiments were performed in duplicate on two or three separate occasions. Binding parameters (Kd and number of binding sites per cell; mean ± S.E.M.) were calculated by Scatchard-plot analysis[29].

4. RESULTS

4.1. Binding of Native Lf and N-Terminally Deleted hLf Species to Jurkat Cells

To delineate the role of $R^2R^3R^4R^5$ of hLf in the binding to Jurkat human lymphoblastic T-cells, we studied the binding of ^{125}I-labeled native hLf and N-terminally deleted hLf species at concentrations ranging from 0 to 80 nM. Results showed that the binding of all hLf species was concentration-dependent and saturable with a single class of binding sites. As shown in Fig. 1, Scatchard analysis revealed that, in the range of hLf concentrations used, the affinity of N-terminally deleted hLf was significantly increased when compared to N-terminal intact hLf. The dissociation constant (Kd) shifted from 69 or 81 nM for hLf or rhLf to 65, 57 and 41 nM for hLf^{-2N}, hLf^{-3N} and hLf^{-4N}, respectively. The highest affinity (Kd of 12 nM) was observed with $rhLf^{-5N}$. In addition, we found that the number of binding sites per cell decreased from 102,000 for intact hLf to 17,000 for both hLf^{-4N} and $rhLf^{-5N}$. Human Lf^{-2N} and hLf^{-3N} bound to about 75,000 and 36,000 binding sites, respectively. These results suggest that the binding of hLf to approximatively 80,000 binding sites on Jurkat cells depends on the presence of $G^1R^2R^3R^4$.

To assess the species specificity of Lf-Jurkat cell interactions, we compared the binding parameters of hLf with that of bLf and mLf. Figure 1 shows that the binding parameters of bLf did not differ from those of hLf; the Kd and number of binding sites were around 60 nM and 100,000 sites per cell for both Lf species. On the other hand, mLf bound to about 8,000 binding sites per cell, with a Kd of 30 nM, which is close to the values obtained with hLf^{-4N} and $rhLf^{-5N}$.

4.2. Effect of Sodium Chlorate Treatment on the Binding of hLf Species to Jurkat Cells

To evaluate to which extent sulfated GAGs, such as heparan sulfate, dermatan- or chondroitin sulfate, determine the binding of hLf to Jurkat cells, we pretreated the cells

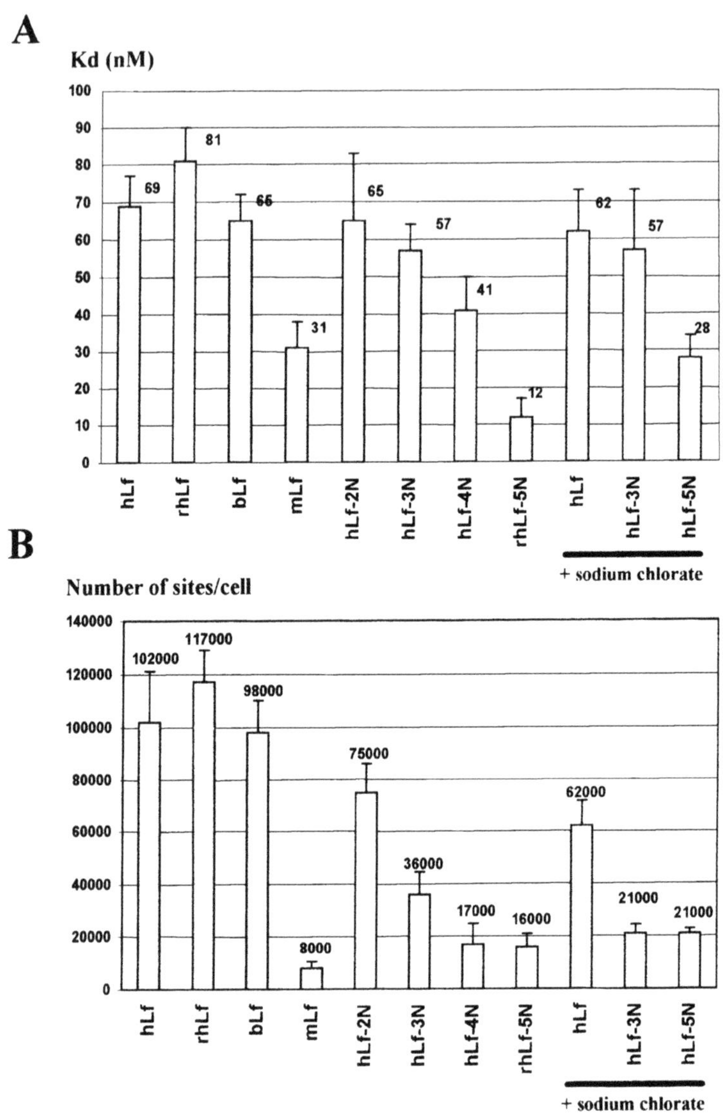

Figure 1. Binding parameters of hLf, hLf variants, bLf and mLf to Jurkat cells treated or not treated with sodium chorate. (A) dissociation constants and (B) number of binding sites per cell. Values are means of two to three separate experiments in duplicate.

with sodium chlorate. Chlorate has been shown to inhibit sulfation of carbohydrate residues on intact cells without interfering with cell growth or protein synthesis[24]. Culture of Jurkat cells in the presence of sodium chlorate did not alter their growth rate nor their morphology. Figure 1 shows that chlorate treatment of Jurkat cells decreased the amount of binding sites for native hLf from 102,000 to 62,000 per cell, whereas the Kd was slightly reduced from 71 to 62 nM. Depletion of cell-associated sulfate groups resulted in about 21,000 binding sites for hLf^{-3N} and $rhLf^{-5N}$ with Kd values of 57 and 28 nM, respectively. This number of binding sites is very close to the number found in untreated cells for both hLf^{-4N} and $rhLf^{-5N}$ (around 17,000 sites/cell). The chlorate treatment reduced the number of binding sites for hLf^{-3N} from 36,000 to 21,000, but did not affect the number of sites for $rhLf^{-5N}$. This suggests that hLf^{-3N}, but not $rhLf^{-5N}$, is still able to interact with sulfated groups exposed on the cell surface of untreated Jurkat cells.

5. DISCUSSION

In the present work, we studied the binding of N-terminally deleted hLf variants lacking either G^1R^2, $G^1R^2R^3$, $G^1R^2R^3R^4$ or $G^1R^2R^3R^4R^5$ to Jurkat cells[26]. The binding parameters of native hLf to Jurkat cells were close to those previously described[28], whereas subsequent removal of the N-terminal arginine residues progressively decreased the number of the binding sites per cell, while slightly increasing the binding affinity. These results indicate that R^2, R^3 and to a lower extent, R^4 of hLf contribute to the binding of hLf to about 80% of the total number of binding sites. Removal of the $G^1R^2R^3R^4R^5$ portion from hLf about six-fold increased the affinity of hLf for a residual number of 20,000 binding sites presumably representing the Lf specific receptor. Comparison of the Kd's of hLf^{-4N} with that of $rhLf^{-5N}$ suggest that R^5 sterically hinders interaction of the hLf lymphocytic receptor with the second basic cluster, $R^{28}K^{29}V^{30}R^{31}$, a region that we previously identified as part of the hLf receptor binding site[20]. The X-ray crystallographic data of hLf indeed indicates that R^5 is linked to the protein core through a hydrogen bond (data not shown). R^5 is thus likely more involved in the structural integrity of hLf than in the interactions of hLf with other molecules.

Chlorate-treatment of Jurkat cells resulted in a 40% decrease in the total number of binding sites for native hLf, suggesting that at least half of the binding sites which interact with the first N-terminal basic cluster of hLf includes sulfate-containing molecules. A recent study suggested that $G^1R^2R^3R^4R^5S^6$ acts with the second basic cluster, $R^{28}K^{29}V^{30}R^{31}$, through formation of a 'cationic cradle' that binds to proteoglycans[10]. Wu *et al.*[11] proposed that a structural motif formed by residues R^5 only together with residues R^{25}, R^{28}, R^{29} and R^{31} represents the proteoglycan-binding domain. Our results suggest that R^2, R^3 and R^4 but not R^5 contribute to the binding of hLf to proteoglycans on Jurkat cells. This is in line with our observation that the N-terminal strech of four consecutive arginines has a decisive role in the interaction of hLF with heparin, Lipid A, human lysozyme and DNA[14].

As shown in Fig. 2, the N-terminal cluster of four consecutive arginine residues is unique for hLf[16]. Nevertheless, the binding of bLf and hLf to Jurkat cells was comparable. Murine Lf displayed relatively high-affinity binding to a much lower number of binding sites (about 8,000 per cell), which resembles the binding parameters of hLf^{-4N} and $rhLf^{-5N}$. This suggests that, in contrast to hLf and bLf, mLf, which contains a single K residue at position 1, does not interact with the sulfated molecules on the Jurkat cells but only with the lymphocyte receptor. Since bLf bound to Jurkat cells in a similar way to hLf, it may be assumed that other basic residues in the N-terminus of bLf contribute to proteoglycan binding. As a matter of fact, both hLf and bLf contain 9 basic amino acids at different positions between residues 1 to 37, whereas mLf only contains 6 basic residues.

In conclusion, our data indicate that $R^2R^3R^4$ but not R^5 of hLf interacts with about 80,000 binding sites which include sulfated cell surface molecules on Jurkat cells. Only about 20,000 binding sites are likely to correspond to the hLf lymphocyte receptor we have previously characterized[6]. Binding of hLf to these binding sites does not require the presence of the first basic cluster. In the light of these results, one could expect that native hLf has no preferential binding to its receptor but mainly interacts with sulfated molecules

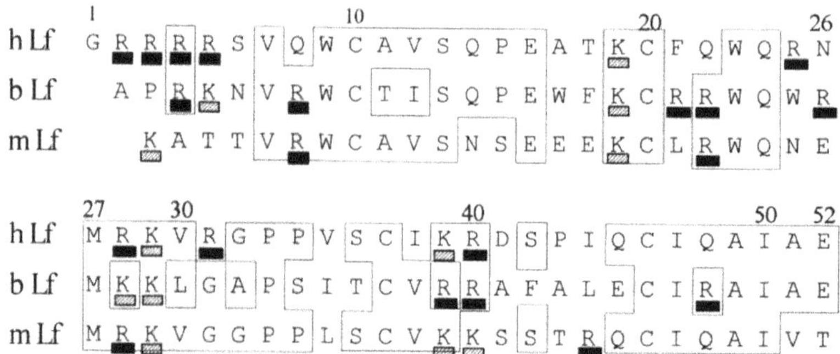

Figure 2. N-terminal protein sequence alignment of hLf, bLf and mLf. Identical amino acids between hLf[16,17], bLf[18] and mLf[19] are boxed. R and K residues are underlined with black and hatched blocks, respectively. Numbering of the sequence is according to[17].

on T-lymphocytes. It was previously shown that proteoglycans modulate Lf uptake by the liver, by modifying the internalization of lactoferrin into rat hepatocytes via the lipoprotein remnant receptor[22]. One could hypothesize that native hLf is bound by sulfated molecules for further presentation to the hLf lymphocyte receptor. In that process, sulfated molecules-bound hLf may be N-terminally degraded by membrane proteases or by extracellular proteases present in secretions or released during inflammation, thus leading to preferential binding of the protein to its receptor.

ACKNOWLEDGMENTS

This work was supported by the Université des Sciences et Technologies de Lille I, the Centre National de la Recherche Scientifique (U.M.R. du CNRS n°111, Director : Prof. A. Verbert) and the Ministère de l'Education Nationale. We would like to thank Prof. A. Tartar and Dr. H. Drobecq (Pasteur Institute, Lille) for the determination of the N-terminal sequence. We are very indebted to Miss I. Duthille for culturing Jurkat cells, to Miss N. Branza for participating to site-directed mutagenesis of hLf cDNA and to Mrs M. Benaïssa for technical assistance.

REFERENCES

1. Mikogami, T., Heyman, M., Spik, G. and Desjeux, J.F. (1994) Am. J. Physiol. 267, G308–G315
2. Rochard, E., Legrand, D., Lecocq, M., Hammelin, R., Crépin, M., Montreuil, J. and Spik, G. (1992) Anticancer Res. 1, 2047–2052
3. McAbee, D.D. and Esbensen, K. (1991) J. Biol. Chem. 226, 23624–23631
4. Ziere, G.J., van Dijk, M.C.M., Bijsterbosch, M.K. and van Berkel, T.J.C. (1992) J. Biol. Chem. 267, 11229–11235
5. Birgens, H.S., Karle, H., Hansen, N.E. and Kristensen, L. (1984) Scand. J. Haematol. 33, 275–280
6. Mazurier, J., Legrand, D., Hu, W.L., Montreuil, J. and Spik, G. (1989) Eur. J. Biochem. 179, 481–487
7. Leveugle, B., Mazurier, J., Legrand, D., Mazurier, C., Montreuil, J. and Spik, G. (1993) Eur. J. Biochem., 213,1205–1211
8. Appelmelk, B.J., An, Y.Q., Geerts, M., Thijs, B.G., de Boer, H.A., MacLaren, D.M. de Graaff, J. and Nuijens, J.H. (1994) Infect. Immun. 62, 2628–2632

9. Elass-Rochard, E., Roseanu, A., Legrand, D., Trif, M., Salmon, V., Motas, C., Montreuil, J. and Spik, G. (1995) Biochem. J. 312, 839–845
10. Mann, D.M., Romm, E. and Migliorini, M. (1994) J. Biol. Chem. 269, 23661–23667
11. Wu, H.F., Monroe, D.M. and Church, F.C. (1995) Arch. Biochem. Biophys. 317, 85–92
12. Hutchens, T.W, Henry, J.F. and Yip, T.T. (1991) Pediatr. Res. 29, 243–250
13. Jorieux, S., Mazurier, J., Montreuil, J. and Spik, G. (1985) Prot. Biol. Fluids 32, 115–118
14. van Berkel, P.H.C., Geerts, M.E.J., van Veen, H.A., Mericskay, M., de Boer, H.A. and Nuijens, J.H., (1997) Biochem. J., 328, 145–151
15. van Berkel, P.H.C., Geerts, M.E., van Veen, H.A., Kooiman, P.M., Pieper, F.R., de Boer, H.A. and Nuijens, J.H. (1995) Biochem. J. 312, 107–114
16. Metz-Boutigue, M.H., Jollès, J., Mazurier, J., Schoentgen, F., Legrand, D., Spik, G., Montreuil, J. and Jollès, P. (1984) Eur. J. Biochem. 145, 659–676
17. Rey, M.W., Woloshuk, S.L., de Boer, H.A. and Pieper, F.R. (1990) Nucleic Acids Res. 18, 5288
18. Pierce, A., Colavizza, D., Benaïssa, M., Maes, P., Tartar, A., Montreuil, J. and Spik, G. (1991) Eur. J. Biochem. 196, 177–184
19. Pentecost, B.T. and Teng, C.T. (1987) J. Biol. Chem. 262, 10134–10139
20. Legrand, D., Mazurier, J., Elass, A., Rochard, E., Vergoten, G., Maes, P., Montreuil, J. and Spik, G. (1992) Biochemistry 31, 9243–9251
21. Zimecki, M., Mazurier, J., Machnicki, M., Wieczorek, Z., Montreuil, J. and Spik, G. (1991) Immun. Lett. 30, 377–382
22. Ziere, G.J., Kruit, J.K., Bijsterbosch, M.K. and van Berkel, T.J.C. (1993) J. Biol. Chem. 268, 27069–27075
23. Huettinger, M., Retzek, H., Hermann, M. and Goldenberg, H. (1992) J. Biol. Chem. 267, 18551–18557
24. Keller, K.M., Brauer, P.R. and Keller, J.M. (1989) Biochemistry 28, 8100–8107
25. Spik, G., Strecker, G., Fournet, B., Bouquelet, S., Montreuil, J., Dorland, L., van Halbeek, H., Vliegenthart, J.F.G. (1982) Eur. J. Biochem. 121, 413–419
26. Legrand, D., van Berkel, P.H.C., Salmon, V., van Veen, H.A., Slomianny, M.-C., Nuijens, J.H. and Spik, G. (1997) Biochem. J. 327, 841–846
27. Salmon, V., Legrand, D., Georges, B., Slomianny, M.-C., Coddeville, B. and Spik, G. (1997) Protein Expression Purif. 9, 203–210
28. Bi, B.Y., Liu, J.L., Legrand, D., Roche, A.C., Capron, M., Spik, G. and Mazurier, J. (1996) Eur. J. Cell Biol. 69, 288–296
29. Scatchard, G. (1949) Ann. N.Y. Acad. Sci. 51, 660–672

GLYCATION LIGAND BINDING MOTIF IN LACTOFERRIN

Implications in Diabetic Infection

Yong Ming Li

The Picower Institute for Medical Research
350 Community Drive
Manhasset, New York 11030

1. SUMMARY

Lactoferrin and lysozyme are two important, naturally occurring antibacterial proteins found in saliva, nasal secretions, milk, mucus, serum and in the lysosomes of neutrophils and macrophages. Both proteins bind specifically to glucose-modified proteins bearing advanced glycation endproducts (AGEs). Exposure to AGE-modified proteins blocks the bacterial agglutination and bacterial killing activities of lactoferrin and also inhibits the bactericidal and enzymatic activity of lysozyme. Peptide mapping by AGE ligand blot revealed two AGE-binding domains in lactoferrin, and a single AGE-binding domain in lysozyme. None of these AGE-binding domains displayed any significant homology in their primary sequences; however, a common 17–18 amino acid cysteine loop motif ($CX_{15-16}C$) was identified among them, which we named an ABCD motif (AGE-Binding Cysteine-bounded Domain). Similar domains are also present in other antimicrobial proteins such as defesins. Hydrophilicity analysis indicated that each of these ABCD loops is markedly hydrophilic. Synthetic peptides, corresponding to these motifs in lactoferrin and lysozyme, exhibited AGE-binding activity. Since diabetes is associated with abnormally high levels of tissue and serum AGEs, the elevated AGEs may inhibit endogenous antibacterial proteins by binding to the conserved ABCD motif, thereby increasing susceptibility to bacterial infections in diabetic individuals. These results may provide a basis for the development of new approaches to prevent diabetic infections.

2. ANTIBACTERIAL ACTIVITY OF LACTOFERRIN

There is a large body of evidence supporting for an important role for lactoferrin in general defense against bacterial and viral infections. Lactoferrin together with lysozyme

Advances in Lactoferrin Research, edited by Spik *et al.*
Plenum Press, New York, 1998.

are first line defense proteins in certain body compartments such as the nasal cavity[1] and the bronchi[2]. Plasma lactoferrin has been recognized as a marker of infection in elderly individuals[3]. Apparently, lactoferrin has a broad spectrum of non-specific defense functions. It can kill bacteria directly[4] or inhibit the entry of bacteria into cells[5]. The defense functions of lactoferrin have been demonstrated on various microorganisms in vitro, including *Candida albicans*[6], *Legionella pneumophila*[7], *E. coli*, *Helicobacter pylori*[8], herpes simplex virus[9], cytomegalovirus[10] and *Toxoplasma gondii* parasites[11]. Lactoferrin binds directly to the bacterial surface and interacts with lipopolysaccharide[12]. The antibacterial activity of lactoferrin is located in the N-terminal region[4,13].

3. DIABETES AND GLUCOSE-MODIFIED PROTEINS

Proteins spontaneously react with reducing sugars such as glucose in non-enzymatic reactions to form AGEs[14]. Although the exact chemical structure of most AGEs remains unknown, they comprise a heterogeneous group of biologically reactive molecules that can be detected and quantitated by immunological methods[15]. AGEs have been implicated in tissue remodeling, the induction of growth factors and cytokines, and atherosclerosis. Both insulin dependent diabetes mellitus and non-insulin dependent diabetes mellitus patients, as well as aging individuals show increased levels of serum AGEs, and tissue AGEs accumulate at an accelerated rate in vascular and renal tissues of diabetic patients[16]. Because of their biochemical properties and distribution, AGEs are considered important factors in the pathogenesis of diabetic and aging complications[17–19].

4. AGE-BINDING PROTEINS

AGE-specific cell-surface binding proteins have been identified on many cell types, including phagocytic and mesenchymal cells. They were first identified on macrophages and characterized to mediate the uptake and degradation of AGEs[20]. Interaction of AGE-modified proteins with the macrophage AGE-binding proteins not only serves to degrade AGE-proteins, but also to induce synthesis and release of cytokines and growth factors, suggesting a dual purpose: disposal of senescent AGE-modified molecules and initiation of tissue repair and protein turnover[21,22]. An 80 kDa AGE-binding protein has also been isolated from pulmonary tissues while searching for AGE receptors, that was further showed to be identical to lactoferrin[23,24]. It has been proposed that lactoferrin together with other AGE-binding proteins may participate in endothelial cell activation and the functional enhancement of procoagulant activity and vascular permeability. Recently, we found that lysozyme binds AGEs with a high affinity[25]. It is interesting to us that the two AGE-binding proteins lactoferrin and lysozyme share a common biological activity—general defense.

5. COMPETITION OF AGE-BINDING BY PROTEOGLYCANS AND SUGAR CONJUGATES

Defense proteins, such as lactoferrin and lysozyme, are a group of anti-microbial proteins present at a high level in saliva, nasal secretions, mucus, serum and in the lysosomes of neutrophils and macrophages. They usually bind to and kill bacterial though a mechanism independent of the immune system. Defense proteins may share a common sugar-conjugate binding domain or motif, which is responsible for the recognition of microbial surface markers. This interaction is critical for bacteriostatic or bactericidal activity. Using ^{125}I-AGE-BSA as a probe, we have examined the specificity of AGE binding to lactoferrin and lysozyme by co-incubation with various proteoglycans and sugar conjugates. We found that both proteins have a similar AGE-specific binding activity[25]. Human lactoferrin binds AGE-BSA with an affinity similar to bovine lactoferrin as reported earlier for bovine lactoferrin[23]. The binding of AGE-BSA to both lactoferrin and lysozyme was time-dependent, reaching a plateau at 30 min, and was saturable with increasing concentrations of ligand. The interaction between AGE-BSA and lactoferrin or lysozyme was AGE-specific and non-covalent. AGE-BSA binding was reversible, and competitive with other AGE-modified proteins, but not with non-glycated carrier protein. Neither early glycation products (propylamine Amadori product), glucose itself nor glucosamine compete AGE-binding to lactoferrin or lysozyme[25]. As is the case for AGE binding to macrophages, certain negatively charged glycosaminoglycans (such as heparin and fucoidan) inhibited AGE binding to lactoferrin and lysozyme. This interaction, however, was not a simple charge-dependent affect, as co-incubation with keratan, chondroitin or poly-glutamic acid did not have the same inhibitory result. Overall, the binding characteristics of AGEs to lactoferrin and lysozyme are very similar as determined by analog compound competition.

6. AGEs INHIBIT DEFENSE PROTEIN-MEDIATED BACTERIAL KILLING

Using lyophilized *Micrococcus lysodeikticus* as a standard substrate, we found that lactoferrin-dependent agglutination of *M. lysodeikticus* was completely blocked by the concurrent addition of AGE-BSA (Figure 1A). Lactoferrin-induced bacterial agglutination could also be reversed by subsequent addition of AGE-BSA. Similarly, the presence of AGE-modified BSA inhibited the enzymatic activity of lysozyme in a dose-dependent manner, whereas the addition of BSA without AGE modification had no such effect. Functional bactericidal activity of lactoferrin was inhibited by AGE treatment, as reflected by a marked increase in the minimal inhibitory concentrations (MIC) required for these host defense proteins after exposure to AGE-modified albumins. AGE-BSA increased the MIC of lactoferrin for *M. lysodeikticus* by 5-fold (Figure 1B). These results suggest that the AGEs in diabetic serum may act as an intrinsic inhibitory factor to endogenous antibacterial proteins.

7. A COMMON AGE-BINDING MOTIF IN LACTOFERRIN, LYSOZYME, AND DEFENSINS

Although lysozyme and lactoferrin do not have primary sequence homology, we found that both proteins exhibit AGE-specific binding activity. Using ^{125}I-labeled AGE-BSA as a ligand, the calculated apparent dissociation constant was 250 nM for lactoferrin. Peptide mapping revealed two AGE-binding domains in lactoferrin and one AGE-binding domain in lysozyme[25]. None of these AGE-binding domains of lactoferrin and lysozyme displayed any distinct interprotein homology in primary sequence. However, a common 17–18 amino acid cysteine-bounded loop motif ($CX_{15-16}C$) was identi-

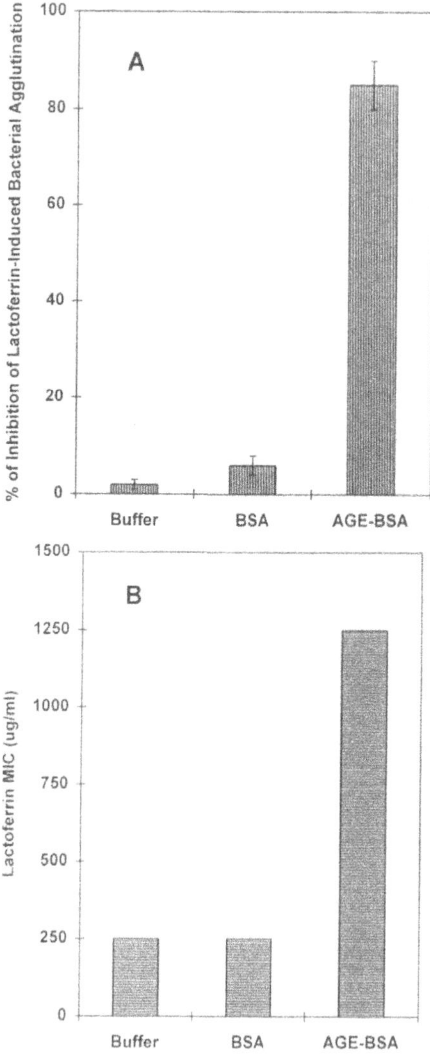

Figure 1. AGE-BSA inhibits bacterial agglutination (**A**) and bactericidal activities (**B**) of lactoferrin[25].

fied among these domains which we named an "ABCD" motif (AGE-Binding Cysteine-bounded Domain). Hydrophilicity analysis indicated that each of these ABCD loops is hydrophilic. Synthetic peptides corresponding to the ABCD motifs of lactoferrin (CysPheGlnTrpGlnArgAsnMetArgLysValArgGlyProProValSerCys) and lysozyme (CysAsnAspGlyArgIleProGlySerArgAsnLeuCysAsnIleProCys exhibited AGE-specific binding activity, as judged by Western ligand blot analyses[25]. These results suggest that the conserved ABCD motif in antibacterial proteins is responsible binding to AGEs. More interestingly, this AGE-binding cysteine loop is also present in other antimicrobial proteins, such as defensins (Figure 2), suggesting a potential general effect of AGEs on anti-microbial proteins.

Defense Proteins	Amino Acid #	Cysteine Loop																	
Lysozyme	62-78	Cys	Asn	Asp	---	Gly	Arg	Ile	Pro	Gly	Ser	Arg	Asn	Leu	Cys	Asn	Ile	Pro	Cys
Lactoferrin (N)	8-25	Cys	Phe	Gln	Trp	Gln	Arg	Asn	Met	Arg	Lys	Val	Arg	Fly	Pro	Pro	Val	Ser	Cys
Lactoferrin (C)	613-630	Cys	Leu	phe	Gln	Ser	Glu	Thr	Lys	Asn	Leu	Leu	Phe	Asn	Asp	Asn	Thr	Gly	Cys
Defensin (HNP-1)	4-19	Cys	Arg	Ile	---	Pro	Ala	Cys	Ile	Ala	Gly	Gly	Arg	Arg	Try	Gly	Thr	---	Cys
Defensin (HNP-2)	3-18	Cye	Arg	Ile	---	Pro	Ala	Cys	Ile	Ala	Gly	Gly	Arg	Arg	Tyr	Gly	Thr	---	Cys

Figure 2. BCD motif in defense proteins, lactoferrin, lysozyme, and defensins.

8. AGE AND DIABETIC INFECTION

Diabetes and aging are associated with abnormally high susceptibility to infection, and the mechanism remains unknown[7]. Early studies failed to reveal a significant effect of diabetes-associated metabolic disturbances such as hyperglycemia on either host defense systems or on bacterial growth. Accumulated evidence, however, indicated that antibacterial activity is significant decreased in diabetes, whereas the defense protein level remains the same compared with normal controls. These results strongly suggest that there are intrinsic defense-inhibitory factors in diabetes. Our data show that AGEs have a strong inhibitory effect on the defense functions of lactoferrin and lysozyme. We hypothesize that elevated levels of AGEs in diabetic patients may mimic the molecules on the bacterial cell wall, bind specifically to a sugar-conjugate recognition motif in endogenous antibacterial proteins, inhibit their bactericidal activity and normal defense, and lead to a high incidence of infections.

In summary, we found that the antibacterial activity of two major antibacterial proteins lactoferrin and lysozyme is inhibited by binding of advanced glycation products to a conserved motif. This finding provides a potential mechanism to explain diabetes- and aging-associated infections and may aid in the development of new therapeutic approaches for infections in diabetes and aging.

ACKNOWLEDGMENTS

This work was supported in part by a Research Award from the American Diabetes Association and a Research Award from the Institute for Advanced Studies in Immunology and Aging.

REFERENCES

1. Raphael GD, Jeney EV, Baraniuk JN, Kim I, Meredith SD, Kaliner MA. Pathophysiology of rhinitis. lactoferrin and lysozyme in nasal secretions. J Clin Invest 1989;84(5):1528–35.
2. Taylor DC, Cripps AW, Clancy RL. A possible role for lysozyme in determining acute exacerbation in chronic bronchitis. Clin Exp Immunol 1995;102(2):406–16.
3. Adeyemi EO, D'Anastasio C, Impallomeni MG, Hodgson HJ. Plasma lactoferrin as a marker of infection in elderly individuals. Aging (Milano) 1992;4(2):135–7.
4. Bellamy W, Takase M, Wakabayashi H, Kawase K, Tomita M. Antibacterial spectrum of lactoferricin b, a potent bactericidal peptide derived from the N-terminal region of bovine lactoferrin. J Appl Bacteriol 1992;73(6):472–9.
5. Longhi C, Conte MP, Seganti L, Polidoro M, Alfsen A, Valenti P. Influence of lactoferrin on the entry process of *Escherichia coli* hb101 (pri203) in HELA cells. Med Microbiol Immunol (Berl) 1993;182(1):25–35.
6. Bellamy W, Wakabayashi H, Takase M, Kawase K, Shimamura S, Tomita M. Killing of *Candida albicans* by lactoferricin b, a potent antimicrobial peptide derived from the N-terminal region of bovine lactoferrin. Med Microbiol Immunol (Berl) 1993;182(2):97–105.
7. Byrd TF, Horwitz MA. Lactoferrin inhibits or promotes *Legionella pneumophila* intracellular multiplication in nonactivated and interferon gamma-activated human monocytes depending upon its degree of iron saturation. Iron-lactoferrin and nonphysiologic iron chelates reverse monocyte activation against *Legionella pneumophila*. J Clin Invest 1991;88(4):1103–12.
8. Miehlke S, Reddy R, Osato MS, Ward PP, Conneely OM, Graham DY. Direct activity of recombinant human lactoferrin against *Helicobacter pylori*. J Clin Microbiol 1996;34(10):2593–4.
9. Marchetti M, Longhi C, Conte MP, Pisani S, Valenti P, Seganti L. Lactoferrin inhibits *Herpes simplex* virus type 1 adsorption to Vero cells. Antiviral Res 1996;29(2–3):221–31.

10. Shimizu K, Matsuzawa H, Okada K, Tazume S, Dosako S, Kawasaki Y, Hashimoto K, Koga Y. Lactoferrin-mediated protection of the host from murine cytomegalovirus infection by a T-cell-dependent augmentation of natural killer cell activity. Arch Virol 1996;141(10):1875–89.
11. Tanaka T, Omata Y, Saito A, Shimazaki K, Igarashi I, Suzuki N. Growth inhibitory effects of bovine lactoferrin to *Toxoplasma gondii* parasites in murine somatic cells. J Vet Med Sci 1996;58(1):61–5.
12. Miyazawa K, Mantel C, Lu L, Morrison DC, Broxmeyer HE. Lactoferrin-lipopolysaccharide interactions. Effect on lactoferrin binding to monocyte/macrophage-differentiated HL-60 cells. J Immunol 1991;146(2):723–9.
13. Bellamy W, Takase M, Yamauchi K, Wakabayashi H, Kawase K, Tomita M. Identification of the bactericidal domain of lactoferrin. Biochim Biophys Acta 1992;1121(1–2):130–6.
14. Cerami A, Vlassara H, Brownlee M. Glucose and aging. Sci Am 1987;256(5):90–6.
15. Bucala R, Vlassara H. Advanced glycosylation endproducts in diabetic renal disease: clinical measurement, pathophysiological significance, and prospects for pharmacological inhibition. Blood Purif 1995;13(3–4):160–70.
16. Li YM, Steffes M, Donnelly T, Liu C, Fuh H, Basgen J, Bucala R, Vlassara H. Prevention of cardiovascular and renal pathology of aging by the advanced glycation inhibitor aminoguanidine. Proc Natl Acad Sci U S A 1996;93(9):3902–7.
17. Vlassara H, Bucala R, Striker L. Pathogenic effects of advanced glycosylation: biochemical, biologic, and clinical implications for diabetes and aging. Lab Invest 1994;70(2):138–51.
18. Li YM, Baviello G, Vlassara H, Mitsuhashi T. Glycation products in aged thioglycollate medium enhance the elicitation of peritoneal macrophages. J Immunol Meth 1997;201:183–8.
19. Li YM, Dickson DW. Enhanced binding of advanced glycation endproducts (AGE) by the ApoE4 isoform links the mechanism of plaque deposition in Alzheimer's disease. Neuroscience Lett 1997;226: 155–158.
20. Vlassara H, Brownlee M, Cerami A. Novel macrophage receptor for glucose-modified proteins is distinct from previously described scavenger receptors. J Exp Med 1986;164(4):1301–9.
21. Yang Z, Makita Z, Horii Y, Brunelle S, Cerami A, Sehajpal P, Suthanthiran M, Vlassara H. Two novel rat liver membrane proteins that bind advanced glycosylation endproducts: relationship to macrophage receptor for glucose-modified proteins. J Exp Med 1991;174(3):515–24.
22. Li YM, Mitsuhashi T, Wojciechowicz D, Shimizu N, Li J, Stitt A, He C, Banerjee D, Vlassara H. Molecular identity and cellular distribution of advanced glycation endproduct receptors: relationship of p60 to ost-48 and p90 to 80k-h membrane proteins. Proc Natl Acad Sci U S A 1996;93(20):11047–52.
23. Neeper M, Schmidt AM, Brett J, Yan SD, Wang F, Pan YC, Elliston K, Stern D, Shaw A. Cloning and expression of a cell surface receptor for advanced glycosylation end products of proteins. J Biol Chem 1992;267(21):14998–5004.
24. Schmidt AM, Vianna M, Gerlach M, Brett J, Ryan J, Kao J, Esposito C, Hegarty H, Hurley W, Clauss M, et al. Isolation and characterization of two binding proteins for advanced glycosylation end products from bovine lung which are present on the endothelial cell surface. J Biol Chem 1992;267(21):14987–97.
25. Li YM, Tan AX, Vlassara H. Antibacterial activity of lysozyme and lactoferrin is inhibited by binding of advanced glycation-modified proteins to a conserved motif. Nat Med 1995;1(10):1057–61.

8

MOUSE LACTOFERRIN GENE

Promoter-Specific Regulation by EGF and cDNA Cloning of the EGF-Response-Element Binding Protein

Christina Teng, Huiping Shi, Nengyu Yang, and Hiroyuki Shigeta

Gene Regulation Group
Laboratory of Reproductive and Developmental Toxicology
National Institute of Environmental Health Sciences
National Institutes of Health
Research Triangle Park, North Carolina 27709

1. SUMMARY

Expression of the lactoferrin gene in a variety of tissues is regulated differentially. We have previously demonstrated that the lactoferrin gene is regulated by estrogen and mitogen in mouse uterus. The mouse lactoferrin gene responded to forskolin, cAMP, TPA and EGF stimulation via two adjacent enhancer elements, the CRE and EGFRE and collectively referred to as the Mitogen Response Unit (MRU). We found that CRE is responsible for forskolin, cAMP and TPA whereas EGFRE is for EGF stimulation. We examined the minimal promoter and enhancer elements of the mouse lactoferrin gene that are required for EGF induced transcriptional activation. We found that the CRE and noncanonical TATA box (ATAAA) are the minimal promoter elements for basal activity of the CAT reporter construct, whereas, the EGFRE is needed for an additional activity induced by EGF in transiently transfected human endometrial carcinoma RL95-2 cells (RL95-2). The EGFRE, however, did not function in heterologous promoters (SV 40 and TK). Therefore, EGF-stimulated lactoferrin gene activity is promoter specific in RL95-2 cells. Mutation made at either elements or insertion of extra nucleotides between the two elements, severely affected EGF-stimulated activity. Nuclear protein prepared from RL95-2 cells protected the EGFRE, CRE and noncanonical TATA from DNAase I digestion in a footprinting analysis. Nuclear protein which interacted with the CRE were previously identified as AP1 and CREB. In this study, we isolated a cDNA clone from an RL95-2 expression library that encodes the EGFRE binding protein. Partial sequence of the cDNA clone revealed 100% nucleotide identity with a GC-box binding protein, BTEB2. Protein-protein interaction among the transcription factors could fine-tune the mouse lactoferrin expression in various tissues.

Advances in Lactoferrin Research, edited by Spik *et al.*
Plenum Press, New York, 1998.

2. INTRODUCTION

Lactoferrin is hormonally regulated in lactating mammary glands and reproductive organs (Green and Pastewka, 1978; Teng et al., 1986; Pentecost and Teng, 1987; Teng et al., 1989). We have previously demonstrated that the expression of lactoferrin gene in mouse uterine tissue is clearly dependent on estrogen (Teng et al., 1986; Pentecost and Teng, 1987; Liu and Teng, 1991, 1992). We identified and characterized an *E*strogen *R*esponsive *M*odule (ERM) present in the mouse lactoferrin gene. This ERM consists of an overlapping COUP-transcription factor binding sequence with estrogen responsive element and confers estrogen action to homologous and heterologous promoters. Both estrogen receptor (ER) and COUP transcription factors (COUP-TF) bind to ERM (Liu and Teng, 1992). We showed that COUP-TF represses the ERM response to estrogen stimulation. Mutation and deletion of the COUP-TF-binding element or reduction of the endogenous COUP-TF increase mouse ERM estrogen responsiveness. Likewise, overexpression of the COUP-TF expression vector blocked the estrogen-stimulated response of ERM in transfected cells. The molecular mechanism of this repression is due to the competition between COUP-TF and the ER for binding at identical contact sites in the overlapping region of the mouse ERM (Liu et al., 1993).

The human lactoferrin gene also contains overlapping COUP-TF- and ER-binding elements similar to the mouse; however, COUP-TF does not bind to this element. Nonetheless, human ERM acted as an enhancer in response to estrogen stimulation in transfected cells (Teng et al., 1992). Upstream from the ERM, there is an extended steroid binding half-site which participates in estrogen responsiveness (Yang et al., 1996). Mutation at Gs in this steroid half-site reduces the estrogen responsiveness of the reporter plasmid containing the intact ERE. Through molecular cloning, we found that the protein which binds to this element is the estrogen-receptor related orphan receptor 1a (hERR1a same as the estrogen-receptor like 1a, hESRL1a). By far-western analysis, we demonstrated that the fusion protein of hERR1a interacts with human ER (Yang et al., 1996). These findings demonstrated that the orphan receptor, hERR1a, modulates the estrogen response of the human lactoferrin promoter activity possibly by binding to a DNA element next to the imperfect ERE and interacts with the estrogen through protein-protein interaction.

The molecular mechanisms that govern lactoferrin gene expression in other tissues or through other signal transduction pathways are currently unknown and would be an interesting question to explore. At least in uterine tissue, the expression of lactoferrin gene could also be stimulated by EGF (Nelson et al., 1991) in addition to estrogen. A convergence between the actions of EGF and estradiol has been demonstrated both *in vivo* and *in vitro* (Ignar-Trowbridge et al., 1992, 1993). However, EGF stimulates transcription of many genes through multiple mechanisms (Elsholtz et al., 1986; Matrisian et al., 1986; Lewis and Chikaraishi, 1987; Fisch et al., 1989). Among the known signalling pathways are the Ras/Raf-1 pathway (Moodie et al., 1993) and phosphorylation of the cytoplasmic STAT transcription factors (Darnell et al., 1994). These signaling pathways are independent of protein synthesis. For some of the EGF-responses, protein synthesis is required (Matrisian et al., 1986; Kerr et al., 1988). There was no consensus EGF response element found among the EGF response genes. Nevertheless, the cis-regulatory element for EGF is GC-rich (Shi and Teng, 1994; Elsholtz et al., 1986; Fisch et al., 1989; Lewis and Chikaraishi, 1987). In some cases, the EGF-mediated transcriptional activation is dependent on the AP-1 element (Elsholtz et al., 1986; Fisch et al., 1989). The mechanism(s) by which EGF activates the lactoferrin gene in mouse uterus is unclear. In this study we identified a cluster of sequence elements that responded to cAMP, TPA and EGF/TGFa in the mouse lactoferrin gene (*M*itogen *R*esponse *U*nit, MRU). To further investigate the minimum regulatory elements of the lactoferrin gene that are required for EGF action, we explored the relationship between the EGFRE and CRE. In addition, we isolated the cDNA which encodes the binding protein to EGFRE. The two clusters of response elements, ERM and MRU, in the mouse lactoferrin gene are located 300 bp apart. The organization of these elements provided an interesting model system to study the "cross-talk" between the steroid hormone and second message signalling pathways.

3. MATERIALS AND METHODS

3.1. Materials

Human endometrial carcinoma cell line RL95–2 and NIH 3T3 cells were obtained from ATCC. Tissue culture components were obtained from Gibco-Bethesda Research Laboratories (Grand Island, NY). 8-bromoadenosine 3:5′-cyclic monophosphate (8-bro-cAMP), forskolin, 12-O-tetradecanoylphorbol 13-acetate (TPA), mouse submandibulary gland epidermal growth factor (EGF) and recombinant transforming growth factor-α (TGF-α) were purchased from Sigma. Radiolabeled compounds were obtained from New England Nuclear Corp. (Boston, MA).

3.2. Plasmid Construction

The mLF0.1-CAT reporter plasmid was constructed as previously described (Shi and Teng, 1994). The ptk-CRE-CAT, ptk-EGFRE-CAT and ptk-MRU1-CAT were constructed by inserting double-stranded synthetic oligonucleotides (Table I) into the BamH I site of the pBLCAT vector (thymidine kinase promoter, ATCC, Rockville, MD). The copy number and orientation of the inserted sequences were determined by the dideoxy chain termination DNA sequencing with a sequenase kit (USB, Cleveland, OH). All plasmid DNAs used for transfection experiments were twice CsCl gradient purified.

3.3. Mutant Constructions

Recombinant PCR (Higuchi, 1990) were used for both site-directed mutagenesis and insertion mutations. The sequence between –75 and –53 of mouse lactoferrin 5′-flanking region was mutated by PCR with the primers presented in Table 1. The insertion mutants were constructed by inserting 4 or 10 bp nucleotides between EGFRE and CRE of the mLF0.1-CAT. The deletion mutants were constructed either by restriction enzyme digestion or direct PCR with the primers presented in Table 1. Deletion mutant 7 (d7) was constructed by digesting the mLF0.1-CAT with Stu I and BstE II , blunting it with mung bean endonuclease and inserting it with EGFRE oligonucleotide. The amplified fragments were cloned into the pCAT-Basic plasmid at the Hind III site.

3.4. Cell Culture, DNA Transfection, and CAT Assay

Human endometrial carcinoma cells (RL95–2) were maintained in Dulbecco's modified Eagle's medium/Ham's F12 (1:1) supplemented with 10% fetal calf serum (Life Technologies, Inc., Gaithersburg, MD). Transient transfections were performed by the calcium

Table 1. Oligonucleotide sequences

The synthetic oligonucleotide sequences used in CAT reporter constructs are as follows:
CRE, 5′-GAT CGC CCG GTG AGG TCA CCC AGC A-3′
EGFRE, 5′-GAT CGG GCA ATA GGG TGG GGC CAG CCC-3′
MRU, 5′-GAT CGG GCA ATA GGG TGG GGC CAG CCC GGT GAG GTC ACC CAG-3′
Primers used in mutagenesis are as follows:
m5, 5-G*TTA*GGTGAGGTCACCCAGCA-3′
m10, 5′-GGG *AAC* TAG GGT GGG GCC AG-3′
m11, 5′-CAA TA*C* *T*GT GGG GCC AGC CCG G-3′
m12, 5′-GGG T*TT* *T*GC CAG CCC GGT GAG G-3′
m15, 5′-ATA *CGT GCA CT*C CAG CCC GGT GAG GTC A-3′
Primers used in deletion mutants are as follows:
d3, 5-TTTAAAGCTTGGCCAGCCCGGTGAGGT-3′
d5, 5′-TAA GCT TGG GCC CTT TAT CTG TGC-3′
d8, 5′-GTT CAA GCT TGC TGG GTG ACC TCA CCG-3′

phosphate method (Cellphect transfection kit, Pharmacia). The cells were transfected with 2.5 μg of plasmid DNA/3.5 cm well. Sixteen hours after transfection, cells were washed, cultured in serum- and phenol red-free medium and different mitogens were added. Forskolin, cAMP and TPA treated cells were incubated for 4 hours whereas EGF and TGF-α were incubated overnight before cell harvesting. CAT enzyme assays were performed in whole cell extracts after normalization for protein concentration (Bio-Rad protein assay system). The CAT reactions were carried out with either 2 or 5 μg of protein for 1 hour at 37°C. The acetylated and nonacetylated forms of [^{14}C]chloramphenicol were separated by thin layer chromatography, autoradiographed, and quantitated by PhosphoImager (Molecular Dynamics, Sunnyvale, CA). All experiments were repeated at least three times and performed in duplicate to ensure reproducibility. Statistical analyses were carried out by paired t-test (Statview program).

3.5. Nuclear Protein Preparation and DNAase I Footprinting Analysis

The RL95–2 cell nuclear extracts were prepared as described previously (Liu et al., 1993). The 104 bp (+1 to −103) fragment of the mouse lactoferrin gene was PCR and subcloned into EcoR I site of the Bluescript SK$^+$ with blunt-end ligation, and a large quantity of the plasmids prepared. The DNA fragment was cut out with either Hind III/Sma I or Hinc II/BamH I and end-labeled with [α-^{32}P]dATP according to the standard protocol. Specific activity of the labeled DNA was 3–6 × 10^7 cpm/μg. Protein binding and DNAase I protection assay were performed according to Roesler et al. (1989). The samples were analyzed on a 6% sequencing gel. Chemical reactions for G and G+A in the same DNA fragment were included as standard (Ausubel et al., 1990).

3.6. Preparation of Probe for Expression cDNA Library Screening

The 24 base pairs of EGFRE oligonucleotide were multimerized with modification in a direct head to tail orientation according to Rosenfeld and Kelly (1986). After the ligation reaction, the larger DNA fragments were isolated from 1.2% agarose gel and subsequently subcloned into the BamH I/Bgl II sites of pSL1180 vector (Pharmacia). A plasmid containing four tandem repeats of the EGFRE sequence was identified by sequencing. This plasmid was used for duplication of the insert. In the third round of duplication, we obtained the plasmid containing 32 head to tail repeats of the EGFRE sequence. This DNA fragment was isolated from the plasmid, purified, labeled with [α-^{32}P]dCTP and used as the probe for screening the expression cDNA library. The expression cDNA library of the RL95-2 cells was constructed as previously described (Yang et al., 1996).

4. RESULTS

4.1. The Mouse Lactoferrin Promoter Region Contains Multiple Nuclear Protein Binding Sites

Analyzing the nucleotide sequence of the 122 bp lactoferrin promoter sequence (+1 to −121), we found three GC-rich regions, in addition to the consensus sequence for the two CAAT boxes, noncanonical TATA box, AP1 and CREB binding sites (Fig. 1). The region between −52 and −41 contained the nucleotide sequence that resembles the CRE and AP1 consensus sequences and we named this region CRE. Two of the GC-rich regions located at distal (GC-I) and proximal (GC-II) to the 5′ and one (GC-III) to the 3′ of the noncanonical TATA element. The two CAAT boxes were present at the 5′ of GC-I and GC-II, respectively. To investigate whether these elements were interacted with the nuclear protein of human endometrial carcinoma RL-95-2 cells, we analyzed the DNA fragment between +1 to −103 by DNAase I footprinting protection assay (Fig. 2). Three regions were found to be protected by protein from the RL95-2 cell nuclear extract. The areas protected were the noncanonical TATA, CRE and GC-II regions (Fig. 2A). A map of various footprint areas in both strands of the DNA fragment is presented in Figure 1B. Protein bound to these regions might play an important role in determining the transcriptional activity of the lactoferrin gene.

4.2. Mouse Lactoferrin Gene Contains a Mitogen Response Unit (MRU)

In order to test whether these putative regulatory elements are indeed functional in the lactoferrin promoter, we transfected the mLF0.1-CAT reporter constructs (+1 to −103 of the mouse lactoferrin gene linked to the basic CAT plamid) into RL95-2 cells to assess its transcriptional activity. We found that the CAT activities in RL95-2 cells were elevated by forskolin, cAMP, TPA, EGF and TGF-α. Therefore, this region of the mouse lactoferrin gene responded to multiple signal transduction pathways.

Stimulation of the reporter constructs by each of the mitogens was dependent on the concentrations (Fig. 3) and time (Fig. 4). It was interesting to find a synergistic effect between TPA and forskolin. This observation was more prominent at 2 hours of treatment (Fig. 4, upper panel). After the longer treatments, the response became additive rather than synergistic. The time courses for EGF and TGF-α induction, however, were much slower (Fig. 4, lower panel). There was no significant change in CAT activity for the first two hours of treatment and a three-fold increase in CAT activity was seen at 4 hours of treatment. The activity continued to rise to five-fold stimulation 24 hours later. These results established that the mLF0.1-CAT reporter, containing the first 104 bp of lactoferrin promoter sequence, could be regulated through second messenger signalling pathways.

4.3. Functional Dissection of the Mitogen Response Elements

To identify the sequence in mLF0.1-CAT that was responsible for the forskolin, TPA and EGF/ TGF-α stimulation, we deleted various putative elements of the mouse lactofer-

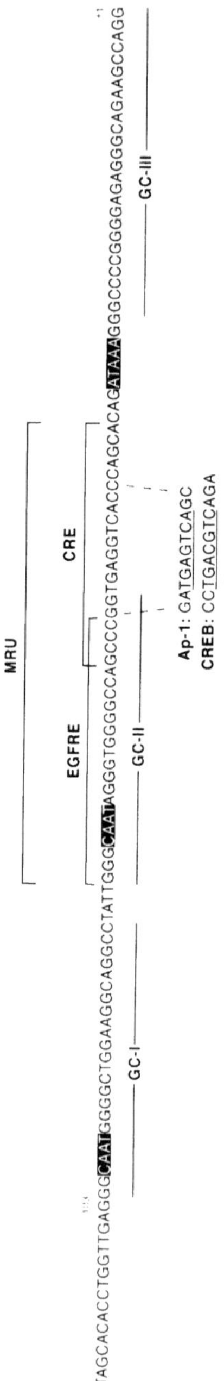

Figure 1. Sequence of the mouse lactoferrin promoter. The response elements and the GC-rich regions are indicated.

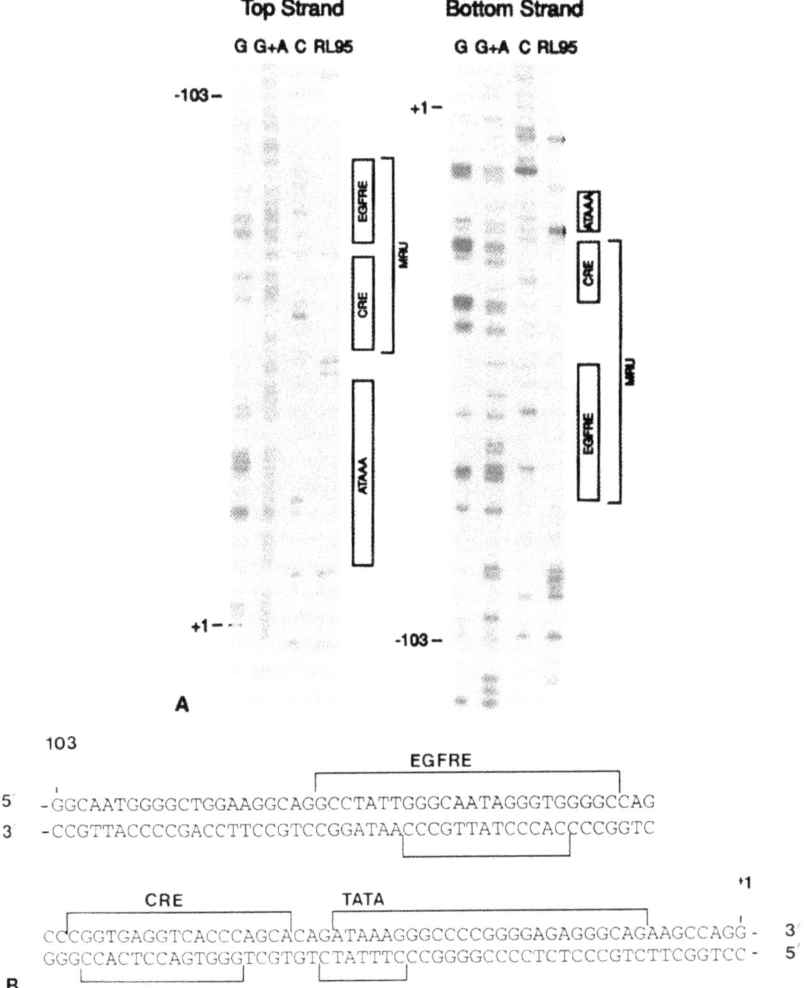

Figure 2. DNAase I footprinting of nuclear protein from RL95-2 cells to the mouse lactoferrin promoter/enhancer region. (A) DNA fragment from −103 to +1 of the lactoferrin gene. Nuclear protein-protected regions were labeled. (B) Diagrammatic presentation of the DNAase I footprint protected regions.

rin gene from the mLF0.1-CAT reporter construct and examined their response to various treatments. Deletion of GC-I (d1) and GC-III regions (d5) did not affect the overall forskolin-, TPA- and EGF-stimulated CAT activity, although, the basal activity was increased (Shi and Teng, 1994). Further deletion of GC-II (del 3), a considerable reduction in both basal and forskolin, or TPA-stimulated activity were found, however, the fold of stimulations by the activators was maintained unless the CRE region was also deleted (Shi and Teng, 1994). The EGF responsiveness was dramatically effected in d3 mutant (Fig. 5). Thus, the element responsible for EGF response resides in the GC-II region, the EGFRE, whereas forskolin, cAMP and TPA reside in the CRE (Fig. 1).

To map the nucleotides that are critical for EGF stimulation, we generated a number of mutations between nucleotides −75 and −54 where the putative EGF response element (EGFRE) is located and the CAT reporter's activities were tested (Fig. 6). Partial mutation

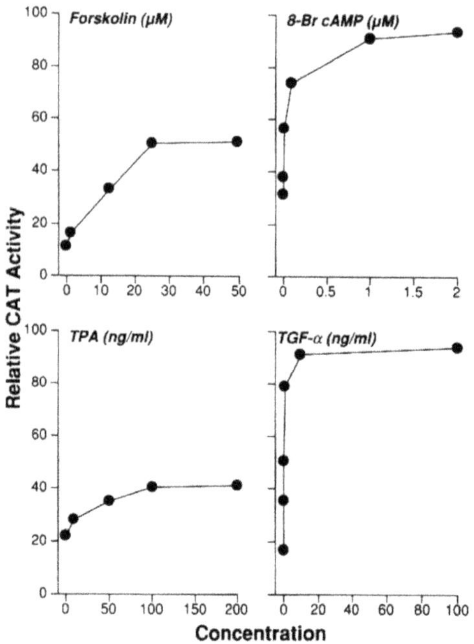

Figure 3. Stimulation of the mouse lactoferrin promoter by mitogen and growth factor are dependent on the concentration.

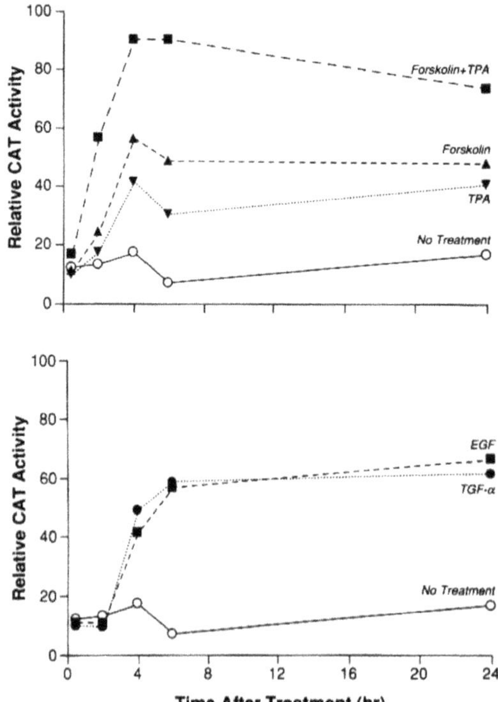

Figure 4. Stimulation of mouse lactoferrin promoter by mitogen and growth factor are time-dependent.

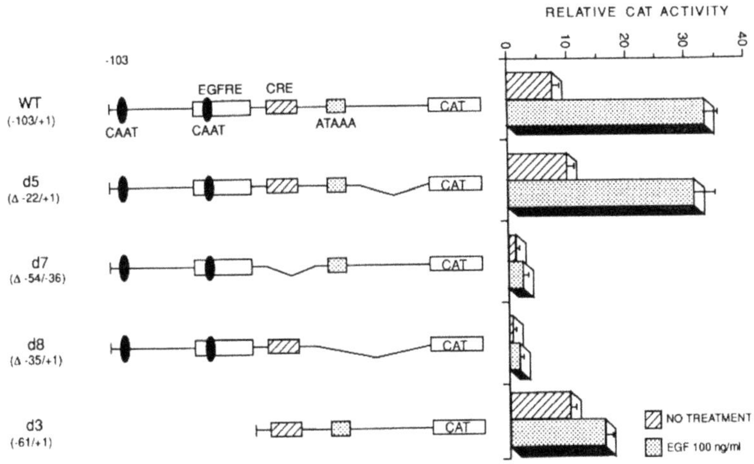

Figure 5. Functional delineation of the response elements in the mouse lactoferrin promoter.

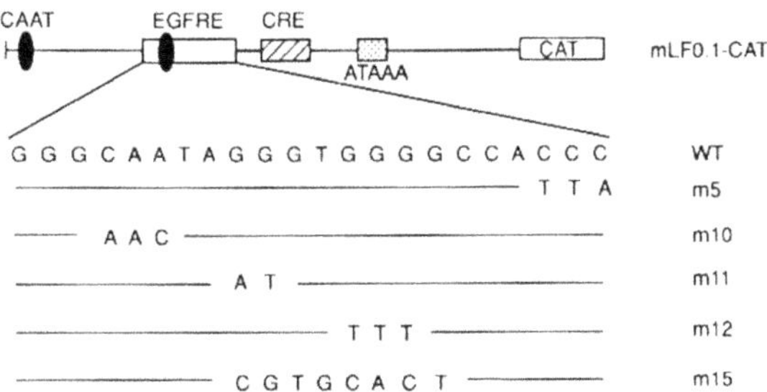

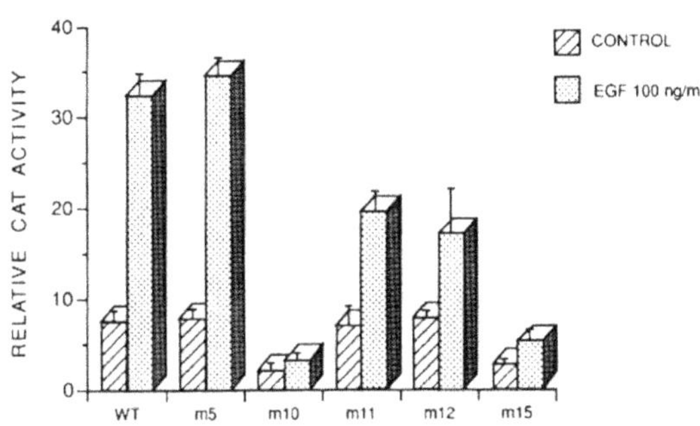

Figure 6. Identification of the critical nucleotides in EGFRE that is responsible for the EGF-stimulation.

at the GGGTGGG region (m11 and m12) reduced, but did not attenuate the EGF responsiveness. As the entire GGGTGGG region was scrambled, basal activity was diminished and EGF-stimulated activity abolished (m15). Mutation of CAAT to AACT in mutant 10 (m10) produced similar effects as the m15. The basal activity was reduced and the EGF responsiveness was completely lost (Fig. 6, m10). Therefore, CAAT and GGGTGGGG in the EGFRE are necessary for its function.

4.4. EGF Induction of Mouse Lactoferrin Gene Requires Both EGFRE and CRE

The importance of CRE and ATAAA box in EGF responsiveness was examined. Deletion of either element impaired the function of basal transcriptional machinery, hence the EGF stimulated activity (Fig. 5, compare wt to d7 and d8). These experiments suggest that both CRE and ATAAA are the minimal promoter of mouse lactoferrin gene. When the EGFRE and CRE were moved 70 bp upstream from the ATAAA box, this reporter plasmid became unresponsive to EGF, and the basal promoter activity was reduced (Shi and Teng, 1996). Thus, both nucleotide sequence of the CRE and the space between the CRE and ATAAA were important in sustaining the mouse lactoferrin promoter activity.

Since EGFRE and CRE were functionally related and physically linked in the mouse lactoferrin promoter, the spatial constraint between these two elements was investigated. Insertion of 4, 10, 30 or 42 nucleotides between CRE and EGFRE reduced the EGF-stimulated activity significantly (Table 2). It was interesting to find that addition of 10 nucleotides between the two elements actually increased the basal activity about three times. The EGF-stimulated activity also increased slightly, however, the fold of induction was decreased. Addition of 4 nucleotides between the EGFRE and CRE severely impaired the EGF-stimulated activity, yet did not affect the basal activity. When 20 or 30 bp was inserted between EGFRE and CRE, EGF-response was reduced.

4.5. The Mouse Lactoferrin MRU Is Promoter Specific

In order to verify whether EGFRE is promoter specific, we have cloned EGFRE into two different heterologous promoters such as SV40- or tk-CAT in either orientations and reporter activity was determined. EGFRE did not confer EGF-stimulated activity in tk-CAT, even in the presence of CRE (Fig. 7, compare fold of induction by EGF in lane 4 to lane 2 and 3). Similar results were also obtained by using SV40 promoter (Shi and Teng, 1996). In contrast, CRE was functional in tk-CAT (fold of induction by FSK in lane 1). It was unexpected to find that forskolin could slightly stimulate the CAT-reporters containing the EGFRE alone in transiently transfected RL95-2 cells (two-fold induction by FSK in lane 2). It was unclear whether any sequence in the EGFRE could mimic CRE's function.

4.6. Isolation and Identification of the cDNA Clone that Binds to EGFRE

We screened an expression library made from poly(A) RNA of the RL95–2 cells with a concatenated EGFRE sequence in order to isolate the nuclear protein that binds to

Table 2. Distance constraint between EGFRE and CRE

Base pair insert	Basal CAT	EGF treatment	Fold induction
0	7.5 ± 1.2	32.5 ± 2.3	4.3
4	5.9 ± 0.4	10.9 ± 1.0	1.8
10	21.8 ± 3.6	41.9 ± 4.3	1.9
30	9.9 ± 0.5	26.7 ± 0.9	2.7
42	5.9 ± 0.8	14.5 ± 2.4	2.5

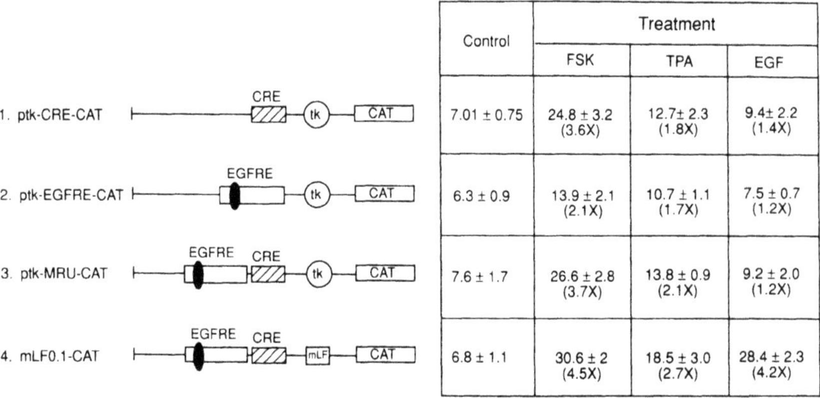

Control	Treatment		
	FSK	TPA	EGF
7.01 ± 0.75	24.8 ± 3.2 (3.6X)	12.7± 2.3 (1.8X)	9.4± 2.2 (1.4X)
6.3 ± 0.9	13.9 ± 2.1 (2.1X)	10.7 ± 1.1 (1.7X)	7.5 ± 0.7 (1.2X)
7.6 ± 1.7	26.6 ± 2.8 (3.7X)	13.8 ± 0.9 (2.1X)	9.2 ± 2.0 (1.2X)
6.8 ± 1.1	30.6 ± 2 (4.5X)	18.5 ± 3.0 (2.7X)	28.4 ± 2.3 (4.2X)

Figure 7. The mouse lactoferrin MRU is promoter specific.

the EGFRE region in human endometrial cells. Among the 1×10^6 clones screened, we identified several positive clones, and the longest, EGFREB, was partially sequenced. The first 149 bp of the EGFREB was identical to a GC-box binding protein, BTEB2, 5′ untranslated region (Fig. 8). Whether the cDNA also encodes the BTEB2 protein needs to be verified.

5. DISCUSSION

We found a DNA sequence within 100 bp upstream from the transcription initiation site of the mouse lactoferrin gene that is capable of conferring EGF/TGF-α , cAMP, forskolin and TPA action. We have delineated the functional role that confers EGF/TGF-α and cAMP/TPA induced regulation into two separate regions, the EGFRE and CRE. These two regions, however, were not totally independent of each other. Deletion at the EGFRE drastically reduced the basal and forskolin- and TPA-stimulated transcriptional activity (Shi and Teng, 1994) even though the CRE was intact. Likewise, mutation or deletion at the CRE region reduced the overall basal and EGF/TGF-α stimulated transcriptional activity (Fig. 5). A similar observation was made with the transferrin receptor gene, in that the CRE element and the GC-rich elements depended on each other for serum or growth factor stimulated activity (Ouyang et al 1993). CRE was also found necessary for basal transcription and for maximal EGF induction of the c-fos promoter (Fisch et al., 1987).

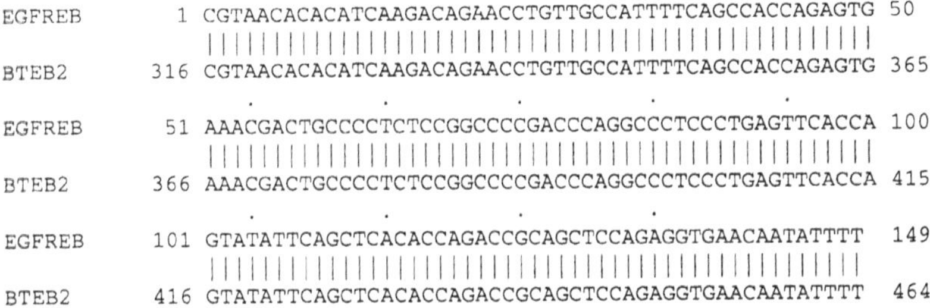

Figure 8. Partial sequence comparison between EGFRE binding protein.

The CRE and ATAAA constituted the basic promoter of mouse lactoferrin gene, because neither one could function independently as the promoter in a reporter construct transiently transfected into RL95–2 cells. The spatial constraint between CRE and ATAAA was also critical (Shi and Teng, 1996). Therefore, CRE became an integral part of the basic promoter machinery in mouse lactoferrin gene. It was interesting to find that EGFRE could not confer EGF action even in the presence of CRE in a heterologous promoter (Fig. 7). Thus, the protein bound to EGFRE must interact directly or indirectly with the initiation complex assembled at the ATAAA and CRE of the mouse lactoferrin gene. Recently, two functionally distinct factors are found which differ from the TATA-binding protein that binds to the rabbit uteroglobin noncanonical TATA-box region at two adjacent sites (Klug et al., 1994). The expression of the uteroglobin gene is confined to the tissues containing these factors. The lactoferrin gene, like the uteroglobin gene was constitutively expressed in one tissue and hormonally regulated in the other (Teng et al., 1989). It is possible that a distinct set of nuclear factors in the RL95–2 cells could recognize the ATAAA element of the mouse lactoferrin gene and interact with it in the same way as the TATA factor interacts with the uteroglobin gene promoter.

Spatial distribution of EGFRE and CRE is essential for EGF-mediated activation of mouse lactoferrin promoter. Insertion of 4 bp between the two elements attenuated the EGF effect (Table 2). As the two elements moved apart for 10 bp, the EGF-mediated activity actually increased slightly. Hence, the protein-protein interaction might be dependent on the stereoalignment of the EGFRE and CRE in the promoter. When 30 or 42 bp was inserted between the EGFRE and CRE, EGF response was reduced. Therefore, it is important that protein bound to these elements interacted with each other. The transcription factor binding sites that exhibit activity in a helical turn-dependent manner similar to our current finding were reported previously by other laboratories (Wu and Berk, 1988; Takahashi et al., 1986; Vilen et al., 1992). Systems, such as the pit-1 activated rat prolactin gene promoter (Smith et al., 1995) and a variety of promoters in E. coli that were activated by cAMP receptor protein (Kolb et al., 1993), were found to exhibit a unique bimodal pattern of position-dependent activation and synergism by the binding of transcription factors to their DNA elements. It will be interesting to determine whether activation of mouse lactoferrin promoter by EGF also exhibited a bimodal pattern of position-dependent transcriptional activity.

The mouse lactoferrin EGF response element resembles the EGF response elements of prolactin (Elsholtz et al., 1986), gastrin (Merchant et al., 1991), the tyrosin hydroxylase gene (Lewis and Chikaraishi, 1987) and transferrin receptor gene (Ouyang et al., 1993). The common feature of these EGF response elements is GC-rich in nature, although no consensus response element for EGF has been derived from these sequences. The EGF response region was present in the GC-II region of the mouse lactoferrin promoter; the GC-II region contained 70% of G and C. Within this area, both CAAT and GGGTGGGG were critical nucleotides for EGF response (Fig. 6). The GGGTGGGG is a consensus binding sequence for several GC-box binding protein including the BTEB2 (Gallarda et al., 1989; Postel et al., 1989; Sogawa et al., 1993). Partial sequence of the 2.4 kbp cDNA that we isolated from the human endometrial carcinoma RL95-2 cell expression cDNA library with EGFRE oligos showed 100% identity with the BTEB2 5′ untranslated region (Fig. 8). Whether this cDNA also encodes the BTEB2 protein needs to be verified by complete nucleotide sequencing of the entire clone.

Studying of lactoferrin gene expression in the mouse uterus suggested that multiple signal transduction pathways are involved. Estrogen confers its action through a composite estrogen response module, the ERM, located in the distal region of the lactoferrin promoter (Liu and Teng, 1992) and EGF confers its action through a composite mitogen response unit, the MRU, located at the proximal region (Shi and Teng, 1994). The organization of the ERM and MRU in the mouse gene provides an interesting model system to study the cross-communication between the steroid/thyroid and second messenger signalling pathways at the gene level. There is evidence that the steroid hormone response element can be further modulated by physical association with additional factors bound at remote sites (Imai et al., 1993; Schild et al., 1993). In the Xenopus vitellogenin B1 promoter, deposition of a nucleosome onto a site between 140 bp and 300 bp upstream from the transcription initiation site further stimulates estrogen induced transcription (Schild et al., 1993). The maximum response of the phosphoenolpyruvate carboxykinase (PEPCK) gene to glucocorticoids required the presence of both the glucocorticoid response unit (GRU) and cyclic AMP response element (CRE) (Imai et al., 1993). Therefore, steroid receptors could interact with additional transcription factors bound at different locations, enhancing their effects. Comparing the ERM and MRU of the mouse lactoferrin gene to the GRU and CRE of the PEPCK, we found these response elements are located and organized similarly in their respective promoters. When the lactoferrin gene is packaged in the chromatin structure in the uterine cells, the two response elements may be brought closer together by looping the DNA with a nucleosome. Therefore, direct interaction of the estrogen receptor with the DNA binding protein of the MRU is likely to occur. We will examine the functional relationship between the estrogen response module and mitogen response unit through the use of mouse lactoferrin gene promoter.

REFERENCES

Ausubel FM, Brent RE, Kingston RE, Moore DD, Seidman JD, Smith JA, Struhl K 1990 Current Protocols in Molecular Biology. PP. a7.5.1–7.5.7. John Wiley & Son, Inc.

Darnell JEJ, Kerr IM, Stark GR 1994 Jak-STAT pathways and transcriptional activation in response to IFNs and other extracellular signaling proteins. Science 264:1415–1421

Elsholtz HP, Mangalam HJ, Potter E, Albert VR, Supowit S, Evans RM Rosenfeld MG 1986 Two different cis-active elements transfer the transcriptional effects of both EGF and phorbol esters. Science 234:1552–1557

Fisch TM, Prywes R, Roeder RG 1987 c-fos sequence necessary for basal expression and induction by epidermal growth factor, 12–0-tetradecanoyl phorbol-13-acetate and the calcium ionophore. Mol. Cell. Biol. 7: 3490–3502

Fisch TM, Prywes R, Roeder RG 1989 An AP1-binding site in the c-fos gene can mediate induction by epidermal growth and 12–0-tetradecanoyl phorbol-13-acetate. Mol. Cell. Biol. 9:1327–1331

Gallarda JL, Foley KP, Yang A, Engel JD 1989 The β-globin stage selector element factor is erythroid-specific promoter/enhancer binding protein NF-E4 Genes & Dev. 3:1845–1859

Green MR, Pastewka JV 1978 Lactoferrin is a marker for prolactin response in mouse mammary explants. Endocrinology 103: 1510–1513

Higuchi R.1990 Recombinant PCR, in M. A. Innis, D. H. Gelfand, J. J. Sninsky, T. J. White. PCR protocols: A guide to methods and applications. (Innis, M.A., Gelfand, D. H., Sninsky, J. J., and White, T. J., eds) pp. 177–183. Academic Press Inc., San Diego, CA.

Imai E, Miner JN, Mitchell JA, Yamamoto K R, Granner D K 1993 Glucocorticoid receptor-cAMP response element-binding protein interaction and the response of the phosphoenolpyruvate carboxykinase gene to glucocorticoids. J. Biol. Chem. 268: 5353–5356

Ignar-Trowbridge DM, Nelson KG, Bidwell MC, Curtis SW, Washburn TF, McLachlan JA, Korach KS 1992 Coupling of dual signaling pathways: epidermal growth factor action involves the estrogen receptor. Proc. Natl. Acad. Sci. USA 89: 4658–4662

Ignar-Trowbridge DM, Teng CT, Ross KA, Parker MG, Korach KS, McLachlan JA 1993 Peptide growth factors elicit estrogen receptor-dependent transcriptional activation of an estrogen-responsive element. Mol. Endocrinol. 7: 992–998

Kerr LD, Holt JT, Matrisian LM 1988 Growth factors regulate transin gene expression by c-fos-dependent and c-fos-independent pathways. Science 242:1424–1427

Kolb A, Busby S, Buc H, Garges S, Adhya S 1993 Transcriptional regulation by cAMP and its receptor protein. Annu. Rev. Biochem. 62: 749–795

Klug J, Knapp S, Castro I, Beato M 1994 Two distinct factors bind to the rabbit uteroglobin TATA-box region and are required for efficient transcription. Mol. Cell. Biol. 14: 6208–6218

Lewis EJ, Chikaraishi DM 1987 Regulated expression of the tyrosine hydroxylase gene by epidermal growth factor. Mol. Cell. Biol. 7:3332–3336

Liu YH, Teng CT 1991 Characterization of estrogen responsive mouse lactoferrin promoter. J. Biol. Chem. 266: 21880–21885

Liu YH, Teng CT 1992 Estrogen response module of the mouse lactoferrin gene contains overlapping chicken ovalbumin upstream promoter transcription factor and estrogen receptor binding elements. Mol. Endocrinol. 6:355–364

Liu YH, Yang N, Teng CT 1993 COUP-TF acts as a competitive repressor for estrogen receptor-mediated activation of the mouse lactoferrin gene. Mol. Cell. Biol. 13:1836–1846

Marchant JL, Demedick B, Brand SJ 1991 A GC-rich element confers epidermal growth factor responsiveness to transcription from the gastrin promoter. Mol. Cell. Biol. 11: 2686–2696

Moodie SA, Willumsen BM, Weber MJ, Wolfman A 1993 Complexes of Ras GTP with Raf-1 and mitogen-activated protein kinase kinase. Science 260:1658–1661

Matrisian LM, Leroy P, Ruhlmann C, Gesnel MC, Breathnach R 1986 Isolation of the oncogene and epidermal growth factor-induced transin gene: complex control in rat fibroblasts. Mol. Cell. Biol. 6:1679–1686

Nelson KG, Takahashi T, Bossert NL, Walmer DK, McLachlan JA 1991 Epidermal growth factor replaces estrogen in the stimulation of female genital-tract growth and differentiation. Proc. Natl. Acad. Sci. USA 88: 21–25

Ouyang Q, Bommakanti M, Miskimins WK 1993 A mitogen-responsive promoter region that is synergistically activated through multiple signalling pathways. Mol. Cell. Biol. 13:1796–1804

Pentecost BT, Teng CT 1987 Lactotransferrin is the major estrogen inducible protein of mouse uterine secretions. J. Biol. Chem. 262: 10134–10139

Postel EH, Mango SE, Flint SJ 1989 A nuclease-hypersensitive element of the human c-myc promoter interacts with a transcription initiation factor. Mol. Cell. Biol. 9: 5123–5133

Rosenfeld PJ, Kelly TJ 1986 Purification of nuclear factor I by DNA recognition site affinity chromatograpny. J. Biol. Chem. 261: 1389–1408

Roesler WJ, Vandenbark GR, Hanson RW 1989 Identification of multiple protein binding domains in the promoter-regulatory region of the phosphoenolpyruvate carboxykinase (GTP) gene. J. Biol. Chem. 264: 9657–9664.

Schild C, Claret FX, Wahli W, Wolffe AP 1993 A nucleosome-dependent static loop potentiates estrogen-regulated transcription from the xenopus vitellogenin B1 promoter in vitro. EMBO J. 12:423–433.

Shi HP, Teng CT 1994 Characterization of a mitogen-response unit in the mouse lactoferrin gene promoter. J. Biol. Chem. 269:12973–12980

Shi HP, Teng CT 1996 Promoter-specific activation of mouse lactoferrin gene by epidermal growth factor involves two adjacent regulatory elements Mol. Endocrinol 10: 732–741

Smith KP, Liu B, Scott C, Sharp SD 1995 Pit-1 exhibits a unique promoter spacing requirement for activation and synergism. J. Biol. Chem. 270: 4484–4491

Sogawa K, Imataka H, Yamasaki Y, Kusume H, Abe H, Fujii-Kuriyama Y 1993 cDNA cloning and transcriptional properties of a novel GC box-binding protein, BTEB2 Nucleic Acids Res. 21:1527–1532

Takahashi K, Vigneron M, Matthes H, Wildeman A, Zenke M, Chambon P 1986 Requirement of stereospecific alignments for initiation from the simian virus 40 early promoter. Nature 319:121–126

Teng CT, Pentecost BT, Chen YH, Newbold RR, Eddy EM, McLachlan JA 1989 Lactotransferrin gene expression in the mouse uterus and mammary gland. Endocrinology 124: 992–999

Teng CT, Walker MP, Bhattacharyya SN, Klapper DG, DiAugustine RP, McLachlan JA 1986 Purification and properties of an oestrogen-stimulated mouse uterine glycoprotein (approx. 70 kDa). Biochem. J. 240: 413–422

Teng CT, Liu YH, Yang NY, Walmer D, Panella T 1992 Differential molecular mechanism of the estrogen action that regulates lactoferrin gene in human and mouse. Mol. Endocrinol. 6: 1969–1981

Vilen BJ, Penta JF, Ting JPY 1992 Structural constraints within a trimeric transcriptional regulatory region. Constitutive and interferon-gamma-inducible expression of the HLA-DRA gene. J. Biol. Chem. 267: 23728–23734

Wu L, Berk A 1988 Constraints on spacing between transcription factor binding sites in a simple adenovirus promoter. Genes & Develop. 2: 403–411

Yang NY, Shigeta H, Shi HP, Teng CT 1996 Estrogen-related receptor, hERR1, modulates estrogen receptor-mediated response of human lactoferrin gene promoter. J. Biol. Chem. 271:5795–5804

CLONING OF HUMAN GENOMIC LACTOFERRIN SEQUENCE AND EXPRESSION IN THE MAMMARY GLANDS OF TRANSGENIC ANIMALS

Sun Jung Kim, Dae-Yeul Yu, Yong-Mahn Han, Chul-Sang Lee, and Kyung-Kwang Lee

Plant and Animal Cell Technology Research Division
Korea Research Institute of Bioscience and Biotechnology
Taejon 305-333, Korea

1. INTRODUCTION

Transgenic animals which express foreign proteins in their mammary glands have been useful tools for the mass production of human proteins[1]. DNA sequences conferring mammary-specific expression to the foreign genes of interest have been derived from the promoters of major milk protein genes such as whey acidic protein[2,3], β-lactoglobulin[4], α-casein[5], β-casein[6,7] and α-lactalbumin[8]. In most studies, the foreign proteins were synthesized and secreted into the milk although in some cases the expression of the transgene was not restricted to the mammary gland. The advent of such technology has prompted researches of using transgenic animals as bioreactors. Pharmaceuticals and other useful human proteins such as human (α1-antitrypsin[4,9], tissue plasminogen activator[2,10], human serum albumin[11], and human hemoglobin[12] were produced in transgenic mouse, rat, sheep, goat, and pig. Although very few transgenic systems have been described where the expression level of the transgene approached that of the endogenous gene, many of them proved to be successful by expressing more than a gram of transgene product per liter of milk.

Lactoferrin (Lf) is an iron binding protein in milk and a member of transferrin family[13]. Human lactoferrin (hLf) is a glycoprotein consisting of a single polypeptide chain (~80 kDa) containing 691 amino acids[14]. The levels of hLf can reach up to 6 g/l in colostrum and decrease to 1 g/l in mature milk. The levels of Lf in bovine milk, however, are lower ranging from 0.02 to 0.2 g/l[15]. Two studies have been reported for the mass production of hLf. One is about the filamentous fungi Aspergillus orizae in which up to 25 mg/l of hLf was expressed[16] The other study exploited transgenic mice system in which regulatory sequences of the bovine S1-casein gene were used to target hLf expression into the mammary glands[17].

Advances in Lactoferrin Research, edited by Spik *et al.*
Plenum Press, New York, 1998.

We have previously reported the generation of transgenic mice carrying human lactoferrin cDNA[7,18]. Regulatory elements of the bovine β-casein gene were used to control the expression of hLf cDNA. A hybrid intron composed of intron 2 of the rabbit-globin gene and intron 1 of bovine β-casein gene was also placed in the hybrid gene to induce higher expression of hLf. It is reported that mice and sheep carrying minigenes composed of genomic sequence of human α1-antitrypsin[4,19] or human serum albumin[20,21] behind the β-lactoglobulin promoter secreted high levels of the corresponding proteins, while no expression was detected in the mammary gland of transgenic mice carrying the cDNA of these genes[20,22].

We have constructed an expression vector comprised of 10 kb bovine β-casein 5′-flanking sequence and 30 kb human genomic sequence, and used it to express the protein in the mammary gland of transgenic mice.

2. CONSTRUCTION OF EXPRESSION VECTOR

For the construction of genomic expression vector, we first cloned genomic lactoferrin sequence from a cosmid library. Figs. 1 and 2 are Southern hybridization data of the human genomic DNA and the positive cosmid clone digested with some restriction enzymes, respectively. The bands appeared in Fig. 2 actually denote the exon-containing restriction fragments because we used hLf cDNA sequence as a probe. This fact helped us to map the exon/intron structure of the gene. Southern hybridization was also performed for the human genomic DNA. The sizes of the enzyme fragments are well matched each other for the same restriction enzyme fragments.

For BamHI, HindIII, and EcoRI restriction enzymes, we mapped their location on the sequence (Fig. 3). The clone has about 40 kb and contained full coding region. The first 10kb was the 5′-flanking sequence including the promoter. The remaining 30 kb frag-

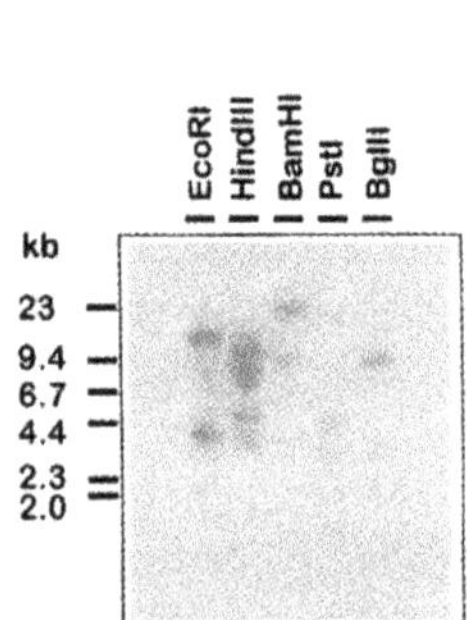

Figure 1. Southern blot analysis of human genomic DNA probed with lactoferrin cDNA.

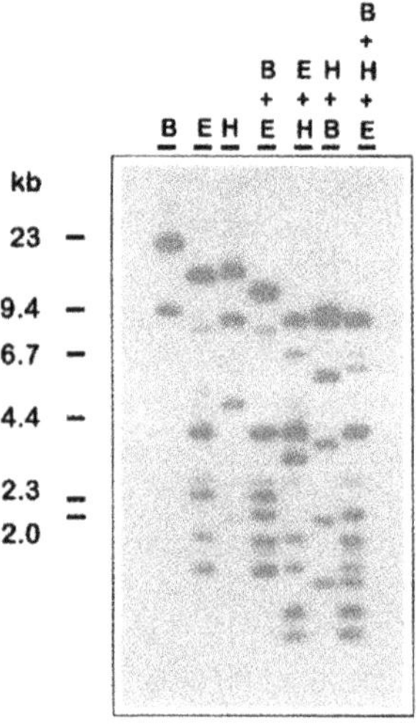

Figure 2. Southern blot analysis of a cosmid clone containing human lactoferrin gene. The restriction enzymes used are B; BamHI, E; EcoRI, and H; HindIII.

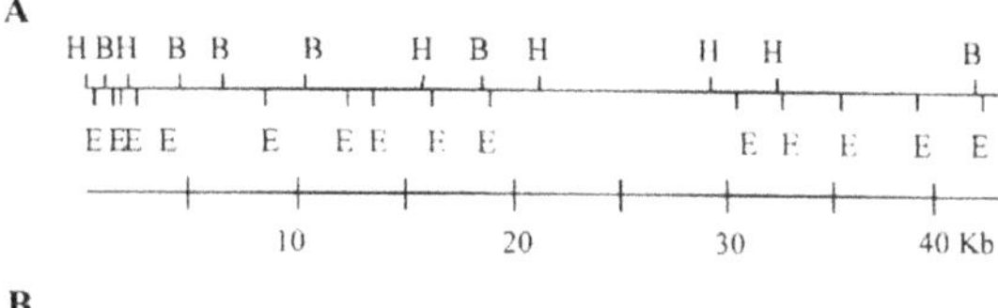

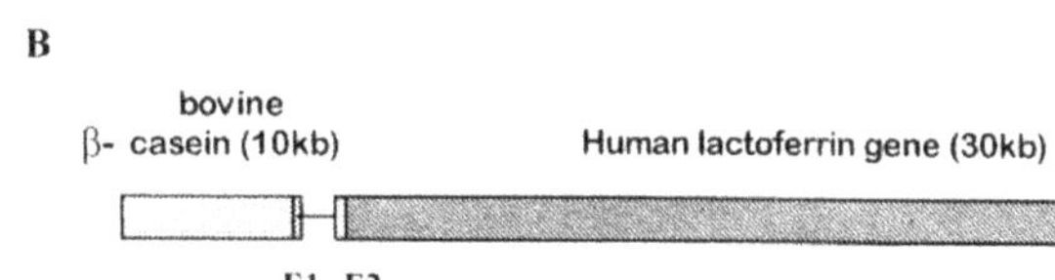

Figure 3. Restriction map of human lactoferrin gene (A) and schematic diagram of the human lactoferrin expression vector (B). Dotted boxes are regulatory sequences of the bovine β-casein gene. Stippled box indicates hLf gene.

ment which has the entire coding region was used for the construction of expression vector by joining with 10 kb of bovine-casein promoter. Because of the long length of the vector, there were some obstacles to construct it using plasmid vectors, so we used a cosmid vector.

3. GENERATION OF TRANSGENIC MICE

For the microinjection of large-sized DNA, some people prepare overlapping fragments of the injection DNA, however there are reports that people microinjected YAC clone of which size is usually hundreds kb long and we also used the traditional microinjection method to get the transgenic mice. Transgenic mice were generated using a standard method[23] as described previously[7]. The mice were an F1 hybrid C57BL/6 × CBA strain. We have obtained 3 transgenic mice from 54 founder mice analyzed and this transgenic efficiency is about 3 times lower than the cDNA mice. We are not sure now whether this lower integration efficiency comes from the long length of the injection DNA.

4. CHARACTERIZATION OF hLf IN MILK OF TRANSGENIC MICE

The 3 transgenic mice, two males and one female, were mated with normal mice and at the lactation stage milk was collected for the assay of expression of lactoferrin. Milk to be assayed was diluted with equal volume of phosphate-buffered saline and centrifuged at 4°C for 30 min at 14,000 × g to separate the whey, casein, and fat fractions. Polyclonal rabbit anti-hLf antiserum (diluted 1:1000 in 1% bovine serum albumin) purchased from Sigma Chemical Co. was usedto detect hLf by the western blot analysis and ELISA. Bound antibody was detected by addition of alkaline phosphatase-conjugated goat anti-rabbit IgG (Bio-Rad). Fig. 4A is western blot, Fig. 4B is the Coomassie stained SDS-PAGE gel. The size of the recombinant Lf is same as the commercial Lf. Its expression level was determined by ELISA and reach up to 400 to 500 μg per ml. This expression level is higher than the highest cDNA expression level, 200 μg/ml. The expression levels of hLf in milk of transgenic mice were 2.5 to 200 μg/ml. The expression level of hLf in four transgenic mice generated in this study, 2.5 to 34 μg/ml, was similar to the results of Platenberg et al.[17] in which its level was 0.1 to 36 μg/ml. One line expressing 200 μg/ml of hLf may be influenced by position effect. There are many reports in transgenic animal that genomic sequences usually express higher level of foreign protein than cDNA[19–21] and we could determine whether the lactoferrin gene is the same case after assaying more lines of mice.

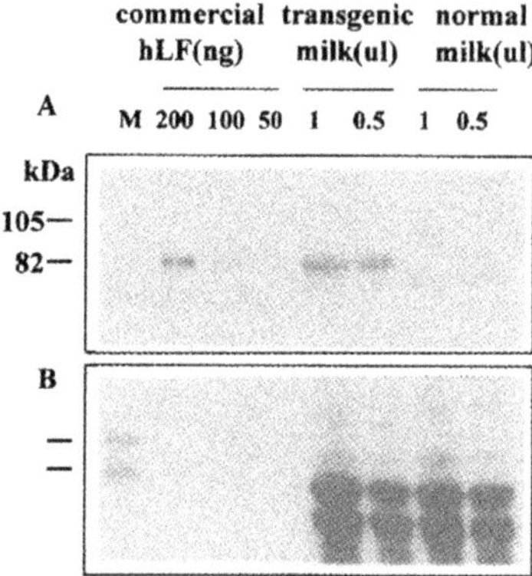

Figure 4. Western blot analysis of milk from transgenic mouse.

5. CONCLUSION

We report the expression of hLf transgene after generating transgenic mice using a bovine β-casein/hLf gene. A lactoferrin expression vector comprised of 10 kb bovine-casein promoter and 30 kb genomic lactoferrin sequence expressed 400–500 μg/ml of human lactoferrin in the mammary glands of transgenic mouse. This expression vector could replace cDNA vector for the efficient expression of the protein.

ACKNOWLEDGMENT

This work is funded by the Ministry of Science and Technology of Korea.

REFERENCES

1. Hennighausen L. The prospects for domesticating milk protein genes. J Cell Biochem 1992; 49: 325–332.
2. Gordon K, Lee E, Vitale JA, Smith AE, Westphal H, Hennighausen L. Production of human tissue plasminogen activator in transgenic mouse milk. Bio/Technology 1987; 5: 1183–1187.
3. Pittius CW, Hennighausen L, Lee E, Westphal H, Nicols E, Vitalc J, Gordon K. A milk protein gene promoter directs the expression of human tissue plasminogen activator cDNA to the mammary gland in transgenic mice. Proc Natl Acad Sci USA 1988; 85: 5874–5878.
4. Archibald AL, McClenaghan M, Hornesy V, Simons JP, Clark AJ. High-level expression of biologically active human 1-antitrypsin in the milk of transgenic mice. Proc Natl Acad Sci USA 1990; 87: 5178–5182.
5. Meade H, Gates L, Lacy E, Lonberg N. Bovine alpha S1-casein gene sequences direct high level expression of active human urokinase in mouse milk. Bio/Technology 1990; 8: 443–446.
6. Lee KF, DeMayo FJ, Aitee SH, Rosen JM. Tissue specific expression of the rat beta-casein gene in transgenic mice. Nucl Acid Res 1988; 16: 1027–1041.
7. Kim SJ, Cho YY, Lee KW, Yu DY, Lee CS, Han YM, Lee KK. Expression of human lactoferrin in milk of transgenic mice using bovine-casein/human lactoferrin cDNA fusion gene. Mol Cells 1994; 4: 355–360.
8. Stinnakre MG, Vilotte JL, Soulier S, L'Harison R, Charlier M, Gaye P, Mercier JC. The bovine alpha-lactalbumin promoter directs expression of ovine trophoblast interferon in the mammary gland of transgenic mice. FEBS Let 1991; 284: 19–22.
9. Carver A, Wright G, Cottom D, Cooper J, Dalrymple M, Temperley S, Udell M, Reeves D, Perey J, Scott A, Barrass D, Gibson Y, Jeffrey Y, Samuel C, Colman A, Garner I. Expression of human 1-antitrypsin in transgenic sheep. Cytotechnology 1992; 9: 77–84.

10. Ebert K, DiTullio P, Barry CA, Schindler JE, Ayres SL, Smith TE, Pellerin LJ, Meade HM, Denman J, Roberts B. Induction of human tissue plasminogen activator in the mammary gland of transgenic goats. Bio/Technology 1994; 12: 699–702.
11. Barash I, Faerman A, Ratovitsky T, Puzis R, Nathan M, Hurwitz DR, Shani M. Ectopic expression of β-lactoglobulin/human serum albumin fusion genes in transgenic mice: hormonal regulation and in situ localization. Transgenic Research 1994; 3: 141–145.
12. Sharma A, Martin MJ, Okabe JF, Truglio RA, Dhanjal NK, Logan JS, Kumar R. An isologous porcine promoter permits high level expression of human hemoglobin in transgenic swine. Bio/Technology 1994; 12: 55–59.
13. Brock JH. Transferrins. In: Harrison PM (ed), Metalloproteins. London: MacMillan Press; 1985: 183–262.
14. Rey MW, Woloshuk SL, deBoer HA, Pieper FR. Complete nucleotide sequence of human mammary gland lactoferrin. Nucleic Acids Res 1990; 18: 5288.
15. Reiter B. Review of nonspecific and microbial factors in milk. Ann Rech Vet 1978; 9: 205–224.
16. Ward P, Lo JY, Duke M, May GS, Headon DR, Conneely OM. Production of biologically active recombinant human lactoferrin in Aspergillus oryzae. Bio/Technology 1992; 10: 784–789.
17. Platenburg GJ, Kootwijk EPA, Kooiman PM, Woloshuk SL, Nuijens JH, Krimpenfort PJA, Pieper FR, de Boer HA, Strijker R. Expression of human lactoferrin in milk of transgenic mice. Transgenic Res. 1994; 3: 99–108.
18. Kim SJ, Lee KK, Yu DY, Han YM, Lee CS, Nam MS, Moon HB, Lee KK. Expression analysis of a bovine β-casein/human lactoferrin hybrid gene in transgenic mice. J. Reproduction and Develop. 1997; 43 (in press)
19. Wright G, Carver A, Cottom D, Scott A, Simons P, Wilmut I, Garner I, and Colman A. High level expression of active human alpha-1-antitrypsin in the milk of transgenic sheep. Bio/Technology 1991; 9: 830–834.
20. Shani M, Barash I, Nathan M, Ricca G, Seafoss GH, Dekel I, Faerman A, Givol D, and Hurwitz DR. Expression of human serum albumin in the milk of transgenic mice. Transgenic Res. 1992; 1: 195–208.
21. Hurwitz DR, Nathan M, Barashi I, Ilan N, and Shani M. Specific combinations of human serum albumin introns direct high level expression of albumin in transfected COS cells and in the milk of transgenic mice. Transgenic Res. 1994; 3:365–375.
22. Clark AJ, Cowper A, Wallace R, Wright G, Simons JP. Rescuing transgene expression by co-integration. Bio/Technology 1992; 10: 1450–1454.
23. Hogan B, Costantini F, Lacy E. Manupulating the Mouse Embryo: Laboratory Manual. New York: Cold Spring Harbor Laboratory Press; 1986: 127–252.

EXPRESSION OF HUMAN LACTOFERRIN IN TRANSFECTED RAT MAMMARY EPITHELIAL CELLS

Haruto Kumura, Yoriko Hiramatsu, Yukako Ukai, Katsuhiko Mikawa, and Kei-ich Shimazaki

Laboratory of Dairy Science
Faculty of Agriculture
Hokkaido University
060 Sapporo, Japan

1. INTRODUCTION

The expression of heterogeneous proteins in mammary gland of transgenic animals is an alternative method of producing recombinant proteins. Several groups have already described initial experiments in different animal systems , mainly in the mouse model system[2]. Although some works reported high expression level with an estimation by electrophoresis, adaptation of protein purification procedure is quite difficult in the mouse model system due to limited availability of mouse milk.

Rats are comparable to mouse as the host with respect to litter size, time to sexual maturity and gestation time and could be expected higher ability to secrete the milk. Since the expression level of genomic rat β-casein in transgenic mouse were low and regulatory sequence elements of bovine β-casein gene seemed to work much better in mammary glands of mouse than that of rat[8], different factors affecting gene expression in rat mammary glands might be involved.

Kim and his co-workers compared expression levels of human lactoferrin (hLf) cDNA under the control of bovine β-casein gene supplemented with some artificial introns by using HC11 cell originated from mouse mammary epithelial cells[6]. They found improvement of expression level by addition of rabbit β-globin intron II and some introns from bovine β-casein gene and finally, produced transgenic mice that gave hLf in their milk at concentration of 150–200 μg/ml[5]. Preliminary studies by using cell culture system are helpful when the expression of a transfected foreign DNA is discussed, however, in the case of cell lines from rat mammary gland, casein gene expression was not confirmed and it remains unclear to availability towards lactogenic hormones[3].

Advances in Lactoferrin Research, edited by Spik *et al.*
Plenum Press, New York, 1998.

In this study, we carried out qualitative investigation of the expression of hLf by using the expression vector described above in primary culture of transfected rat mammary epithelial cells.

2. MATERIALS AND METHODS

2.1. Construction of Bovine β-Casein/hLF Fusion Genes

Expression vector named pBL1 was kindly provided by Dr. Dae-Yeul Yu[6]. In brief, upstream of the hLf coding region was connected to 2.0 kb of 5′-flanking region of bovine β-casein gene and rabbit β-globin intron II. At the 3′ end of the hLf cDNA, exon 8 and 9 of bovine β-casein gene followed by SV-40 poly (A) signal was fused (Fig. 1). Control vector, pBLO was obtained by elimination of hLf cDNA from pBL1.

2.2. Preparation, Culture, and Transfection of Primary Rat Mammary Epithelial Cells

Primary rat mammary epithelial cells (PRME) were isolated by enzymatic dissociation and differential centrifugation according to Richards et al.[7] with some modifications. Mammary glands were isolated from 14-day pregnant Wistar rats and minced with a razor blade. The tissue pieces were dissociated in 199 medium containing 0.1% collagenase (type III, Worthington) at 37°C for 15 hours with gentle shaking. Incubation was continued in 0.05% collagenase and 1.5 U/ml of dispase (Sigma) dissolved in 199 medium for an additional one hour. After collection of cells by centrifugation, cells were washed three times with phosphate buffer saline (PBS) and then, suspended in 0.05% pronase (Kaken Chem.) dissolved in 199 medium. Incubation was achieved at 37°C for 30 min with rotary shaker at a speed just sufficient to move the solution slowly. After filtration through 150-μm filter cloth and centrifugation, cells were resuspended in 199 medium containing few drops of Dnase (Sigma) solution and put on a prepared Percoll (Pharmacia) gradient tube to perform centrifugation at 800 × g for 15 min. Layers enriched with epithelial cells were collected and diluted with 199 medium. After centrifugation, cells were suspended in Dulbecco's modified Eagle's medium containing penicilin at 100 U/ml, streptomycin at 0.1 mg/ml (basal medium). The cells were plated and cultured on a 35-mm plastic culture dishes in 2 ml of basal medium supplemented with 5% fetal bovine serum, 5 μg/ml of bovine insulin (Sigma) and 10 ng/ml of human recombinant epidermal growth factor (EGF; Sigma) at 37°C in 95% air/5% CO_2.

After 2-day incubation, medium was changed and the following day, cells were transfected with 10μg of DNA per dish using calcium phosphate-precipitation procedure described by Chen and Okayama[1]. After transfection, cells were washed with PBS and incubation in basal medium in the presence of fetal bovine serum, insulin and EGF were

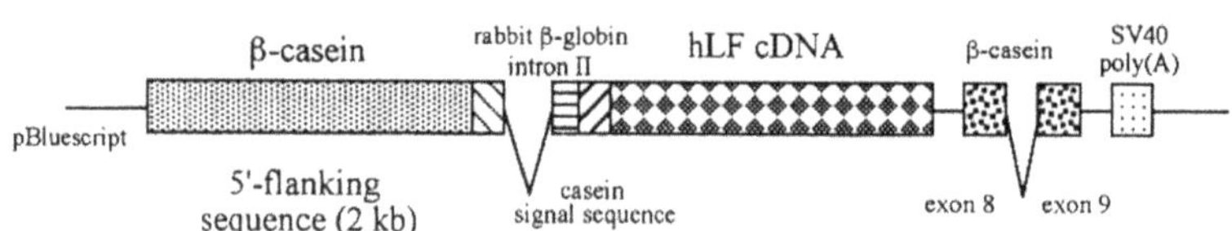

Figure 1. Construction of a bovine β-casein/hLf expression vector.

carried out overnight. Cells were washed by PBS and briefly trypsinized. Cells were seeded on rat tail collagen gel that had been layered onto culture dishes at 1 ml/ 35 mm dish and incubated for 2 days in the same medium.

Hormonal induction was achieved by addition of 5 μg/ml of insulin, 5 μg/ml of ovine prolactin (Sigma) and 1 μg/ml of hydrocortisone (Sigma) in basal medium. Medium was changed every other day and incubation was carried out for 4-days. Media were recovered and used for enzyme-linked immunosorbent assay (ELISA) described below.

3. DETECTION OF EXPRESSED PROTEIN

3.1. Immunofluorescence

Cells on collagen gel were washed with PBS three times and fixed with 4% paraformaldehyde in PBS. Polyclonal rabbit anti-hLf antiserum (Cappel, Division of Copper Diagnostics, Cochranville, PA, USA, diluted 0:1000 in 1% BSA) mouse anti-rat α-casein antiserum (prepared in this laboratory, after purification of α-casein according to procedure of Kaetzel and Ray[4], (diluted 1:3000 in 1% BSA) and rabbit anti-mouse keratin antiserum (Kindly gifted by Dr. F. Nakamura, Hokkaido University, diluted 1 :20 in 1% BSA) were used to detect the corresponded proteins. Bound antibodies were detected by addition fluorescein isothiocyanate-conjugated (FITC) goat anti-rabbit or mouse IgG.

3.2. Detection of hLf and α-Casein by ELISA

Human lactoferrin and α-casein in recovered medium were analyzed by ELISA. Bound antibodies were detected by horseradish peroxydase-conjugated goat anti-rabbit IgG or rabbit anti-mouse IgG. As the substrate, 2,2′-azino-bis(3-ethyl-benzthioazoline-6-sulfonic acid) was used.

3.3. Cellular Localization of β-Galactosidase Activity

To confirm efficiency of the transfection procedure, primary cells were transfected with pZeoSV LacZ. After transfection, cells were incubated for 48 h in basal medium supplemented with 5% fetal bovine serum, 5 μg/ml of insulin and 10 ng/ ml of EGF. Cells were fixed with paraformaldehyde and stained with 4mM $K_3Fe(CN)_6$, 4 mM $K_4Fe(CN)_6$, 2 mM $MgCl_2$ and 1 mg/ml X-Gal in PBS.

4. RESULTS AND DISCUSSION

PRME's were efficiently prepared as clumps from fresh mammary glands by the procedures in this study. The phase contrast microscopic observation of the cells is shown in Fig. 2. Just after the preparation, the cells appeared as clumps, so-caled "organoids" (Fig. 2A). The cells grew well and showed the characteristic cobblestone morphology of epithelial cells (Fig. 2B).

Efficiency of transfection procedure by calcium phsophate precipitation method was confirmed by monitoring with a β-galactosidase expressing construct. Blue signal was observed in the nucleus of epithelial cells and contaminated nerve cells irrespective of transfection, however, the signal was more intense after transfection and some of the epithelial

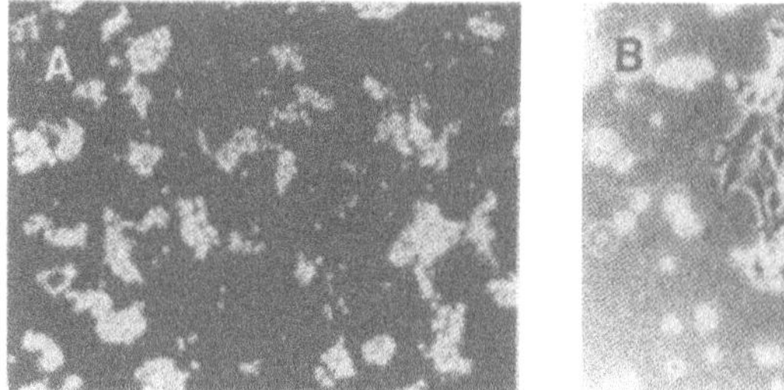

Figure 2. Phase contrast microscopic observation of rat mammary epithelial cells cultured on plastic dishes. (A) acinous fragments (day 0 after seeding); (B) adhering, spreading, and growing cells (day 4 after seeding).

cells were stained entirely in cytoplasm (data not shown). Figure 3A shows fluorescent signal derived from anti-mouse keratin antiserum, which implies colonies of epithelial cells on collagen gel. After transfection, the cells were still able to respond to prolactin, cortisol and synthesize α-casein (Fig. 3B).

Expression level in mammary glands of transgenic animals is frequently much higher when genomic gene is used as the reporter construct compared to cDNA[2]. Despite the pBL1 comprises cDNA of hLF, fluorescent signal derived from hLF-antiserum was more frequently scattered when pBL1 was transfected compared of pBLO (Fig. 4). This

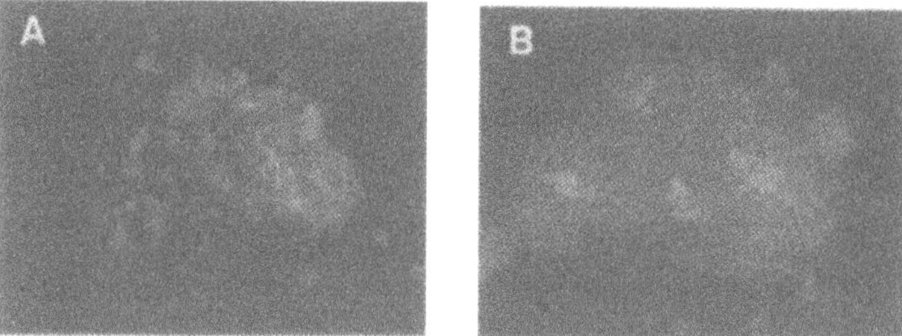

Figure 3. Immunofluorescence showing expression of keratin (A) and α-casein (B) in transfected primary culture of rat mammary epithelial cells.

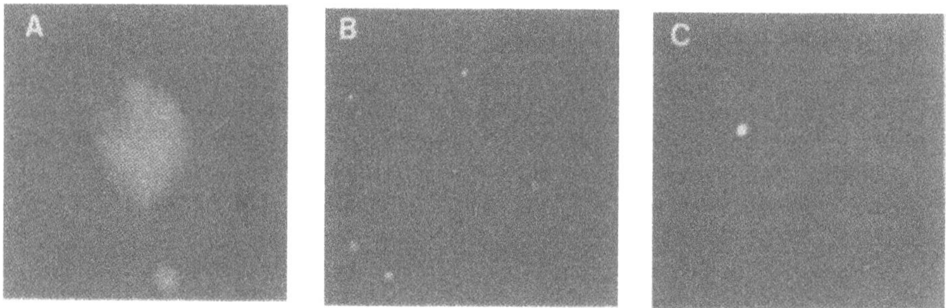

Figure 4. Immunofluorescence showing expression of human lactoferrin in pBL1 (A) or pBL0 (B) transfected primary culture of rat mammary epithelial cells. Untransfected culture was also given (C).

result suggested expression of hLF in transfected rat mammary primary culture. These proteins were thought to be secreted into the culture medium, however, it was failed to quantify expression levels of α-casein and hLF by ELISA, probably due to very low population of the epithelial cells and an expression level under the limit of detection.

Our experimental design required 2-days to adhere trypsinized cells on collagen gel and 4-days induction by lactogenic hormones. Since we observe transient expression, if modification to exclude the 2-days before the onset of induction, it could allow not only favorable duration period of transfected construct, but also sufficient induction effects by lactogenic hormones.

5. CONCLUSION

The expression of recombinant human lactoferrin under the regulation of bovine β-casein gene was studied by using primary culture of rat mammary epithelial cells. Two artificial introns were introduced to expression vector. The cells were prepared from pregnant Wistar rats and transfection was performed by precipitation method of calcium phosphate. After the transfection with the vector containing lactoferrin cDNA, induction of the cell on floating collagen gel was carried out by addition of insulin, hydrocortisone and prolactine. Positive signal against human lactoferrin antiserum was detected. Elimination of the cDNA from the vector resulted in poor response against the antiserum. Intensive reponse against rat α-casein antiserum was recognized irrespective of transfection procedure. These results suggested that isolated mammary epithelial cells sustained its ability to synthesize endogenous milk protein and the vector scheme was suitable for the expression in rat mammary epithelial cells.

ACKNOWLEDGMENTS

We thank Dr. K. Sayama in Shizuoka University, Japan for technical advises and Dr. E. Devinoy in INRA, Jouy en Josas, France for helpful discussion.

REFERENCES

1. Chen C, Okayama H (1987) Mol. Cell. Biol. 7, 2745–2752
2. Houdebine LH (1994) J. Biotechnol. 34, 269–287.
3. Huynh HT, Robitaille G, Turner JD (1991) Exp. Cell. Res. 197, 191–199.
4. Kaetzel CS, Ray DB (1984) J. Dairy Sci. 67, 64–75.
5. Kim SJ, Cho YY, Lee KW, Yu DY, Lee CS, Han YM, Lee KK (1994) Mol. Cells 4, 355–360.
6. Kim SJ, Yu DY, Lee KW, Cho YY, Lee CS, Han YM, Lee KK (1995) J. Biochem. Mol. Biol. 28, 57–61.
7. Richards J, Larson L, Yang J, Guzman R, Tomooka Y, Osborn R, Imagawa W, Nandi S (1983) J. Tissue Cult. Meth. 8, 31–36.
8. Rijnkels M, Kooiman PM, Krimpenfort PJA, De Boer HA, Pieper FR (1995) Biochem. J. 311, 929–937.

11

RESTRICTED SPATIOTEMPORAL EXPRESSION OF LACTOFERRIN DURING MURINE EMBRYOGENESIS

Pauline P. Ward,[1] Marisela M. Mendoza,[1] Odila Saucedo-Cardenas,[1] Christina T. Teng,[2] and Orla M. Conneely[1]

[1]Department of Cell Biology
Baylor College of Medicine
One Baylor Plaza, Houston, Texas 77030
[2]National Institutes of Environmental Health Sciences
National Institutes of Health
Research Triangle Park, North Carolina 27709

1. SUMMARY

Lactoferrin is a member of the transferrin family of iron-binding proteins to which several physiological functions have been ascribed. While there is a wealth of evidence about the distribution and function of this protein in the adult, the expression and function, if any, of lactoferrin during embryogenesis has not been investigated. In the current study, the spatiotemporal distribution of lactoferrin was analyzed during normal murine embryonic development. This analysis demonstrated that lactoferrin is expressed in three distinct patterns during embryogenesis. First, lactoferrin is expressed at the 2-cell stage in the preimplantation embryo where it continues to be expressed until the blastocyst stage when expression ceases. The second phase of lactoferrin expression is not detected until the latter half of gestation when the protein is detected in the myeloid cells, beginning in the fetal liver at embryonic day 11 and later in the spleen and bone marrow coinciding with the onset and diversification of myelopoiesis in these organs during embryogenesis. Finally, lactoferrin is detected in a variety of glandular epithelial cells and/or their secretions, including respiratory and oral epithelia which is consistent with the expression pattern observed for this protein in the adult where it plays an important role in host defense at the mucosal surface. Taken together, these analyses indicate that the role of lactoferrin in the developing embryo is restricted to the preimplantation stage and development of first and second line host defense systems.

Advances in Lactoferrin Research, edited by Spik *et al.*
Plenum Press, New York, 1998.

2. INTRODUCTION

Lactoferrin (Mr = 80 kDa) is a monomeric member of the transferrin family of non-heme iron-binding glycoproteins[1]. The three dimensional structure for lactoferrin has been resolved and it has been shown that the protein is bilobal with a high degree of homology between the N- and C-terminal halves[2]. Each lobe of lactoferrin has the capacity to bind one ferric ion with high affinity, but reversibly. Lactoferrin is a secretory protein and is localized primarily at the mucosal surface where it is secreted by glandular epithelial cells[3,4]. Highest levels of lactoferrin have been detected in the lactating mammary gland where levels up to 6g/l have been detected in milk[5]. Lactoferrin has also been detected in other secretions including nasal, salivary, bronchial, pancreatic, intestinal and genital[3,4,6,46]. Lactoferrin is also expressed in immune cells where it is detected in the secondary granules of neutrophils[7,8].

Several physiological functions for lactoferrin have been proposed based on both *in vitro* and *in vivo* evidence. These functions include antimicrobial activity exerted by bacteriostatic[9,10], bactericidal[11,12] and antiendotoxin[13,14] mechanisms, iron absorption[15,16], cellular growth promotion[17,18] and differentiation[19,20], immunomodulatory activity[21–23] and regulation of myelopoiesis[24–26]. Specific and saturable receptors for lactoferrin have been identified on immune and non-immune cell types and these receptors may play an important role in modulating the pleiotropic functions of this protein[16,27–29].

While much is known about the distribution and function of lactoferrin in the adult, the ontogeny and function, if any, of this protein during embryogenesis remains to be established. In the present study, we have used a combination of in situ hybridization and immunocytochemical analyses to determine the temporal and spatial pattern of lactoferrin mRNA and protein expression during murine embryogenesis. These analysis have shown that lactoferrin is expressed in a restricted spatiotemporal pattern during fetal development. The protein is first detected in the preimplantation embryo at the 2-cell stage and continues to be expressed until the blastocyst stage when expression ceases. The second wave of lactoferrin expression is seen in hematopoietic cells where lactoferrin expression follows the onset and diversification of myelopoiesis during embryogenesis. Finally, the third wave of expression of lactoferrin is detected in glandular epithelial cells and/or their secretions. Our data indicate that lactoferrin may play an important role in both preimplantation embryo development as well as development of first and second line host defense systems.

3. MATERIALS AND METHODS

3.1. Animals

ICR mice were purchased from Harlan Sprague Dawley (Houston, TX). The mice were housed in a 12 hour light/12 hour dark cycle and in compliance with NIH and institutional guidelines.

3.2. Isolation of Preimplantation Mouse Embryos

Mice were superovulated by an intraperitoneal injection of 5 IU pregnant mare serum gonadotrophin (PMSG, Diosynth, Chicago, IL) followed 48 hours later with 5 IU of human chorionic gonadotrophin (hCG, Pregnyl, Organon, Inc, West Orange, NJ) and placed with males. The superovulated females were sacrificed on day 1.5–4.5 of pregnancy and the oviducts and/or uterus were flushed with M2 media (Sigma,) to obtain 1-cell to blastocyst stage embryos.

3.3. Isolation of Postimplantation Mouse Embryos

Matings were set up between ICR mice and the morning of detection of the vaginal plug was designated day 0.5 of pregnancy. At embryonic days E7-E18, pregnant mice were sacrificed and the embryos were isolated.

3.4. Immunohistochemical Localization of Lactoferrin in Postimplantation Embryos

Embryos were fixed in Bouins solution for 2–24 hours depending on the size of the embryo[30]. After fixation, the embryos were washed repeatedly in 70% ethanol and embedded in paraffin. Tissue sections (5 μm) were mounted on microscope slides and were deparaffinized and rehydrated through a graded series of ethanol washes. Endogenous peroxidase activity was quenched in a solution of 3% hydrogen peroxidase in methanol. The sections were incubated for 1 hour in blocking solution (10% normal goat serum) followed by incubation for 3 hours with rabbit polyclonal antiserum directed against mouse lactoferrin (1:400 or 1:2000). Sections were then incubated with a biotinylated goat anti-rabbit secondary antibody (1:500 or 1:2000). Immunoreactivity was detected using streptavidin-peroxidase (Zymed) and using DAB as substrate chromogen (Sigma). Sections were counterstained using hematoxylin and eosin.

3.5. In Situ Hybridization

Preimplantation embryos were fixed directly onto slides using 4% paraformaldehyde. Postimplantation embryos were fixed in 4% paraformaldehyde and embedded in paraffin prior to sectioning. In situ hybridization analysis was performed on 7 μm transverse tissue sections and on embryo whole mounts as previously described[31]. ^{35}S-labeled antisense mouse lactoferrin probe contained nucleotides 1356–2223 of the mouse lactoferrin cDNA[32]. Sense probes displayed no hybridization signal above background.

4. RESULTS

4.1. Expression of Lactoferrin during Preimplantation Development

During the preimplantation phase of embryogenesis, the fertilized egg divides and gives rise to equipotent blastomeres until the 16 cell stage of development. At this stage, the first overt differentiation occurs resulting in the appearance of two distinct lineages, an inner cell mass and an outer trophoectodermal layer. Subsequent cell divisions result in the development of an expanded blastocyst that hatches from the zona pellucida and implants into the uterine wall at E3.5[33,34]. To determine whether lactoferrin is expressed during this period, in situ hybridization was performed using a specific antisense mouse lactoferrin riboprobe. The results from this analysis are illustrated in Figure 1. Lactoferrin mRNA is absent from the developing oocyte (Fig. 1A) but expression begins in the fertilized embryo at the 2–4 cell stage (Fig. 1B) where expression is uniform between the blastomeres until the 16–32 cell stage where staining for lactoferrin mRNA appears to be concentrated mainly in the inner cell mass (Fig. 1C–D). However, lactoferrin message continues to be detected until the blastocyst stage when expression ceases and no mRNA was detected in the hatched blastocyst (Fig. 1E). This restricted pattern of lactoferrin expression suggests that this protein may have an important functional role in the preimplantation embryo.

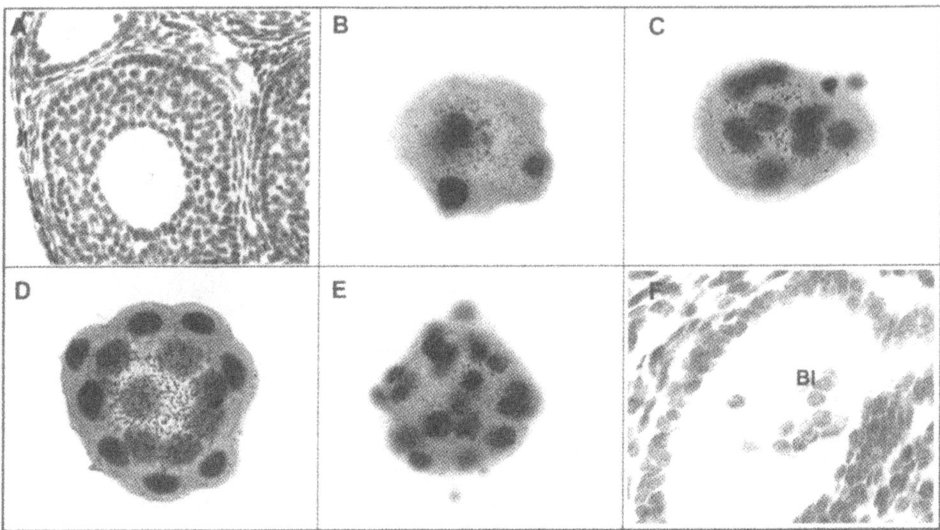

Figure 1. In situ hybridization analysis of lactoferrin expression in the developing oocyte and preimplantation embryo. Embryos were hybridized with specific ^{35}S-labeled antisense mouse lactoferrin probe. Panel A; developing oocyte. Panel B-E; 2–4, 8, 16 and 32 cell respectively. Panel F; hatched blastocyst. Images were obtained using a Zeiss 40x/0.75 objective.

4.2. Lactoferrin Expression Coincides with the Onset and Diversification of Myelopoiesis in the Postimplantation Embryo

Hematopoiesis is a dynamic process during fetal development that begins at embryonic day 7 in the extraembryonic yolk sac and later in the embryonic aorta gonad mesenephrous region (AGM). At this stage, hematopoiesis consists mainly of primitive erythropoiesis and stem cell development. Definitive hematopoiesis does not begin until the latter half of gestation when stem cells from the yolk sac and AGM region colonize the fetal liver at embryonic day 11[35–37]. At later stages of embryonic development, hematopoiesis diversifies from the liver to other organs, mainly the thymus, spleen, and bone marrow, the latter two being major sites of myelopoiesis[38]. To determine whether lactoferrin is likely to play a role in embryonic hematopoiesis, we examined the expression of lactoferrin mRNA and protein during hematopoietic cell development. The results of these analysis are shown in Figures 2, 3 and 4. In situ hybridization analysis demonstrated that lactoferrin was not expressed in the post implantation embryo or extraembryonic yolk sac (Fig. 2A) during the first half of gestation and that expression does not resume until the latter half of gestation when mRNA transcripts are detected in the fetal liver (Fig. 2B).

To gain further insights into the specific cell types where lactoferrin was localized, immunohistochemical analysis was performed using a specific polyclonal IgG directed against murine lactoferrin. The results of this analysis are shown in Figure 3. Consistent with mRNA results, a suppopulation of cells immunoreactive for lactoferrin were detected in the liver at E11 (Fig. 3A). The staining increased as development proceeded reaching a maximum at embryonic day 15 coinciding with the time of maximal hematopoietic activity of this organ. Hepatocytes, erythrocytes and megakaryocytes were clearly devoid of immunoreactivity (Fig. 3B). Analysis of immunoreactive cells at higher

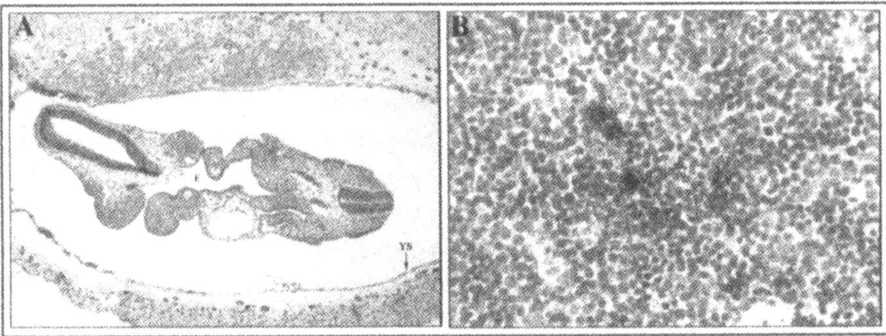

Figure 2. In situ hybridization analysis of lactoferrin expression in hematopoietic sites in the postimplantation embryo. Sagittal sections were obtained from postimplantation embryos and hybridized with a specific ^{35}S-labeled antisense mouse lactoferrin probe. Images were obtained using a Zeiss 40x/0.75 objective. Panel A. Embryo (E) and Yolk sac (YS) at day 8.5 of gestation (20x). Panel B. Fetal liver obtained from day 12.5 embryo (40x).

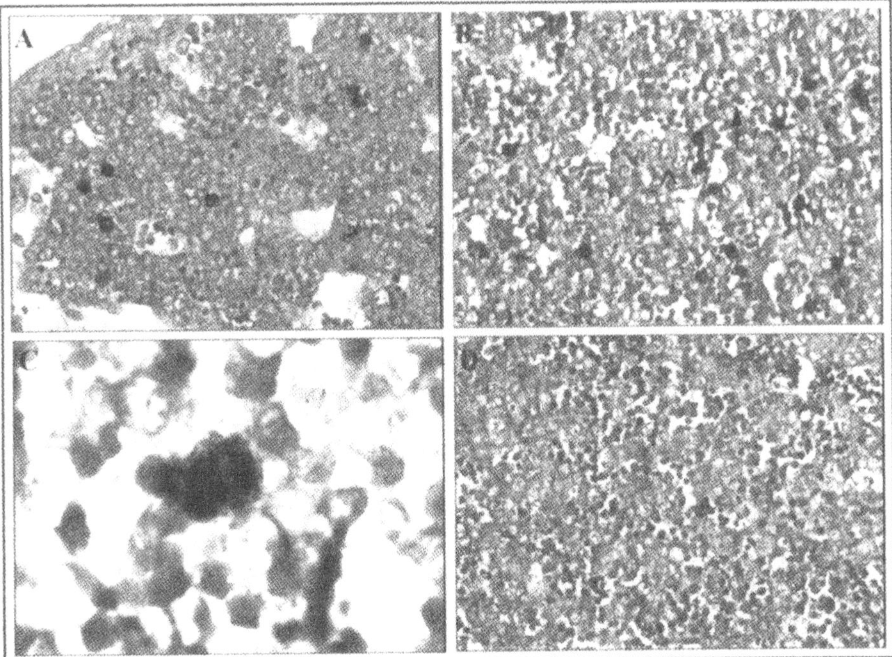

Figure 3. Immunohistochemical localization of lactoferrin in the fetal liver during embryogenesis. Sagittal sections were obtained from Embryos from day 11.5 to 15.5 of gestation. Sections were incubated with a polyclonal antibody directed against mouse lactoferrin Panel A. Liver at E11.5 (40x) showing a few immunoreactive cells staining positive for lactoferrin. Panel B. Liver at E15.5 (40x) showing an increase in lactoferrin staining Note the hepatocytes (*), erythrocytes (↑) and megakaryocytes (arrowhead) are clearly devoid of immunoreactivity. Panel C. Liver at E15.5 (100x) showing a neutrophil with multilobed nucleus which is staining positive for lactoferrin. Panel D. Control section of liver at 15.5 incubated with non-immune serum (40x).

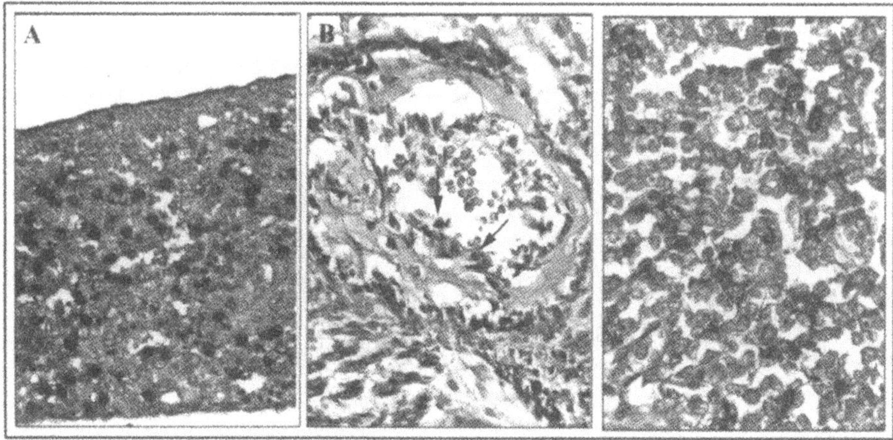

Figure 4. Immunohistochemical localization of lactoferrin in the fetal spleen, bone marrow and thymus. Sections were obtained from embryos at day 16.5 of gestation. Sections were incubated with a polyclonal antibody directed against mouse lactoferrin Panel A. Fetal Spleen, (40x) Panel B, Fetal bone marrow (40x) with arrows pointing to immunoreactive cells. Panel C. Fetal thymus (40x).

magnification (Fig. 3C) showed characteristic neutrophil morphology with a multilobed nucleus and granular cytoplasm. Control sections incubated with nonimmune serum did not show any staining (Fig. 3D).

Consistent with myelopoietic diversification, lactoferrin was also detected in the neutrophils in the bone marrow and spleen during later stages of embryonic development (Figure 4A,B). In the thymus, which is the site of lymphoid cell development, staining for lactoferrin was observed in the supporting epithelial cells. However, a few random lymphoid cells also appeared immunoreactive for lactoferrin although further analysis will be required to conclusively determine the identity and developmental status of these cells.

4.3. Lactoferrin Localization in the Digestive System

Immunohistochemical analysis was used to examine the temporal and spatial pattern of lactoferrin localization in the developing embryonic digestive system. Lactoferrin was detected throughout the developing digestive tract with expression detected in the salivary glands, oral cavity and stomach (Fig. 5A–C). At embryonic day 15, coinciding with the development of the crypt/villi architecture in the intestine, a punctate staining pattern was observed for lactoferrin with specific cells in both the crypt and villi staining strongly for this protein (Fig. 5D). This pattern of localization continued throughout the remainder of embryogenesis (Data not shown).

4.4. Lactoferrin Localization in the Respiratory System

Previous studies have shown that lactoferrin is present in the secretions of the respiratory tract in the adult[3,45]. The immunohistochemical localization of this protein in the embryonic respiratory system was now examined. Consistent with the expression pattern in the adult, lactoferrin was detected in the respiratory system during late embryonic development in the nasal cavities, pharynx and in the bronchioles in the lung (Fig. 6A–C).

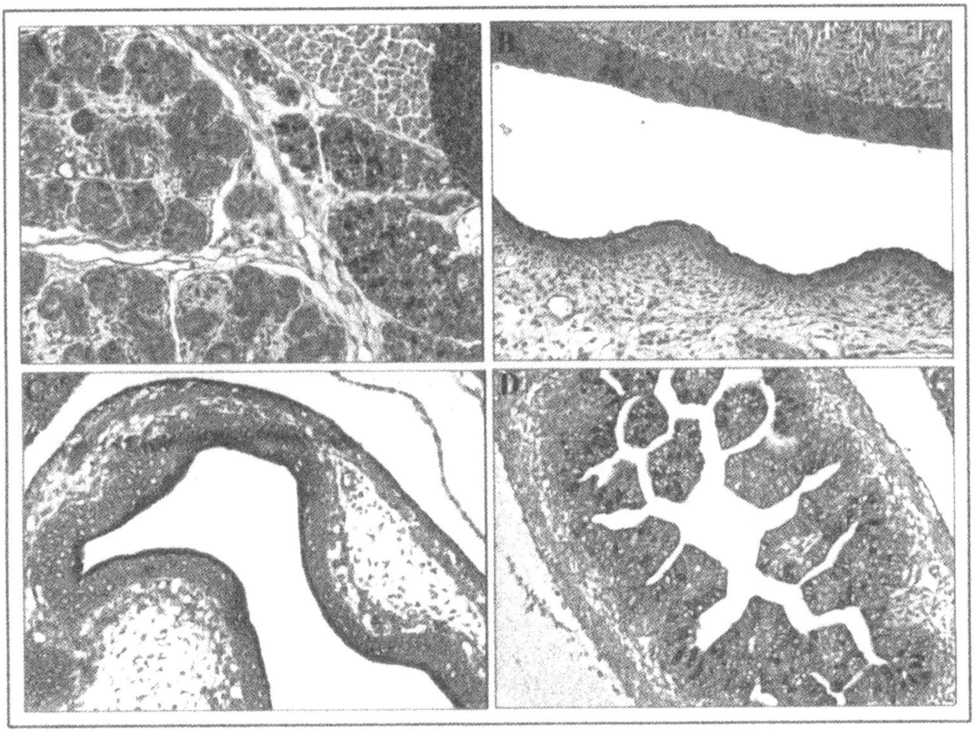

Figure 5. Immunohistochemical localization of lactoferrin in the digestive tract. Sagittal sections were obtained from embryos and incubated with a polyclonal antibody directed against mouse lactoferrin Panel A. E16.5 salivary gland (40x). Panel B. E16.5 oral cavity (40x). Panel C. E16.5 stomach (20x). Panel D. E15.5 gastrointestinal tract (20x).

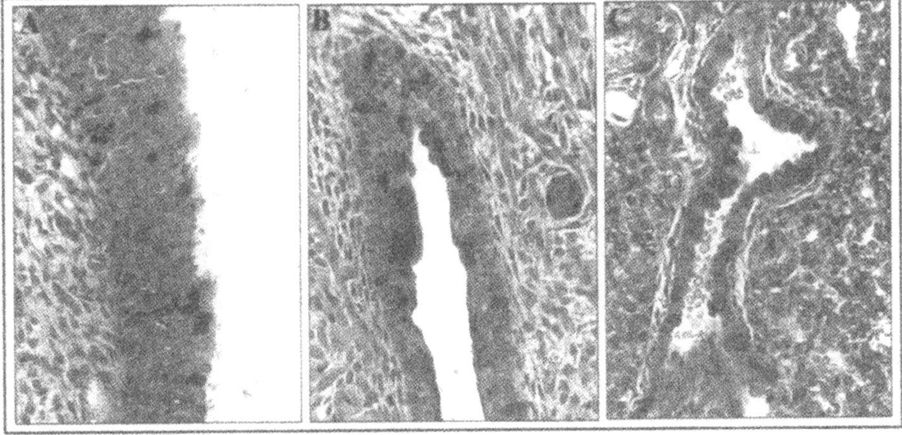

Figure 6. Immunohistochemical localization of lactoferrin in the respiratory system. Sagittal sections were obtained from embryos and incubated with a polyclonal antibody directed against mouse lactoferrin Panel A. E16.5 nasal cavity (40x). Panel B. E16.5 oropharynx (40x). Panel C. E16.5 lung.

5. DISCUSSION

In the present study, the expression and localization of lactoferrin during murine embryogenesis was determined using a combination of in situ hybridization and immunohistochemical techniques. This analysis has identified three distinct spatial and temporal patterns of lactoferrin expression during mouse embryonic development. Lactoferrin expression is first detected in the preimplantation embryo at the 2-cell stage and continues to be expressed until the blastocyst stage when expression ceases. Interestingly, lactoferrin mRNA transcripts are distributed evenly throughout the cells until the 16–32 cell stage. At this stage, two phenotypically distinct type of cells arise in the embryo, the outer polar trophoectodermal cells which mainly give rise to extraembryonic tissue and an inner cell mass which gives rise to the embryo proper aswell as extraembryonic tissue[33,34]. Coincident with this first differentitive step, the lactoferrin mRNA appears to be synthesized primarily by the inner cell mass where it continues to be detected until the blastocyst stage when expression ceases. This tightly restricted expression pattern suggests that lactoferrin may play an important role in the development of the preimplantation embryo.

Definitive hematopoiesis, including myelopoiesis begins in the fetal liver at embryonic day 11 as this organ becomes colonized with stem cells originating from both the yolk sac and aorta gonad mesenephrous region (AGM)[35–37]. Consistent with its detection in the neutrophils in the adult[7], lactoferrin is detected at both the mRNA and protein level in the neutrophils of the fetal liver at E11 where the number of positive cells increases as the hematopoietic activity of this organ expands. At embryonic day 16, myelopoiesis diversifies to the fetal spleen and bone marrow and consistent with the timing of this diversification, lactoferrin is detected at E16 in both of these organs. Interestingly, lactoferrin protein was also detected in the developing thymus. In this organ, staining was detected in cells with characteristic epitheliocyte morphology while random lymphoid cells also appeared positive for lactoferrin. These data suggest that lactoferrin may have a role in thymic lymphocyte development either through the paracrine effect of epitheliocyte derived lactoferrin on a subpopulation of lymphocyte cells or the protein may be expressed in a subpopulation of these cells. However, further analysis is warranted to identify the exact cell types which express and/or uptake this protein.

During mouse embryogenesis, the primitive gut begins as a simple endodermal tube at E9.5. At E15, major differentiation is observed in this organ with the formation of crypt/villi structures characteristic of adult morphology. Coinciding with this first differentiation step, a subset of the cells in the villi and crypts stained strongly for lactoferrin. This punctate staining has been observed in other tissues by us[39] and previously by others[40–42] although the significance remains to be established. In the adult, there is considerable *in vitro* and *in vivo* evidence to suggest that lactoferrin may play an important role in cellular growth and/or differentiation in enteric[17,43] and immune cells[18–20] which may be mediated, in part, by signaling through lactoferrin receptors identified on these cell types[16,27–29]. In the embryo, lactoferrin receptors have been detected on the surface of gastrointestinal cells[44]. Lactoferrin is a secretory protein and while the origin of this protein remains to be established, it is tempting to speculate that the onset and specific localization and/or uptake of this protein in certain cells may be consistent with this protein playing an important role in embryonic gut development.

Finally, lactoferrin was also widely detected in embryonic epithelial cells and/or their secretions in a pattern similar to that previously observed in the adult[3,45,46]. The presence of lactoferrin in the exocrine secretions of late stage embryos is consistent with the hypothesis that lactoferrin is an important regulator of the postnatal non-specific immune system at the mucosal surface. Future studies are now warranted to determine if lactoferrin mRNA and protein colocalize in the same cells which should provide valuable information to decipher possible autocrine and/or paracrine roles of this protein during embryogenesis.

REFERENCES

1. Aisen, P. & Listowsky, I. (1980) Ann. Rev. Biochem. 49, 357–393.
2. Anderson, B. F., Baker, H. M., Norris, G. E., Rice, D.W. & Baker, E.N. (1989) J. Mol. Biol. 209, 711–734.
3. Masson, P. L., Heremans, J. F. & Dive, C. (1966) Clin. Chim. Acta. 14, 735–739.
4. Pentecost, B. T. & Teng, C. T. (1987) J. Biol. Chem. 262,10134–10139.
5. Masson, P. L. & Heremans, J. F. (1971) Comp. Biochem. Physiol. 39, 119–129.
6. Yu, L.-C., & Chen, Y.-H. (1993) Biochem. J. 296, 107–111.
7. Masson, P. L., Heremans, J. F. & Schonne, E. (1969) J. Exper. Med. 130, 643–658.
8. Rado, T. A., Bollekens, J., St. Laurent, G., Parker, L. & Benz Jr., E. J. (1984) Blood 64, 1103–1109.
9. Oram, J. D. & Reiter, B. (1968) Biochim. Biophys. Acta 170, 351–365.
10. Bullen, J. J., Rogers, H. J. & Griffiths, E. (1978) Current Topics in Microbiology and Immunology 80, 1–35.
11. Arnold, R. R., Cole, M. F. & McGhee, J. R. (1977) Science 197, 263–265.
12. Ellison, R. T. & Giehl, T.-J. (1991) J. Clin. Investig. 88, 1080–1091.
13. Applemelk, B. J., An, Y., Geerts, M., Thijs, B. G., DeBoer, H. A., MacLaren, D. M., DeGraaff, J. & Nuijens, J. H. (1994) Infect. Immunity 62, 2628–2632.
14. Elass-Rochard, E., Roseanu, A., Legrand, D., Trif, M., Salmon, C., Motas, C. Montreuil, J., & Spik, G. (1995) Biochem. J. 312, 839–846.
15. Cox, T. M., Mazurier, J., Spik, G., Montreuil, J. & Peters, T. J. (1979) Biochim. Biophys. Acta 558, 129–141.
16. Iyer, S. & Lonnerdal, B. (1993) Eur. J. Clin. Nutri. 47, 232–241.
17. Nichols, B. L., McKee, K., Henry, J. F., & Putman, M. (1987) Pediat. Res. 21, 563–567.
18. Hashizume, S., Kuroda, K. & Murakami, H. (1987) Biochem. Biophys. Res. Commun. 763, 377–382.
19. Zimecki, M., Mazurier, J., Machnicki, M., Wieczorek, Z., Montreuil, J. & Spik, G. (1991) Immunology Lett. 30, 119–124.
20. 20. Zimecki, M., Mazurier, J., Spik, G., & Kapp, J. A. (1995) Immunology 86, 122–127.
21. Zagulski, T., Lipinski, P., Zagulska, A. Broniek, S. & Jarzabek, Z. (1989) Brit. J. Exper. Pathol. 70, 697–704.
22. Crouch, S.P.M., Slater, K. J., & Fletcher, J. (1992) Blood 80, 235–240.
23. Machnicki, M., Zimecki, M. & Zagulski, T. (1993) Intern. J. Exper. Pathol. 74, 433–439.
24. Zucali, J. R., Broxmeyer, H. E., & Ulatowski, J. A. (1979) Blood 54, 951–954.
25. Broxmeyer, H. E., Smithyman, A., Eger, R. R., Meyers, P. A. & deSousa, M. (1978) J. Exper. Med. 148, 1052–1067.
26. Sawatzki, G. & Rich, C. (1989) Blood Cells 15, 371–375.
27. Mazurier, J. Legrand, D., Hu, W.-L., Montreuil, J., & Spik, G. (1989) Eur. J. Biochem. 179, 481–487.
28. Birgens, H. S., Karle, H., Hansen, N. E. & Kristensen, L. O. (1984) Scand. J. Haematol. 33, 275–280.
29. Van Snick, J. L., & Masson, P. L. (1976) J. Exper. Med. 144, 1568–1580.
30. Kaufman, M. H. (1995) The Atlas of Mouse Development, Academic Press.
31. Warburton, D. & Fraser, F. C. (1964) Am. J. Hum. Genet. 16, 1–15.
32. Pentecost, B. T. & Teng, C. T. (1987) J. Biol. Chem. 262, 10134–10139.
33. Johnson, M. H., McConnell, J., & Van Blerkom, J. (1984) J. Embryol. Exp. Morphol 83, 197–231.
34. Rossant, J. (1986) in 'Experimental Approaches to Mammalian Embryonic Development', eds. Rossant, E. J., & Pedersen, R. A. (Cambridge University Press: New York) pp. 97–120.
35. Dzierzak, E &. Medvinsky, A. (1995) TIG 11, 359–366
36. Medvinsky, A. &. Dzierzak, E. (1996) Cell 86, 897–906.
37. Yoder, M. C., Hiatt, K., Dutt, P., Mukherjee, P., Bodine, D. M. & Orlic, D. (1997) Immunity 7, 335–344.
38. Morrison, S. J., Uchida, N., & Weissman, I. L. (1995) Annu. Rev. Cell Dev. Biol. 11, 35–71.
39. Tibbets, T. Personal communication. Lactoferrin immunolocalization in the mouse uterus.
40. Inoue, M., Yamada, J., Kitamura, N., Shimazaki, K.-I., Andren, A. & Yamashita, T. (1993) Tissue and Cell 25, 791–797.

41. McMaster, M. T., Teng, C. T., Dey, S. K. & Andrews, G. K. (1991) Mol. Endocrinol. 5, 101–111.
42. Wichmann, L., Vaalasti, A., Vaalasti, T. & Tuohimaa, P. (1989) Int. J. Androl. 12, 179–186.
43. Heird, W. C., Schwarz, S. W. & Hansen, I. H. (1984) Pediat. Res. 18, 512–515.
44. Kawakami, H. & Lonnerdal, B. (1991) Am. J. Physiol. 261, G841–G846.
45. Masson, P. L., Heremans, J. F., Prignot, J. & Wauters, G. (1966) Thorax 21, 538–544.
46. Miyauchi, J. (1984) Acta Histochem. Cytochem. 17, 177–189

12

CONSTRUCTION OF RECOMBINANT CHIMERIC HUMAN LACTOFERRIN/BOVINE TRANSFERRINS

Henry Wong and Anthony B. Schryvers

Department of Microbiology and Infectious Diseases
Faculty of Medicine
University of Calgary
Calgary, Alberta, Canada

1. INTRODUCTION

One of the contributing factors to bacterial pathogenesis is the ability to acquire iron (Fe) in the extremely Fe scarce environment of the host. The families Neisseriaceae and Pasteurellaceae, which include a number of human and veterinary pathogens, express outer membrane receptors that specifically bind and remove Fe from the hosts' Fe binding proteins, transferrin (Tf) and lactoferrin (Lf)[5]. The bacterial Lf receptor was initially identified by affinity methods with immobilized human Lf (hLf) as the affinity matrix, using high salt and pH conditions to minimize nonspecific interactions[11]. These studies identified a single Fe-repressible outer membrane protein, lactoferrin binding protein (Lbp), in constrast to the two transferrin binding proteins, transferrin binding protein A (TbpA) and transferrin binding protein B (TbpB), that were isolated with a Tf affinity column[11]. Lbp shared several properties with TbpA and once the gene encoding Lbp was cloned[9] it was apparent that the predicted protein sequence was homologous to TbpA.

The obvious parallels between the transferrin and lactoferrin receptors prompted attempts at identifying a TbpB homologue. When lower stringency (low salt and pH) conditions were employed a second putative Lf receptor protein of approximately 75 to 85 kDa named LbpB was isolated from *Neisseria meningitidis, Moraxella catarrhalis* and *M. bovis*[3]. However, recent experiments with *M. catarrhalis* have demonstrated that the 84 kDa protein isolated under low stringency conditions is in fact, another Fe-regulated protein, CopB, and that the authentic LbpB actually comigrates with LbpA and can be isolated under high stringency conditions[4].

A characteristic of the bacterial receptor proteins is their exquisite specificity for host glycoprotein[5]. Thus the human pathogens, *N. meningitidis* and *M. catarrhalis* specifically bind human Lf (hLf) while *M. bovis* will only bind bovine Lf (bLf)[3]. Studies aimed

Advances in Lactoferrin Research, edited by Spik *et al.*
Plenum Press, New York, 1998.

at identifying the region(s) of hLf that is responsible for interacting with the bacterial receptors were attempted. Proteolytically derived N- and C-lobe subfragments of hLf demonstrated that both lobes are capable of binding to the Lf receptors[14]. Obviously, these conclusions are tainted by the lack of understanding of the receptor composition at that time. Furthermore, the approach of using proteolytically derived subfragments of Lf has many inherent limitations, such as the increasing difficulty in isolation and purification of small peptides and the potential loss of tertiary structure that may be required for avid binding interactions. Thus it is apparent that alternative approaches will be required to fully delineate the receptor-ligand interaction.

The ability of the bacterial receptor proteins to distingish between different Lfs initially seems surprising when one considers the high degree of amino acid sequence homology that hLf and bLf share[8]. Crystallographical data for both hLf and bLf are available[1,6,13] yet little is known about their topographical interactions with the bacterial receptors. Nonetheless, the varying amino acids are largely found on the surface regions Lf, particularly in the surface loops connecting the internal alpha-helical segments, and these regions likely dictate the binding specificity to Lbps. The relatively conserved amino acid sequence of the internal folding domains in transferrins and lactoferrins suggest that it should be possible to generate hybrid proteins that will maintain their overall tertiary structure. This approach has been successfully employed to generate chimeric Tfs for mapping the binding epitopes of hTf to Tbps[10]. Thus we are using this innovative approach for delineating the regions on hLf that are responsible for interacting with the bacterial receptors.

2. RESULTS

2.1. Design and Preparation of Hybrid hLf/bTf cDNAs

The objective of this research program was to generate a series of chimeric proteins for mapping the regions of hLf involved in binding to the bacterial receptors. Since the basic N-terminal region of Lf[2] may contribute to its binding properties, we decided to use a bovine Tf (bTf) as a partner for chimeric protein construction, because it does not possess a predominance of basic amino acids in this region. The chimeric protein approach was used in a previous study where a series of hTf/bTf chimeras were produced and evaluated for their interaction with Tf receptors from human pathogens[10]. The PCR-based SOEing (splicing by overlap extension) approach[7] was used to construct the hybrid genes in that study. The SOEing approach has the advantage of enabling the investigator to generate junctions anywhere in the DNA sequence and thus provides tremendous flexibility in construction of the hybrids. However, there are also several disadvantages to this approach. Each specific hybrid needs to be constructed by one or more SOEing reactions and since there is no functional tests to ensure that the original gene segments are not modified, it becomes necessary to completely sequence each hybrid gene that is generated.

The disadvantages of the SOEing approach prompted us to utilize an alternative approach for generating our first generation of hybrid genes. In this approach unique restriction sites are introduced into the two genes of interest at desired junctions without altering the amino acid sequence of the respective genes. The SILMUT (silent mutation) program[12] is used to select the appropriate restriction sites. Once identically placed restriction sites are present in the two genes, it is relatively simple to exchange segments of the individual genes by conventional cloning techniques, and thus generate the desired hybrid genes. Since it is possible to introduce all of the desired restriction sites into the two assembled "parent" genes, one only needs to resequence the parent genes or test for the production of functional protein by these modified genes to ensure that no problematic errors were introduced.

For the construction of the first set of hybrid genes we decided to introduce junctions at interdomain or interlobe regions so that exchange of individual domains would be possible. After aligning the hLf and bTf sequences we analyzed the junctional (interdomain) sequences for the possibility to introduce unique restriction sites. The restriction sites chosen are shown in Figure 1a. It is evident that it was not possible to design unique restriction sites at two of the junctions and that SOEing was necessary to accomplish the desired exchanges. The individual PCR amplified segments were first cloned in the pGEMT vector (Promega), where they can readily be sequenced, and hybrid genes were sequentially assembled in pGEMT. The set of hybrid *hLf/bLf* genes that we have constructed to date is illustrated in Figure 1b.

2.2. Production of Recombinant Chimeric hLf/bTf Proteins with Bac-to-Bac Baculovirus Expression Vector System (BEVs)

The hybrid *hLf/bTf* genes were subcloned into the pFASTBAC1 plasmid and used to prepare recombinant baculovirus using the BEVs (GIBCO BRL) as described in the meth-

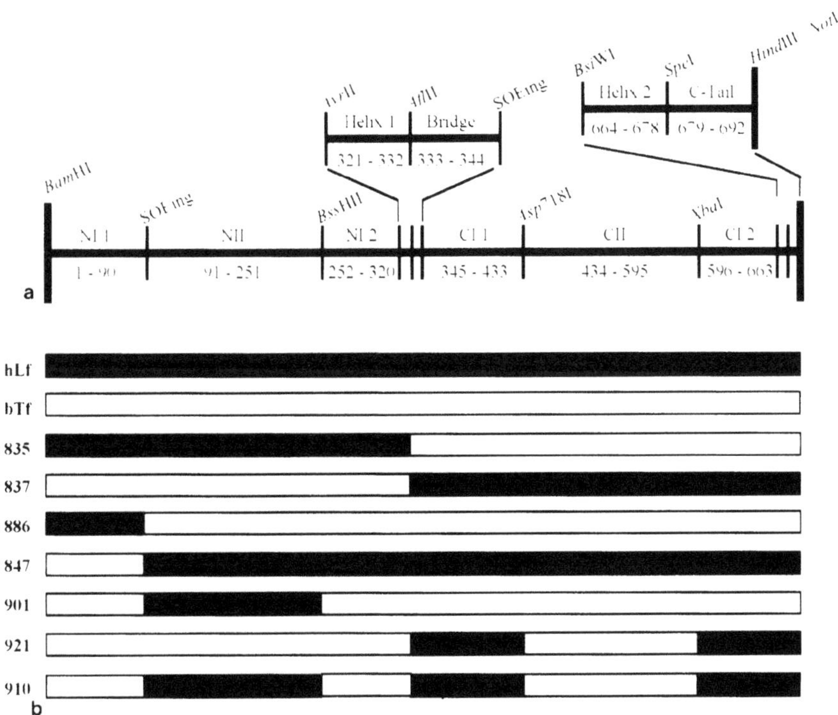

Figure 1. a. Schematic of the hLf/bTf polypeptide. The restriction sites introduced at the interdomain/lobe regions (*slanted*) and the residue numbers, based on [1], corresponding to each region are shown. **b.** Schematic of chimeric hLf/bTfs. The various chimeric *hLf/bTf* constructs are depicted. Regions derived from *hLf* are shown in *grey* and the ones from *bTf* are shown in *white*. hLf: native hLf; bTf: native bTf; 835: hLf N-lobe/bTf C-lobe; 837: bTf N-lobe/hLf C-lobe; 886: hLf N1.1/bTf rest; 847: bTfN1.1/hLf rest; 901: hLf N2/bTf rest; 921: hLf C1/bTf rest; 910: hLf N2+C1/bTf rest.

ods section. Culture supernatants from insect cells infected with the recombinant baculovirus were resolved on 10% SDS-polyacrylamide gels, electroblotted and were analyzed for reactivity against anti-hLf and anti-bTf antisera. The results from this analysis illustrate that recombinant protein of the appropriate size and antibody reactivity was produced (Figure 2). More refined analysis awaits additional production and purification of the individual chimeras.

3. DISCUSSION

The recombinant chimeric hLf/bTf constructs will have a wide range of applications. Binding epitopes of both hLf and bTf to bacterial receptors can be mapped. Once the binding epitopes are identified, synthetic peptides resembling the epitopes could serve as the antigen for generating antibodies with the antigen-binding sites resembling the bacterial receptors' binding domain. These antibodies would in turn be used as the antigen for raising a population of anti-idiotypic antibodies that bear the specificity for the bacterial receptors and be used for vaccine purposes. It is also possible to generate a hybrid that lack the binding epitopes to the bacterial receptors but still retain the attributes of hLf. Such a hybrid would be a safe additive in infant milk formula to counteract the Lf receptor-mediated Fe acquisition by invading pathogens.

4. METHODS

4.1. Incorporation of Restriction Sites in bTf and hLf cDNAs

The various restriction sites were incorporated into the cDNAs by oligonucleotide directed PCR mutagenesis. Different regions of *bTf* and *hLf* were amplified with oligonucleotide primers incorporating the appropriate restriction sites on both 5′- and 3′-termini of the PCR products. Standard PCR conditions used are as follows: 30 cycles of 1 min at 94°C, 1 min at 50°C, and 1–2 min at 72°C. Following the last cycle, a 10 min incubation at 72°C was added to ensure that all the amplified fragments were completely extended to the 3′-terminus. Taq^{+} DNA polymerase (Sangon), containing Pfu, was used to reduce the possibility of misincorporations. The amplified gene fragments were ligated into the pGEMT vector according to the manufacturer's instructions. Clones harboring the inserts in the same orientation were isolated, the hybrid genes were assembled by digestion of plasmids pair that contain the adjacent subfragments of interest (*upstream* and *downstream*). Using the 3′ restriction site of the *upstream* insert, which is the same as the 5′ site

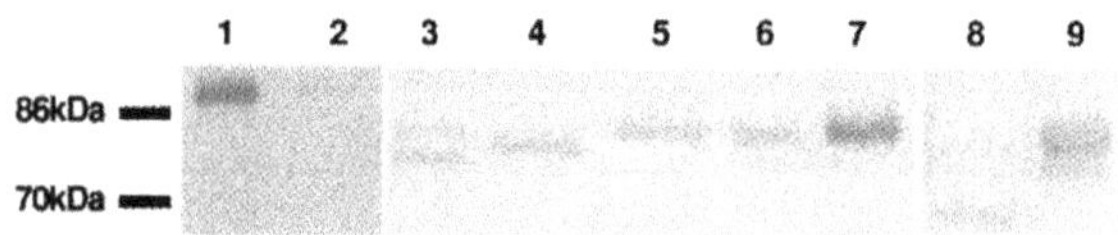

Figure 2. Western blot analysis of recombinant chimeric hLf/bTf constructs. Samples were boiled in the presence of 100mM DTT for 10 minutes prior to being resolved on a 10% SDS-polyacrylamide gel for electroblotting onto an Immobilon membrane. Proteins were detected using anti-hLf and anti-bTf antisera. *Lane 1*: commercial hLf, *lane 2:* construct 832, *lane 3*: commercial bTf, *lane 4*: construct 840, *lane 5*: construct 835, *lane 6*: construct 837, *lane 7*: construct 847, *lane 8*: construct 901, *lane 9*: construct 910.

of the *downstream* insert, and one of the unique vector site available, the appropriate DNA bands were gel purified and ligated together.

4.2. Production of Recombinant Chimeric hLf/bTf Proteins with BEVs

After assembly of the hybrid genes, the different constructs were subcloned from the pGEMT vector into pFASTBAC1. Plasmid DNA from the pFASTBAC1 clones was isolated for transforming *E. coli* strain DH10BAC. White colonies were selected and the high molecular weight plasmid DNA (Bacmid) was isolated for transfection of *Spodoptera frugiperda* cells, grown in serum-free medium (SF900II-SFM, GIBCO BRL), by liposome-mediated transfection according to the BEVs instruction manual. Recombinant proteins were seeked for by analyzing clarified supernatant on western blots probed with anti-hLf and anti-bTf antisera 3 days post-transfection. Since *the S. frugiperda* cells were not co-transfected with wild type *Autographa californica* nuclear polyhedrosis virus, viral plaque purification was unnecessary. Instead, supernatant containing the recombinant virus was used to infect mid-log phase *S. frugiperda* cells to generate high titer viral stocks. Production of the recombinant chimeras involves infecting 5×10^7 *S. frugiperda* cells at mid-log phase, at a concentration of 1×10^7 cells/mL. After a 1 hour incubation at room temperature with gentle rocking, cells were diluted to 1.2×10^7 cells/ml with SF900II-SFM in a T-175 flask (NUNC) and incubated at 27°C for 5 days. *S. frugiperda* cells were clarified from the supernatant by centrifugation for 15 min at 2000 rpm followed by centrifugation for 30 min at 12,000 rpm to remove cellular debris.

ACKNOWLEDGMENTS

This work is supported by the Medical Research Council of Canada grant number MT10350. We thank Dr. J.W. Tweedie for generously providing the clone of hLf. H. Wong and A.B. Schryvers are supported by an Alberta Heritage Foundation for Medical Research Studentship and Fellowship, respectively.

REFERENCES

1. Anderson, B.F., H.M. Baker, G.E. Norris. D.W. Rice and E.N. Baker. 1989. Structure of human lactoferrin: Crystallographic structure analysis and refinement at 2×8 Å resolution. J.Mol.Biol. 209:711–734.
2. Bellamy, W., M. Takase, K. Yamauchi, H. Wakabayashi, K. Kawase and M. Tomita. 1992. Identification of the bactericidal domain of lactoferrin. Biochim.Biophys.Acta Protein Struct.Mol.Enzymol. 1121:130–136.
3. Bonnah, R.A., R.-H. Yu and A.B. Schryvers. 1995. Biochemical analysis of lactoferrin receptors in the Neisseriaceae: Identification of a second bacterial lactoferrin receptor protein. Microbial Pathogenesis 19:285–297.
4. Bonnah, R.A., R.-H. Yu, H. Wong and A.B. Schryvers. 1997. Biochemical and immunological properties of lactoferrin binding protein B from *Moraxella (Branhamella) catarrhalis*. Microb.Pathog. in press.
5. Gray-Owen, S.D. and A.B. Schryvers. 1996. Bacterial transferrin and lactoferrin receptors. Trends in Microbiol. 4:185–191.
6. Haridas, M., B.F. Anderson, H.M. Baker, G.E. Norris and E.N. Baker. 1994. X-ray structural analysis of bovine lactoferrin at 2.5 A resolution. Adv.Exp.Med.Biol. 357:235–238.
7. Ho, S.N., H.D. Hunt, R.M. Horton, J.K. Pullen and L.R. Pease. 1989. Site-directed mutagenesis by overlap extension using the polymerase chain reaction. Gene 77:51–59.
8. Metz-Boutigue, M.H., J. Jolles, J. Mazurier, F. Schoentgen, D. Legrand, G. Spik, J. Montreuil and P. Jolles. 1984. Human lactotransferrin: amino acid sequence and structural comparisons with other transferrins. Eur.J.Biochem. 145:659–676.

9. Pettersson, A., A. Maas and J. Tommassen. 1994. Identification of the *iroA* gene product of *Neisseria meningitidis* as a lactoferrin receptor. J.Bacteriol. 176:1764–1766.
10. Retzer, M.D., A. Kabani, L. Button, R.-H. Yu and A.B. Schryvers. 1996. Production and Characterization of Chimeric Transferrins for the Determination of the Binding Domains for Bacterial Transferrin Receptors. J.Biol.Chem. 271:1166–1173.
11. Schryvers, A.B. and L.J. Morris. 1988. Identification and characterization of the human lactoferrin-binding protein from *Neisseria meningitidis*. Infect.Immun. 56:1144–1149.
12. Shankarappa, B., D.A. Sirko and G.D. Ehrlich. 1992. A general method for the identification of regions suitable for site-directed silent mutations. Biotechniques 12:382–384.
13. Shongwe, M.S., C.A. Smith, E.W. Ainscough, H.M. Baker, A.M. Brodie and E.N. Baker. 1992. Anion binding by human lactoferrin: Results from crystallographic and physicochemical studies. Biochemistry 31:4451–4458.
14. Yu, R.-H. and A.B. Schryvers. 1993. Regions located in both the N-lobe and C-lobe of human lactoferrin participate in the binding interaction with bacterial lactoferrin receptors. Microb.Pathog. 14:343–353.

13

THE LDL-RECEPTOR FAMILY

Lactoferrin and Lipid Metabolism

M. Huettinger,[1] M. Meilinger,[1] Ch. Gschwentner,[1] and H. Lassmann[2]

[1]Institut Medical Chemistry
University Vienna
Währingerstr. 10, A-1090 Vienna, Austria
[2]Institute Neurology
University Vienna
Schwarzspanierstr. 17, A-1090 Vienna, Austria

1. INTERSECTION OF LACTOFERRIN AND LIPOPROTEIN METABOLISM

1.1. Lactoferrin Inhibits Chylomicron Remnant Metabolism in Vivo

Recently we described the specific inhibitory action of lactoferrin on chylomicron remnant (CR) uptake into the intact rat liver. Injection of lactoferrin but not transferrin (7 mg/100 g animal weight) together with radiolabelled CR inhibits in vivo uptake into liver by 50%. No inhibition of uptake into spleen was observed[1]. We demonstrated, that inhibition of uptake is directly connected to endocytosis, since endosomes purified by zonal rotor sucrose gradient centrifugation lack CR-radioactivity almost completely, when prepared from livers of rats that have been injected lactoferrin prior to labelled CR. The radioactivity associated with the liver in the presence of lactoferrin is exclusively located in compartments with a higher density, presumably plasma membranes or associated with endothelial cells. CR do not interact with lactoferrin and are processed normally by lipoprotein lipase in the presence of lactoferrin. Inhibition therefore takes place at an early step of the endocytotic uptake mechanism of CR located on the liver cell plasma membrane. Endocytosis is not blocked totally, as lactoferrin is taken up into endosomes. In addition, in vivo as well as in tissue culture we observed inhibition of CR uptake only and no effect was seen on receptor mediated LDL uptake or uptake of asialoorosomucoid[2].

Advances in Lactoferrin Research, edited by Spik *et al.*
Plenum Press, New York, 1998.

1.2. Structural Homologies Suggest Competition of Lf and Apolipoprotein E for One Receptor

We next determined the structural features underlying this sequel. Here we noted that the effect was lost upon minute disturbances in the protein structure, but was not affected by removal of the carbohydrate chain. As can be deduced from X-ray crystallography analysis data from lactoferrin, positively charged groups are arranged in an alpha helical turn very much alike the arrangement of the positive charges in ApoE, the candidate ligand for the CR-receptor(s). These results suggest that the CR uptake system recognises positive charged aminoacids arrayed like the sequence on ApoE. A discrete, not yet explored, structural feature, on lactoferrin presumably prevents interaction with another ApoE recognising receptor, the LDL receptor.

2. THE LDL-RECEPTOR FAMILY AND THE LIGAND FAMILY

Receptor mediated endocytosis is the mechanism to concentrate plasma circulating substances for highly effective uptake into cells. A vast number of receptors serving this purpose have been described in great detail, for example LDL-receptor and transferrin-receptor. It seemed a rule that receptors recognise and internalise the ligands with high specificity and affinity.

2.1. LDL-Receptor Related Protein (LRP) Is the Chylomicron Remnant Receptor

Since the description of an LDL-receptor related protein (LRP) bearing many apoE binding sites, this protein was assumed to serve removal of apoE rich chylomicron remnant particles via receptor mediated endocytosis into liver with high affinity and specificity. The finding of a second ligand for the putative chylomicron remnant receptor, LRP, the metabolically unrelated protease inhibitor alpha-2-macroglobulin, produced some irritation. By now, some seven ligands are ascribed to be bound by this plasma membrane receptor: apoE, alpha-2-macroglobulin, lactoferrin, PA and/or PA/PAI-1 (plasminogen-activator, PA-Inhibitor-1), RAP (receptor associated protein), Pseudomonas Exotoxin (PE), LPL , tissue factor inhibitor and vitellogenin.

In an attempt to find a unifying explanation for the existence of diverse ligands for one receptor we focused on their binding behaviour to the receptor. By finding groups of ligands that use the same binding site, we expected to see more clearly the method of functions. We have previously reported that lactoferrin competes internalisation of CR but not internalisation of alpha-2-macroglobulin. We have tested whether alpha-2-macroglobulin and CR interfere with each others internalisation and whether lactoferrin competes PA/PAI-1, LPL, RAP or PE internalisation and found interference for all but not with alpha-2-macroglobulin.

The LDL receptor family consists of a number of endocytotically active plasma membrane receptors[3]. They share cysteine rich DE containig type A binding repeats, EGF precursor homology repeats and internalisation signal sequences in varying multiplicity. The different manifolds of repeats determine the variety of ligands they recognise. Thus beside apoE containing lipoprotein mass transport a variety of other functions may be fulfilled by these receptors.

We have described LRP mediated endocytosis of lactoferrin that competes with the apoE mediated uptake of chylomicron remnants. Lactoferrin and apoE seem to compete because they bear the same binding portions for LRP. Lactoferrin is an iron binding protein secreted by many glandular epithelia providing an unspecific immune barrier on surfaces. This function is ascribed to the iron scavenging potency that deprives bacteria of a necessary growth factor. In addition, a proteolytic breakdown product of Lf, a peptide named lactoferricin has bacteriostatic potency different from iron scavenge. Lactoferricin includes the sequence portion we have proposed to be the receptor binding determinant because of its homology to the apoE sequence portion that mediates binding to LRP, which results in competition of Lf and chylomicron remnant uptake in vivo.

2.2. Binding of Ligands to LRP Is Preceded by Reaction with Heparan Sulfate Proteoglycans (HSPG)

Most, if not all ligands of LRP exhibit sequence portions of homology with heparin binding sequences described by Cardin and Weintraub. For apoE and lactoferrin, Glycosaminoglycan binding was investigated in detail[4–6]. It was found that heparin binding homology sequences converge with the putative receptor binding regions. We searched for the occurrence of such regions in a variety of ligands of LRP, the largest member of the LDL receptor family. Peptides synthesised according to these regions in lactoferrin, RAP and PAI-I were tested for their heparin binding ability and their ability to compete binding of the respective holo-ligand. For lactoferrin this region was also described to represent the bactericidal principle of proteolytic digest fragments of lactoferrin (lactoferricin).

3. FUNCTION OF LRP IN MICROENVIRONMENT

3.1. Besides Lipid Mass Transport, LRP Functions as a Multiligand Receptor in Cell Migration/Tissue Repair

LRP is believed to function in lipoprotein mass transport. Its abundance in cells not involved in this process is unclear.

LRP is constitutively expressed on hepatocytes, many epithelia, neurons, astrocytes, smooth muscle cells, fibroblasts and in macrophages/monocytes and their precursors as well as erythroblasts. It is not found on granulocytes and their precursors nor in lymphocytes, indicating a strict differentiation dependency of expression. It was also studied extensively in chicken[7–12]. Lucid concepts, introduced recently, assume a barrier carrier function for transport of classes of molecules or, as we propose, a function in balancing inflammatory activation.

Work in our laboratory has elucidated some features of this receptor. Of great help was the detection the new ligand lactoferrin. We demonstrated binding of the granulocyte derived lactoferrin at low pH[2] and in addition association of lactoferrin with senile plaques in Alzheimer's disease as well as florid foci within brain tissue of multiple sclerosis postmortem specimen—further pointing towards a role in the inflammatory process and the ability of macrophages and astrocytes to metabolise lactoferrin.

3.2. Alzheimer's Disease and Lactoferrin

Lf may be involved: competition with apolipoprotein E[13]. Insoluble Lf deposits in Alzheimer's plaques underline the necessity of intact Lf metabolism in inflammatory foci[13,14].
Next to heart disease, cancer and stroke, Alzheimer's disease (AD) is the fourth leading cause of death in the developed world. Alzheimer's disease is a neurodegenerative disorder

that results in progressive dementia. The disease is classified into three general clinical categories: early-onset familial disease, which occurs before 60 years of age; late onset familial disease, which occurs after 60 years of age; and sporadic late onset disease, with no apparent familial association. Interestingly, epidemiological studies in subjects with late onset Alzheimer's disease demonstrated a high frequency of apoE 4 allele. In biochemical studies a role of apoE 4 in the pathogenesis of the Alzheimer disease also has been suggested[15–19]. The E4 allele is a better ligand to LRP and consequently will be able to displace Lf more effectively from binding to LRP.

Definitive diagnosis of the disease can be obtained only by post-mortem microscopic examination of the cortex. The two histopathological hallmarks of the disorder are intracellular neurofibrillary tangles, consisting largely of abnormally phosphorylated tau protein, and the senile, neuritic plaque, a largely extracellular lesion. The principal component of the neuritic plaques is the 39–43 amino acid β-amyloid protein derived from a much larger integral membrane glycoprotein, termed amyloid precursor protein (APP). The β-amyloid peptide, a proteolytic fragment of APP, encompasses 28 amino acids of the APP ectodomain and the adjacent 11–15 residues of the membrane spanning domain.

The senile plaques also contain a variety of additional elements: the protease inhibitors, antichymotrypsin and alpha-2-macroglobulin, as well as complement factors, apolipoprotein E, lactoferrin and cellular elements including astrocytes and microglia. Our group could also confirm by immunocytochemistry the presence of lactoferrin in senile plaques in sections from hippocampus of AD patients. In addition we show that LRP is expressed in astrocytes and neurons making them capable of regulating Lf metabolism in interdependency with apoE as soon as cell damage caused by βA deposition induces tissue repair.

Lactoferrin is thought to be responsible for primary defence against microbial infection. Many other functions have been attributed to lactoferrin, including immunomodulation and cell growth regulation. It was reported that lactoferrin expression was greatly up-regulated in both neurons and glia in post-mortem brain of AD patients[14].

Last year intriguing new studies appeared providing evidence for glycation of amyloid-β protein, and suggesting that it may play a role in the pathogenesis of AD. Interestingly, a lactoferrin-like molecule on the surface of the endothelial cells and macrophages has been shown to act as a receptor for AGE-proteins (advanced glycation endproducts)[20]. Recently free lactoferrin molecule itself has been confirmed to be able to bind AGE-proteins. Taken together, the molecular interactions between AGE-β-amyloid protein and lactoferrin might provide one of the possible explanations of the colocalisation of these molecules in the senile plaques and elucidate the complex pathogenesis of AD.

4. CONCLUSION

Lactoferrin metabolism mediated by LRP is particularly meaningful when:

- Lactoferrin can be bound by extracellular matrix heparan-sulfate-proteoglycans.
  - The existence of high capacity and low affinity binding sites has been reported.
  - Sequences on lactoferrin responsible for heparin binding are evident.
- Activation of tissue repair processes induce conformational changes in lactoferrin.
  - Cleavage of heparin binding sites on lactoferrin occurs by several proteases.
  - Molecular shape and function changes after this modification.
- Internalisation of modified lactoferrin signals to the cell the inflammatory events occurring outside and prepares the cell for proper response in concert with the other ligands of LRP, which are all molecules of acute phase reaction.

ACKNOWLEDGMENT

This work was funded in part by OENB project 5508.

REFERENCES

1. Huettinger M, Retzek H, Eder M, Goldenberg H. Characteristics of chylomicron remnant uptake into rat liver. Clin Biochem. 1988;21:87–92
2. Huettinger M, Retzek H, Hermann M, Goldenberg H. Lactoferrin specifically inhibits endocytosis of chylomicron remnants but not alpha-macroglobulin. J. Biol. Chem. 1992;267:18551–18557
3. Brown M, Herz J, Goldstein J. Calcium cages, acid baths and recycling receptors. Nature 1997;388:629–630
4. Cardin AD, Demeter DA, Weintraub HJ, Jackson RL. Molecular design and modeling of protein-heparin interactions. Meth. Enzymol. 1991;203:556–583
5. Mann DM, Romm E, Migliorini M. Delineation of the glycosaminoglycan-binding site in the human inflammatory response protein lactoferrin. J. Biol. Chem. 1994;269:23661–23667
6. Ji ZS, Fazio S, Lee YL, Mahley RW. Secretion-Capture Role for Apolipoprotein-E in Remnant Lipoprotein Metabolism Involving Cell Surface Heparan Sulfate Proteoglycans. J. Biol. Chem. 1994;269:2764–2772
7. Morwald S, Yamazaki H, Bujo H, et al. A novel mosaic protein containing LDL receptor elements is highly conserved in humans and chickens. Arter. Thromb. Vas. Biol. 1997;17:996–1002.
8. Nimpf J, Stifani S, Bilous PT, Schneider WJ. The Somatic Cell-Specific Low Density Lipoprotein Receptor-Related Protein of the Chicken - Close Kinship to Mammalian Low Density Lipoprotein Receptor Gene Family Members. J. Biol. Chem. 1994;269:212–219
9. Bujo H, Hermann M, Lindstedt KA, Nimpf J, Schneider WJ. Low density lipoprotein receptor gene family members mediate yolk deposition J. Nutr. 1997;127 Supplement:S801–S804.
10. Bujo H, Hermann M, Schneider WJ, Nimpf J. A new branch of the LDL-receptor family tree: VLDL-receptors. Z. Gastroenterol. 1996;34 Suppl 3:124–126
11. Novak S, Hiesberger T, Schneider WJ, Nimpf J. A new low density lipoprotein receptor homologue with 8 ligand binding repeats in brain of chicken and mouse. J. Biol. Chem. 1996;271:11732–11736
12. Hiesberger T, Hermann M, Jacobsen L. et al. The chicken oocyte receptor for yolk precursors as a model for studying the action of receptor-associated protein and lactoferrin. J. Biol. Chem. 1995;270:18219–18226
13. Bellosta S, Nathan BP, Orth M, Dong LM, Mahley RW, Pitas RE. Stable expression and secretion of apolipoproteins E3 and E4 in mouse neuroblastoma cells produces differential effects on neurite outgrowth. J. Biol. Chem. 1995;270:27063–27071
14. Rebeck GW, Harr SD, Strickland DK, Hyman BT. Multiple, diverse senile plaque-associated proteins are ligands of an apolipoprotein E receptor. the alpha 2-macroglobulin receptor/low-density-lipoprotein receptor-related protein. Ann. Neurol. 1995;37:211–217
15. Weisgraber KH, Roses AD, Strittmatter WJ. The role of apolipoprotein E in the nervous system. Curr. Opin. Lipidol. 1994;5:110–116
16. Lippa CF, Smith TW, Saunders AM, Hulette C, Pulaski Salo D, Roses AD. Apolipoprotein E-epsilon 2 and Alzheimer's disease: genotype influences pathologic phenotype. Neurology 1997;48:515–519
17. Goedert M, Strittmatter WJ, Roses AD. Alzheimer's disease: Risky apolipoprotein in brain. Nature 1994;372:45–46
18. Corder EH, Saunders AM, Strittmatter WJ, et al. Gene dose of apolipoprotein E type 4 allele and the risk of Alzheimer's disease in late onset families. Science 1993;261:921–923
19. Strittmatter WJ, Saunders AM, Schmechel D, et al. Apolipoprotein E: high-avidity binding to beta-amyloid and increased frequency of type 4 allele in late-onset familial Alzheimer disease. Proc. Natl. Acad. Sci. (U S A) 1993;90:1977–1981
20. Schmidt AM, Mora R, Cao R, et al. The endothelial cell binding site for advanced glycation end products consists of a complex: an integral membrane protein and a lactoferrin-like polypeptide. J. Biol. Chem. 1994;269:9882–9888

14

IDENTIFICATION AND ANALYSIS OF A CA^{2+}-DEPENDENT LACTOFERRIN RECEPTOR IN RAT LIVER

Lactoferrin Binds to the Asialoglycoprotein Receptor in a Galactose-Independent Manner

Douglas D. McAbee, David J. Bennatt, and Yuan Yuan Ling

Department of Biological Sciences
University of Notre Dame
Notre Dame, Indiana 46556

1. SUMMARY

We identified a 45 kDa Ca^{2+}-dependent Lf binding protein on rat hepatocytes. Dithiobis(sulfosuccimidylproprionate) (DTSSP)-crosslinked ^{125}I-Lf to a 45 kDa adduct in a Ca^{2+}-dependent manner on intact cells. The ^{125}I-labeled crosslinked complexes were absent when either surface-bound ^{125}I-Lf was stripped prior to crosslinking or an excess of unlabeled Lf was included in the DTSSP reaction. Triton X-100 extracts of hepatocyte membrane ghosts were chromatographed on Lf-agarose, and a 45 kDa polypeptide (p45) was eluted by EGTA. Anti-p45 sera blocked vigorously ^{125}I-Lf endocytosis to intact rat hepatocytes, confirming that p45 functions as the Ca^{2+}-dependent Lf receptor on hepatocytes. Two tryptic fragments of p45 showed 100% identity with internal sequences (Leu^{121}→Lys^{126} and Phe^{198}→Lys^{220}) of the major subunit (RHL-1) of the rat asialoglycoprotein receptor. Antisera against p45 and RHL-1 crossreacted equally well with each protein, and asialoorosomucoid blocked the binding of ^{125}I-Lf to hepatocytes. We did not detect the minor subunits (RHL-2/3) of the rat asialoglycoprotein receptor in p45 preparations from Triton X-100-extracts of hepatocytes, and ^{125}I-Lf bound to immobilized RHL-1 but not to RHL-2/3. Exoglycosidases were used to remove terminally-exposed NeuNAc and α- and β-Gal from bovine Lf glycans, and lectin blotting confirmed that glycosidase-treated Lfs lacked detectable terminal Gal. Unexpectedly, deglycosylated Lf exhibited no loss in its ability to compete with unmodified Lf for binding to isolated hepatocytes. Moreover, β-lactose but not sucrose competed vigorously for ^{125}I-Lf endocytosis by hepatocytes, indicating that Lf binds at or near

Advances in Lactoferrin Research, edited by Spik *et al.*
Plenum Press, New York, 1998.

the carbohydrate-recognition domain of RHL-1. We conclude that RHL-1 is the Ca^{2+}-dependent Lf receptor on hepatocytes and that it binds Lf in a Gal-independent manner.

2. INTRODUCTION

The molecular basis for hepatic metabolism of blood lactoferrin (Lf) has received considerable scrutiny over the last several years. The chylomicron remnant receptor can mediate removal of human Lf injected into rat circulation[1, 2], and hepatocyte surface proteoglycans can bind human Lf[3]. We found that isolated rat hepatocytes bind bovine Lf by both high-affinity (K_d ~ 20 nM), Ca^{2+}-dependent sites and low-affinity (K_d ~ 1 μM), Ca^{2+}-independent sites; only the high-affinity sites are endocytically competent[4]. Because human Lf binds to isolated rat hepatocytes at 4°C with lower affinity (K_d ~ 10 μM) than does bovine Lf, and binding is not enhanced by Ca^{2+} but is blocked by polyanions[5], the chylomicron remnant receptor probably does not constitute the Ca^{2+}-dependent Lf receptors on rat hepatocytes. In this study, we identified a 45 kDa polypeptide (p45) that functions as the Ca^{2+}-dependent Lf receptor on hepatocytes. p45 shares amino acid sequence homology and immunocrossreactivity with the major subunit (RHL-1) of the asialoglycoprotein receptor (ASGP-R). Moreover, Lf binds to RHL-1 at or near RHL-1's carbohydrate-recognition domain, yet in a Gal/GalNAc-independent manner.

3. MATERIALS AND METHODS

3.1. Cells and Protein Preparation

Hepatocytes were prepared from male Sprague-Dawley rats by a modification of a collagenase perfusion procedure as described elsewhere[6]. Viability was determined microscopically by trypan blue exclusion. ASOR was prepared as described previously[7]. Bovine colostrum Lf (commercial preparation >90% pure electrophoretically) was further purified by ion-exchange chromatography as described previously[6]. ^{125}I-Lf had specific activities of 10–62 dpm/fmol. RHL subunits of the asialoglycoprotein receptor (ASGP-R) were purified from Triton X-100 extracts of digitonin-permeabilized isolated rat hepatocytes by AF-agarose affinity chromatography[8].

3.2. Crosslinking Experiments

Hepatocytes prebound with ^{125}I-Lf at 4°C were incubated with 2 mM DTSSP for 60 min at 4°C. The cells were washed twice in EGTA/dextran sulfate-containing buffer supplemented with 0.1% glycine to strip non-crosslinked ^{125}I-Lf from hepatocytes and block unreacted DTSSP, then permeabilized with 0.06% (w/v) digitonin to generate hepatocyte ghosts in the presence of 1 mM PMSF. Hepatocyte ghosts were solubilized in Triton X-100-containing buffer at 4°C followed by sedimentation of detergent-insoluble material; supernatants were subjected to SDS-PAGE and autoradiography.

3.3. Electrophoresis and Western, Lectin, and Ligand Blotting

Samples were subjected to denaturing electrophoresis (SDS-PAGE) on discontinuous Tris-HCl slab gels; polypeptides were visualized by fixation staining in Coomassie

Brilliant Blue R250. In some cases, polypeptides fractioned by SDS-PAGE were electrophoresed to nylon-supported nitrocellulose. Western analysis of protein replicas was done by the method of Burnette[9]. Carbohydrates were detected using a DIG Glycan Differentiation Kit (Boehringer Mannheim) according to the manufacture's instructions. Digoxigenin-labeled lectins used included SNA to detect NeuNAc(α2,6)x and RCA to detect Gal(β1,n)x. Polypeptide renaturation and ligand blotting were done according to the method of Zeng et al.[10] Nitrocellulose was incubated with 1 μg ^{125}I-Lf/mL in the presence of 20 mM $CaCl_2$ for 1 h at 22°C. ^{125}I-Lf was detected by exposure of papers to a phosphor screen and developed on a Molecular Dynamics Storm 840 image analyzer.

4. RESULTS

4.1. Crosslinking of ^{125}I-Lf to Hepatocytes

To identify candidate membrane proteins on isolated hepatocytes that interact with Lf in a Ca^{2+}-dependent manner, we employed the homobifunctional crosslinker DTSSP, a membrane-impermeant reagent that forms amide bonds with primary amines on adjacent proteins positioned within 12 Å of each other. We prebound hepatocytes with ^{125}I-Lf at 4°C and washed the cells in the presence of Ca^{2+} and dextran sulfate to remove ^{125}I-Lf bound to Ca^{2+}-independent low-affinity sites[4]. Non-covalently bound ^{125}I-Lf was removed from the cells by EGTA/dextran sulfate wash, and ^{125}I-Lf-adduct complexes solubilized from hepatocytes with Triton X-100 were detected by SDS-PAGE and autoradiography. Autoradiography detected ^{125}I-Lf-containing bands at M_r of 80, 125 and 260 (Fig. 1A). The 80 kDa ^{125}I-polypeptide associated with DTSSP-treated cells comigrated with ^{125}I-Lf monomer (Fig. 1A). The 125 and 260 kDa bands were reduced or absent when ^{125}I-Lf binding was done in the presence of a 50-fold molar excess of unlabeled Lf (Fig. 1A), when DTSSP was omitted from the incubation (Fig. 1A), or when cells were washed free of bound ^{125}I-Lf prior to addition of DTSSP (Fig. 1B). We reported previously that low amounts of bovine Lf dimers (160 kDa) and trimers (240 kDa) are artifactually generated

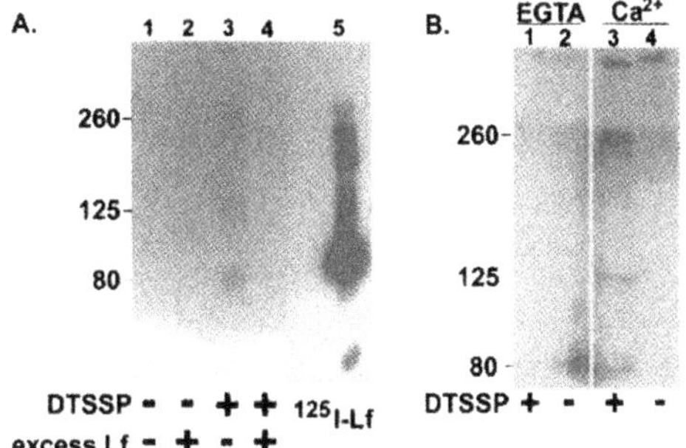

Figure 1. DTSSP crosslinking of Lf to hepatocyte polypeptides. **A.** Isolated rat hepatocytes were prebound with ^{125}I-Lf/mL at 4°C in the presence (lanes 2, 4) or absence (lanes 1, 3) of a 50-fold molar excess of unlabeled Lf, incubated with (lanes 3, 4) or without (lanes 1, 2) DTSSP, and analyzed for ^{125}I-Lf crosslinked complexes as described in Materials and Methods. ^{125}I-Lf incubated with DTSSP similarly but without cells was also analyzed (lane 5). **B.** Hepatocytes prebound with ^{125}I-Lf were washed in the presence of EGTA (lanes 1, 2) or $CaCl_2$ (lanes 3, 4) prior to incubation in the presence (lanes 1, 3) or absence (lanes 2, 4) of DTSSP. ^{125}I-Lf crosslinked complexes were detected as described in Materials and Methods.

during sample preparation for SDS-PAGE and detected when gels are overloaded[6]. While these multimers were detected when ^{125}I-Lf was incubated alone with DTSSP (Fig. 1A), these species did not comigrate with the DTSSP-dependent 125 and 260 kDa bands. In other experiments we have found that the appearance of the 260 kDa species- unlike the 125 kDa species - is not strictly dependent on the presence of DTSSP or Ca^{2+}. The nature of the 260 kDa crosslinked species is unclear. Because monomeric Lf migrates at 80 kDa on SDS-PAGE, these findings suggest that the adduct crosslinked to ^{125}I-Lf to form the 125 kDa species was ~ 45 kDa.

4.2. Purification and Analysis of a 45 kDa Lf-Binding Protein

To purify Ca^{2+}-dependent Lf-binding proteins, we chromatographed solubilized hepatocyte membrane preparations on Lf-agarose under conditions that reduced non-specific binding of Lf to hepatocytes. Briefly, hepatocyte ghosts were washed extensively in Ca^{2+}-free buffers prior to solubilization with Triton X-100 at pH 7.5. The detergent extract was passed sequentially over ethanolamine-agarose and Lf-agarose in the presence of Ca^{2+}. The Lf-agarose was washed with 10 bed volumes in the presence of Ca^{2+} and 0.2% dextran sulfate to remove anionic polypeptides bound non-specifically to Lf-agarose prior to column elution with EGTA. SDS-PAGE of EGTA-eluted fractions revealed a major polypeptide of 45 kDa and a minor band at 90 kDa (Fig. 2, lane 2). We have found that reduction of electrophoresed samples decreases substantially the 90 kDa band with concomitant enrichment of the 45 kDa species, suggesting that the 90 kDa species is an oxidized form of the 45 kDa polypeptide (p45). No other major polypeptide species were purified on Lf-agarose under these conditions. We have also found that p45 is labeled using a membrane-impermeant NHS-ester of biotin (data not shown), indicating that it is a cell-surface protein.

If p45 functions as a Ca^{2+}-dependent Lf receptor on hepatocytes, then antibodies against p45 could bind to the extracellular domain of p45 present on intact cells and prevent Lf binding and endocytosis. Monospecific antiserum against p45 was generated and assayed for its ability to block the interaction of ^{125}I-Lf to isolated rat hepatocytes. Titration of anti-p45 serum progressively reduced total bound (surface and intracellular) ^{125}I-Lf associated with hepatocytes up to 75% compared to pre-immune sera (Fig. 3). Anti-p45 sera also blocked specifically the binding of ^{125}I-Lf to hepatocytes at 4°C (data not shown). Thus, antibodies specific for p45 blocked the binding and endocytosis of Lf by isolated rat hepatocytes. We conclude from all the above findings that p45 is the Ca^{2+}-dependent hepatocyte Lf receptor.

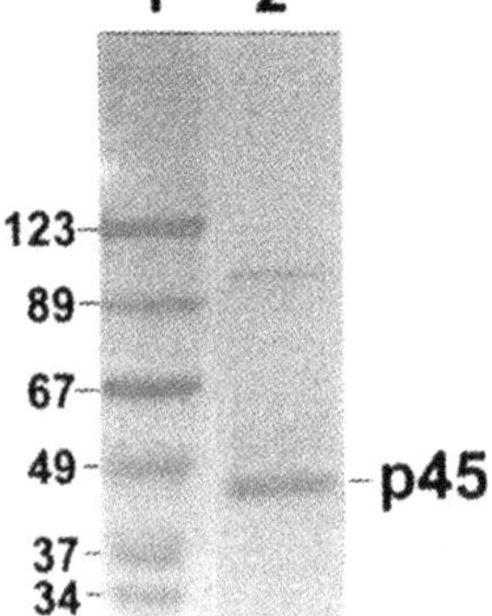

Figure 2. Ligand affinity chromatography of hepatocyte membrane proteins on Lf-agarose. Triton X-100-solubilized membrane proteins from digitonin-permeabilized hepatocyte ghosts were chromatographed sequentially on ethanolamine-agarose and Lf-agarose. Lane 1: molecular weight standards. Lane 2: proteins eluted from Lf-agarose EGTA.

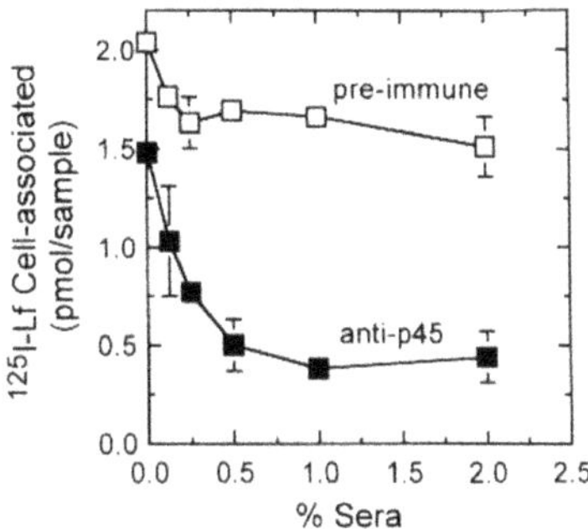

Figure 3. Anti-p45 serum blocks Lf endocytosis by isolated hepatocytes. Hepatocytes (2×10^6 cells/mL) were incubated with ^{125}I-Lf (2 μg/mL) with or without pre-immune serum (□) or anti-p45 serum (■) at 37°C for 45 min. The cells were chilled, washed free of unbound ligand, and assayed for cell-associated radioactivity. Values represent the means of duplicate samples; error bars reflect standard deviations from the means.

4.3. p45 Is Identical to the RHL-1 Subunit of the ASGP-R

To determine the novelty of p45, we electrophoretically purified p45 by SDS-PAGE followed by limited trypsinization and automated NH_2-terminal sequencing of multiple tryptic fragments. Analysis of two fragments gave the sequences LLLHVK and FVQQHMGPLNTWIGLTDQNGPWK which showed 100% identity with $Leu^{121} \rightarrow Lys^{126}$ and $Phe^{198} \rightarrow Lys^{220}$, respectively, of the RHL-1 subunit of the ASGP-R[11]. RHL-1 is a 42 kDa type II membrane protein that is the major component of the rat ASGP-R which consists of three subunits (RHL 1, RHL-2, RHL-3)[12] and binds specifically to Gal- and GalNAc-terminated glycoconjugates[13]. If RHL-1 functions as the Ca^{2+}-dependent Lf receptor, then ligands specific for the ASGP-R should compete with Lf for binding to intact hepatocytes. We found that molar excesses of ASOR - a desialylated form of orosomucoid possessing multiple Gal-terminated, triantennary Asn-linked N-acetyllactosaminyl complex glycan chains-reduced ^{125}I-Lf binding to the Ca^{2+}-dependent Lf binding sites > 90%, comparable to the reduction observed with a 50-fold molar excess of unlabeled Lf (87%; Fig. 4). Also, a molar excess of Lf reduced ^{125}I-ASOR binding to hepatocytes by ~40% (data not shown).

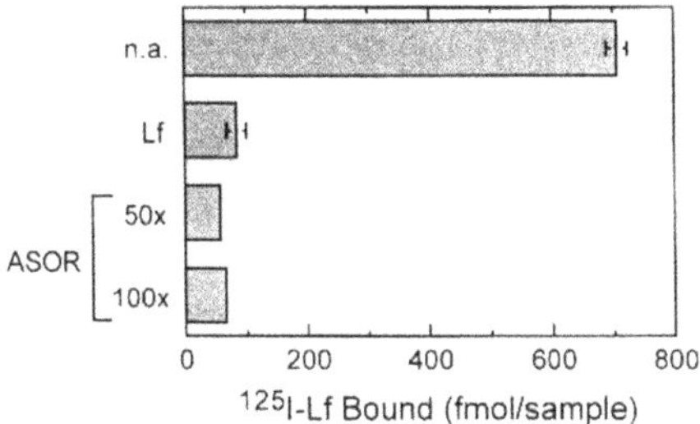

Figure 4. Competition binding of Lf and ASOR on isolated rat hepatocytes. Hepatocytes (2×10^6 cells/mL) were incubated with ^{125}I-Lf (2 μg/mL) at 4°C in the absence (n.a.) or presence of a 50-fold molar excess of unlabeled Lf (Lf) or a 50 and 150-fold molar excess of ASOR. Cells were washed free of unbound ligand and assayed for cell-associated radioactivity. Values are means of duplicate samples; error bars reflect standard deviation from the mean.

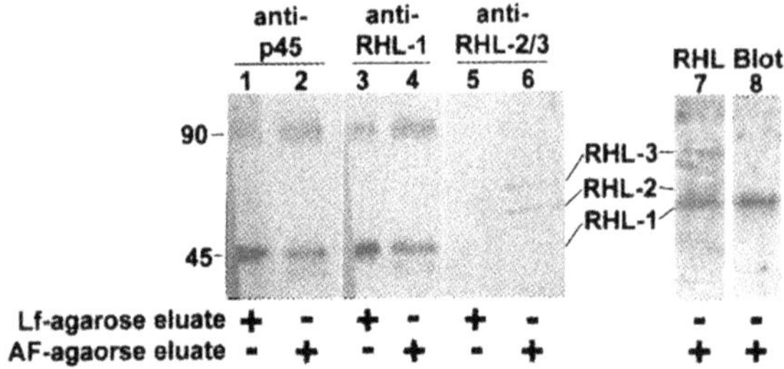

Figure 5. Immunocrossreactivity of p45 and ASGP-R and ^{125}I-Lf binding to immobilized RHL-1. p45 and ASGP-R were purified, subjected to SDS-PAGE, and transferred to nitrocellulose. p45 (Lf-agarose eluate, lanes 1, 3, 5) and ASGP-R (AF-agarose elute, lanes 2, 4, 6–8) were transferred to nitrocellulose and probed with rabbit antisera against p45 (anti-p45, lanes 1 and 2), RHL-1 (anti-RHL-1; lanes 3 and 4), or RHL-2/3 (anti-RHL-2/3; lanes 5 and 6). Bound IgG was detected goat anti-rabbit IgG-alkaline phosphatase conjugate. Immobilized ASGP-R was stained with amido black (lane 7) or incubated with ^{125}I-Lf and analyzed as described in Materials and Methods.

We then examined p45 and RHL subunits for immunocrossreactivity. p45 and RHL subunits were purified by ligand-affinity chromatography, then probed for immunoreactivity using mono-specific antibodies against p45, RHL-1 (carbohydrate-recognition domain), and RHL-2/3. Both anti-p45 and anti-RHL-1 antibodies each detected p45 and RHL-1 in an identical manner (Fig. 5, compare lanes 1, 2 with 3, 4). Antibodies specific for RHL-2/3 detected these proteins in the fractions eluted from AF-agarose (Fig. 5, lane 6), but RHL-2/3 were not detected in fractions eluted from Lf-agarose (Fig. 5, lane 5). We also examined the ability of ^{125}I-Lf to bind to purified RHL subunits immobilized on nitrocellulose. Affinity-purified RHL was subjected to SDS-PAGE, transferred to nitrocellulose, and blocked with BSA in the presence of Ca^{2+} and Tween-20 overnight. Such treatment has been shown to renature the ligand-binding activity of RHL subunits following denaturing electrophoresis[10]. We found that immobilized RHL-1, but not RHL-2/3, bound ^{125}I-Lf (Fig. 5, lane 8), consistent with the conclusion that RHL-1, but not RHL-2/3, interacts with Lf.

4.4. Role of Lf's Glycan Structure in Its Interaction with RHL-1

Ligands specific for various mammalian hepatic lectin systems, when co-injected with Lf, do not reduce Lf clearance from the circulation[14]. In view of these findings, we investigated what role Lf's glycan chains played in its interaction with RHL-1. Bovine Lfs contain up to two biantennary N-acetyllactosaminyl oligosaccharides[15] of which one arm terminates invariably in NeuNAc-(α2,6)-Gal-(β1,4) while the other arm terminates predominately in either NeuNAc-(α2,6)-Gal-(β1,4), Gal-(β1,4)-GlcNAc-(β1,2), or Gal-(α1,3)-Gal-(β1,4)[15,16]. To determine the importance of Lf's glycans in its interaction with RHL-1, we first treated bovine Lf with neuraminidase, α-galactosidase, and β-galactosidase and examined for the presence of terminal α-NeuNAc, β-Gal, and β-GlcNAc by lectin blotting (Fig. 6A). Glycosidase-treated Lf lost all detectable terminal a-NeuNAc (Fig. 6A, lane 4) and β-Gal (Fig. 6A, lane 6) compared to mock-treated Lf (Fig. 6A, lanes 3, 5). Second, we found that ≤ 80-fold molar excess of glycosidase- and mock-treated Lf competed vigorously with ^{125}I-Lf for binding to cells at 4°C (Fig. 6B). We observed no differences in the competition isotherms between the two forms of Lf. Similar results were been obtained using glycosidase-treated human Lf (data not shown). In other experiments,

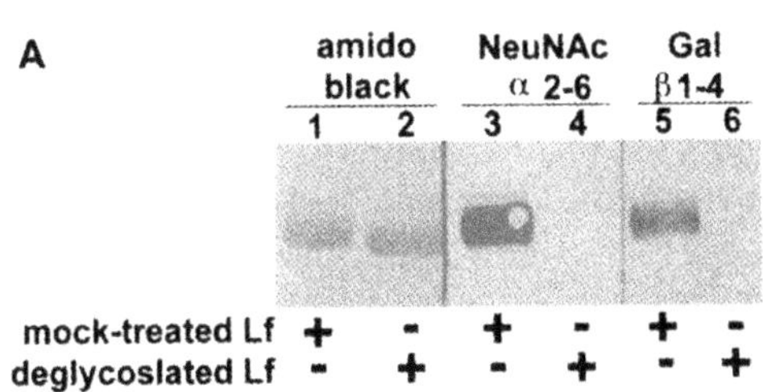

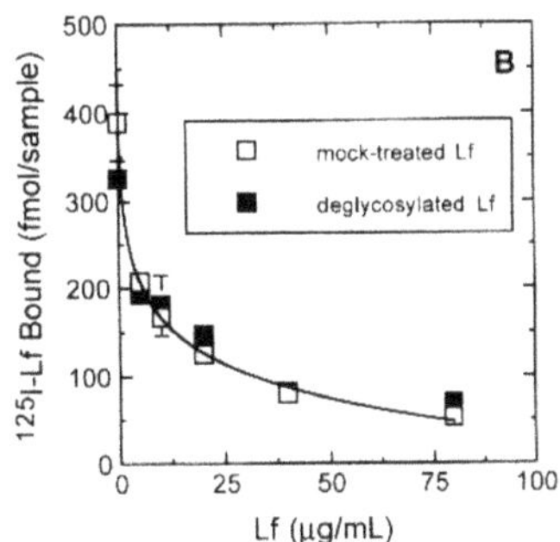

Figure 6. Deglycosylation of bovine Lf and titration of its ligand competition binding activity. **A.** Bovine Lf was treated with neuraminidase, α-galactosidase, β-galactosidase, and N-acetylhexosaminidase. Deglycosylated Lf and mock-treated Lf were subjected to SDS-PAGE and transferred to nitrocellulose. Fractioned Lfs were either stained with amido black (lanes 1, 2) or probed with digoxigenin-labeled SNA to detect terminal NeuNAc (SA a2–6; lanes 3, 4), or digoxigenin-labeled RCA to detect terminal β-Gal (Gal b1–4; lanes 5, 6). Digoxigenin-labeled lectins were detected with anti-digoxigenin IgG. **B.** Hepatocytes (2×10^6 cells/mL) were incubated with ^{125}I-Lf (2 μg/mL) for 90 min at 4°C with or without mock-treated (□) or deglycosylated (■) Lf. Cells were washed and assayed for cell-associated radioactivity. Values represent means of duplicate samples; error bars reflect standard deviations of means.

we purified RHL-1 on affinity columns derivatized with neuraminidase- and β-galactosidase-treated Lf (data not shown). We conclude from all these data, therefore, that RHL-1 binds to Lf in a Ca^{2+}-dependent manner but by a mechanism independent of the presence of terminal Gal or GalNAc moieties on Lfs' glycan structures.

To determine whether Lf bound to the carbohydrate-recognition domain or some other site(s) on ASGP-R, we examined the ability of β-lactose to block the interaction of ^{125}I-Lf to hepatocytes. We found that at large molar excesses, β-lactose blocked hepatocyte interaction of ^{125}I-Lf by ≤ 77% (Fig. 7). An equivalent amount of sucrose had little or no effect on the interaction of ^{125}I-Lf with cells. These findings strongly suggest that Lf interacts with the carbohydrate-recognition domain of ASGP-R.

5. DISCUSSION

We have provided evidence that the Ca^{2+}-dependent Lf receptor on rat hepatocytes is the major subunit (RHL-1) of the ASGP-R. First, ^{125}I-Lf can be crosslinked to an ad-

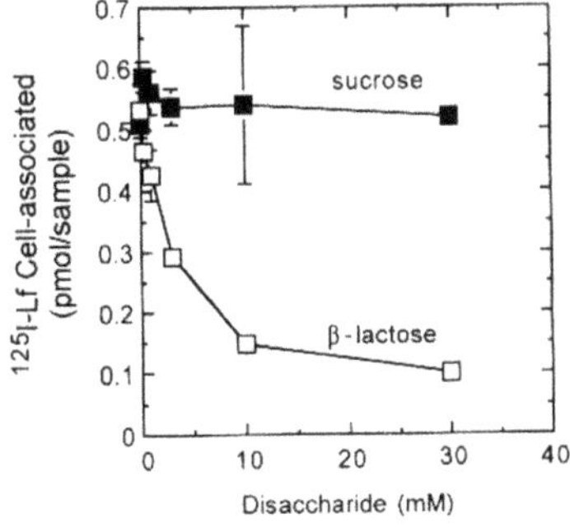

Figure 7. Effect of β-lactose and sucrose on Lf binding to hepatocytes. Hepatocytes (2×10^6/mL) were incubated at 4°C with ^{125}I-Lf (2 μg/mL) with or without β-lactose (□) or sucrose (■) for 90 min. Cells were washed free of unbound ligand and assayed for cell-associated radioactivity. Values represent means of duplicate samples; error bars reflect standard deviations of means.

duct with a molecular weight (45 kDa) close to that reported for RHL-1. Second, a 45 kDa polypeptide was isolated by ligand-affinity chromatography from hepatocyte membrane extracts which shared amino acid sequence identity and immunocrossreactivity with RHL-1. Third, Lf and desialylated ligands competed with each other for binding to hepatocytes. Fourth, ^{125}I-Lf binds to RHL-1 immobilized on nitrocellulose in a Ca^{2+}-dependent manner. We have also found that Lf's interaction with RHL-1 does not require the presence of Gal-terminated glycans even though it binds to RHL-1 at or near the carbohydrate-recognition domain.

Several lines of evidence argue that Lf binds to RHL-1 in a Gal-independent manner. (i) Lf's biantennary complex-type oligosaccharides- in which only one arm terminates in Gal - predicts Lf should bind to the ASGP-R via its terminal Gal sugars with affinity constants in the mM or μM range[17], yet bovine Lf binds to isolated rat hepatocytes with K_d ~ 20–80 nM[6]. (ii) Lf devoid of Gal-terminated oligosaccharides competed as well as unmodified Lfs for binding to cells. (iii) We have expressed two truncated variants of bovine Lf in *E. coli*, one lacking the first 70 NH_2-terminal amino acids, and the other lacking the first 270 N-terminal amino acids. *E. coli* extracts enriched in either of these variant proteins, but not extract from mock-transformed cells, specifically block the interaction of native Lf to the Ca^{2+}-dependent sites on hepatocytes[18]. We conclude, therefore, that Lfs bind to RHL-1 by a mechanism that is independent of Gal- or GalNAc-terminated glycans. To our knowledge, this is the first report showing that the major subunit of the ASGP-R binds to a protein ligand with high affinity in a carbohydrate-independent manner.

Previous reports suggest the ASGP-R is not involved in clearance of plasma Lf based on the observations that ASOR and Lf do not block removal of each other from murine circulation[14, 19]. While ASOR and Lf compete for binding to isolated rat hepatocytes, the two ligands appear to bind ASGP-Rs by different mechanisms. ASOR's interaction with the ASGP-R via its lectin properties involves all three receptor subunits[20, 21] whereas Lf binds to RHL-1 alone. In addition, Lf-unlike ASOR-does not bind to the large intracellular pool of ASGP-R in hepatocytes[22, 23]. It remains to be determined conclusively, therefore, that the lack of ASOR competition for Lf clearance reflects the removal of these two ligands from the circulation by entirely different receptor systems.

One implication of these findings is that Lf clearance from blood and non-blood fluids may be mediated in part by RHL-1-like receptors present on cells of the reticuloendothelial system. For instance, rat peritoneal macrophages possess a 42 kDa Gal/GalNAc receptor[24] that has high homology with RHL-1[25] and is endocytically competent when expressed in COS cells[26]. Notably, Lf has been shown to interact with a 42 kDa polypeptide present on promonocytic U937 cells[27], but it is not known if this polypeptide shares identity with the macrophage 42 kDa ASGP-R. Non-parenchymal liver cells internalize Lf [28], and Kupffer cells exhibit two known Gal-specific lectins, one a 61 kDa polypeptide that binds fucosyl- or Gal-terminated glycoconjugates and shows partial homology with RHL-1[29] and the other, a 30 kDa Gal-particle receptor which is a peripheral membrane protein that shows no immunological crossreactivity to RHL-1 but is homologous to C-reactive protein[30]. Based on our findings here, examination of monocyte/macrophage RHL-1-like receptors for binding and endocytosis of Lf is warranted. Also, studies on plasma Lf content in M2 knockout mice which are deficient in M1 and ASGP-R functional expression[31] should confirm whether or not hepatocyte ASGP-Rs are involved in regulation of hepatic clearance of plasma Lf.

ACKNOWLEDGMENTS

We thank Dr. Paul Weigel for his gift of antibodies to RHL-1 and RHL-2/3. This work was supported by a grant from the National Institutes of Health DK-43355.

REFERENCES

1. Huettinger, M, Retzek H, Hermann M, Goldenberg H. (1992) J. Biol. Chem. 267:18551–18557.
2. Meilinger, M, Haumer M, Szakmary KA, Steinbock F, Scheiber B, Goldenberg H, Huettinger M. (1995) FEBS Lett. 360:70–74.
3. Hu, WL, Regoeczi E, Chindemi PA, Bolyos M. (1993) Am. J. Physiol. 264:G112–G117.
4. McAbee, DD, Nowatzke W, Oehler C, Sitaram M, Sbaschnig E, Opferman JT, Carr J, Esbensen K. (1993) Biochemistry 32:13749–13760.
5. Ziere, GJ, Vandijk MCM, Bijsterbosch MK, Vanberkel TJC. (1992) J. Biol. Chem. 267:11229–11235.
6. McAbee, DD, Esbensen K. (1991) J. Biol. Chem. 266:23624–23631.
7. Schachter, H, Jabbal I, Hudgin RL, Pinteric L, McGuire EJ, Roseman S. (1970) J. Biol. Chem. 245:1090–1100.
8. Ray, DA, Weigel PH. (1985) Anal. Biochem. 145:37–46.
9. Burnette, WN. (1981) Anal. Biochem. 112:195–203.
10. Zeng, FY, Oka JA, Weigel PH. (1996) Glycobiology 6:247–255.
11. Leung, JO, Holland EC, Drickamer K. (1985) J. Biol. Chem. 260:12523–12527.
12. Drickamer, K, Mamon JF, Bins G, Leung JO. (1984) J. Biol. Chem. 259:770–778.
13. Ashwell, G, Harford J. (1982) Annu. Rev. Biochem. 51:531–554.
14. Imber, MJ, Pizzo SV. (1983) Biochem. J. 212:249–257.
15. Coddeville, B, Strecker G, Wieruszeski J-M, Vliegenthart JFG, van Halbeek H, Peter-Katalinic J, Egge H, Spik G. (1992) Carbohydrate Res. 236:145–164.
16. Spik, G, Coddeville B, Montreuil J. (1988) Biochimie 70:1459–1469.
17. Lee, YC, Townsend RR, Hardy MR, Lonngren J, Arnarp J, Haraldsson M, Lonn H. (1983) J. Biol. Chem. 258:199–202.
18. Sitaram, MP, Maloney B, McAbee DD. (1996) FASEB J. 10:1263a.
19. Moguilevsky, N, Retegui LA, Courtoy PJ, Castracane CE, Masson PL. (1984) Lab. Invest. 50:335–340.
20. Herzig, MCS, Weigel PH. (1990) Biochemistry 29:6437–6447.
21. Sawyer, JT, Sanford JP, Doyle D. (1988) J. Biol. Chem. 263:10534–10538.
22. Weigel, PH, Oka JA. (1983) J. Biol. Chem. 258:5095–5102.
23. McAbee, DD, Ling YY. (1997) J. Cell. Physiol. 171:75–86.
24. Kelm, S, Schauer R. (1988) Biol. Chem. Hoppe-Seyler 369:693–704.
25. Ii, M, Kurata H, Itoh N, Yamashina I, Kawasaki T. (1990) J. Biol. Chem. 265:11295–11298.
26. Ozaki, K, Ii M, Itoh N, Kawasaki T. (1992) J. Biol. Chem. 267:9229–9235.
27. Britigan, BE, Ratcliffe HR, Buettner GR, Rosen GM. (1996) Biochim. Biophys. Acta 1290:231–240.
28. Courtoy, PJ, Moguilevsky N, Retegui LA, Castracane CE, Masson PL. (1984) Lab. Invest. 50:329–334.
29. Hoyle, GW, Hill RL. (1991) J. Biol. Chem. 266:1850–1857.
30. Kempka, G, Roos PH, Kolb-Bachofen V. (1990) J. Immunol. 144:1004–1009.
31. Ishibashi, S, Hammer RE, Herz J. (1994) J. Biol. Chem. 269:27803–27806.

15

BACTERIAL LACTOFERRIN RECEPTORS

Anthony B. Schryvers, Robert Bonnah, Rong-hua Yu, Henry Wong, and Mark Retzer

Department of Microbiology and Infectious Diseases
University of Calgary
Calgary, Alberta, Canada

1. SUMMARY

Lactoferrin is thought to play a pivotal role in prevention of infection in the host and its ability to sequester iron from potential pathogens has been considered an important component of its antimicrobial function. A number of bacterial species in the *Neisseriaceae* have developed a mechanism for acquiring iron directly from this host glycoprotein which involves surface receptors capable of specifically binding lactoferrin. Initial attempts at identifying the receptor proteins in *Neisseria* and *Moraxella* species using affinity isolation with immobilized lactoferrin under high stringency conditions presumptively identified a single 100 kDa receptor protein, LbpA (formerly Lbp1). Under modified affinity isolation conditions a second 84 kDa lactoferrin binding protein was isolated and had been presumptively identified as LbpB. This protein was not isolated from a CopB -ve isogenic mutant of *Moraxella catarrhalis,* indicating that it was in fact CopB. However, another lactoferrin binding protein isolated under high stringency conditions, that comigrated with LbpA in most, but not all, *M. catarrhalis* strains, was identified by convalescent antisera. Its biochemical properties suggested that it indeed was LbpB. The identity of these proteins was confirmed by preparing isogenic mutants with the *lbpA* and *lbpB* genes. Growth studies with isogenic mutants deficient in LbpB, LbpA, CopB or FbpA were performed to evaluate their role in iron acquisition from lactoferrin. LbpA and FbpA were essential for this process, supporting prior models of the iron acquisition pathway. LbpB was not essential which is remniscent of studies with the bacterial transferrin receptors. The isogenic CopB -ve isogenic mutants were deficient in iron acquisition from both transferrin and lactoferrin, suggesting that it is a key component in both pathways. A model providing an alternate explanation of the data is presented. The role and surface accessibility of the lactoferrin receptor proteins suggests that they might be useful vaccine antigens and the preferentially reactivity of convalescent antisera with LbpB suggests that it may be the prime candidate.

Advances in Lactoferrin Research, edited by Spik *et al.*
Plenum Press, New York, 1998.

2. INTRODUCTION

The host iron-binding glycoprotein, lactoferrin (Lf), is present on mucosal surfaces and at sites of infection where it is actively released by neutrophils. Lf is proposed to primarily have an antimicrobial role due to iron (Fe) sequestration and to direct interaction with the bacterial surface, properties that have been well established in *in vitro* studies, but for which there is little or no direct evidence *in vivo*[1]. However, the observation that certain bacterial species have evolved mechanisms that are capable of specifically utilizing Lf as a source of iron for growth[2], suggests that the iron withholding properties of Lf has physiological relevance.

The presence of specific mechanisms of iron acquisition from lactoferrin in bacteria that involve surface receptor proteins has largely been based on three demonstrated features; (i) utilization of Fe-Lf as a source of iron for growth, (ii) demonstration of Lf-binding by bacterial cells or isolated membranes, and (iii) identification of Lf-binding proteins by affinity isolation experiments or by binding assays with proteins resolved by SDS-PAGE. However, due its high pI, Lf has the tendency to bind to a number of bacterial surface components such as lipopolysaccharide[3] and porins[4] under 'standard' buffer conditions, indicating that the demonstration of Lf-binding properties of cells, membranes or proteins have to be cautiously intrepreted. The demonstrated exquisite specificity of the *Neisseria* and *Moraxella* receptors for host lactoferrin[2], even under high pH and salt conditions, remniscent of the specificity of the bacterial transferrin receptors[5], clearly distinguishes these interactions from potentially 'nonspecific' interactions under physiological conditions. The presence of functionally relevant lactoferrin receptors in members of the Neisseriaceae has also been conclusively demonstrated by the use of isogenic receptor-deficient strains[2,6,7]. Although there is preliminary evidence suggesting the presence of Lf receptors in other species such as *Helicobacter pylori*[8] and *Staphylococcus aureus*[9], they possess different properties from the receptors found in the Neisseriaceae and their functional significance needs to be established.

3. THE LACTOFERRIN BINDING PROTEINS

3.1. Characterization of LbpA

The initial affinity isolation experiments designed to identify Lf binding proteins used high pH and salt conditions in the binding and wash buffers[10] to avoid potential nonspecific interactions that are notorius for Lf, primarily as a consequence of its high pI[1]. SDS-PAGE analysis in these preliminary studies identified a single 100 kDa protein band (LbpA, formerly Lbp1) in a number of *Neisseria* and *Moraxella* species[11] (Figure 1, lanes 1 & 5, clear arrowhead). Cloning of the gene encoding LbpA from *Neisseria* provided the opportunity for obtaining direct evidence for its role in Lf binding and iron acquisition by expression of recombinant protein and construction of LbpA-ve mutants[2,6,7,12,13]. The predicted protein sequence of LbpA had a high degree of identity with the transferrin receptor protein, TbpA, and moderate homology with other TonB-dependent outer membrane proteins. TonB-dependence is a term applied to outer membrane proteins involved in cell-uptake processes for nutrients such as iron and vitamin B12. The energy for this process is supplied by TonB, an inner membrane protein which is hypothesized to transverse the periplasmic space and directly interact with these uptake proteins. Homology of LbpA to TonB-dependent proteins suggests that it is a transmembrane protein that might ultimately serve as the channel for transport of Fe across the outer membrane[2].

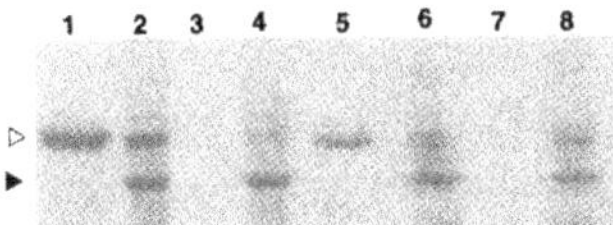

Figure 1. Lactoferrin binding proteins in *M. catarrhalis* and *N. meningitidis*. Lf binding proteins were isolated from iron-deficient membranes from *M. catarrhalis* strain N141 using hLf-Sepharose (lanes 1 and 2) or bLf-Sepharose (lanes 3 and 4) under high stringency (high salt and pH, lanes 1 and 3) or low stringency (low salt and pH, lanes 2 and 4) conditions as described previously[14]. Similarly, Lf binding proteins were isolated from iron-deficient membranes from *N. meningitidis* strain M982 using hLf-Sepharose (lanes 5 and 6) or bLf-Sepharose (lanes 7 and 8) under high stringency (high salt and pH, lanes 5 and 7) or low stringency (low salt and pH, lanes 6 and 8) conditions as described previously[14].

The predicted protein sequence of the meningococcal LbpA has been used to generate a model for its structure which includes 26 transmembrane β-strands, 13 external loops and 12 loops exposed to the periplasmic space[12]. The only experimental evidence currently supporting this model is the reactivity of a monoclonal antibody against a peptide region in predicted external loop 4 with intact cells. We subsequently initiated proteolytic susceptibility studies with *M. catarrhalis* in an attempt to evaluate the surface accessibility of the Lf binding proteins. When intact cells or solublized membranes from different *M. catarrhalis* strains were exposed to proteases and then used for affinity isolation experiments with immobilized hLf, only two prominant peptides of 51 and 37 kDa, respectively, were detected[15]. Reactivity with anti-LbpA antisera and N-terminal amino acid sequence analysis of these peptides revealed that they were derivatives of LbpA. The N-terminal amino acid sequence of the 51 kDa polypeptide, DASQAL..., localizes to the predicted external loop 5 when alignments between the LbpAs from *N. meningitidis* and *M. catarrhalis* are made (data not shown). Thus these results support the proposed model for LbpA structure. The N-terminal sequence of the 37 kDa polypeptide, AAPSAPV..., aligns to a region very near the N-terminus of LbpAs and TbpAs from other bacterial species that is in the periplasm according to the proposed LbpA model[15]. Although these inconsistencies with the proposed model could be attributable to deficiencies in the experimental methodologies, it strongly suggests that the proposed structure of the N-terminal region should be reevaluated.

3.2. Characterization of LbpB

Due to the propensity of Lf to bind to other molecules by virtue of its high pI[1], the initial experiments for isolating bacterial Lf receptor proteins by affinity methods with immobilized Lf involved high stringency conditions[10]. Standard SDS-PAGE analysis of lactoferrin receptor proteins isolated under these conditions revealed what appeared to be a single protein band of approximately 100,000 molecular weight. Since the bacterial transferrin (Tf) receptors were shown to consist of two receptor proteins, Tf binding protein A (TbpA), the homologue of LbpA, and Tf binding protein B (TbpB), a lipoprotein[5], the existence of a second lactoferrin receptor protein was anticipated. Therefore affinity isolation experiments were performed under modified conditions, particularly under lower salt concentrations and lower pH conditions, to determine if additional Lf-binding proteins could be isolated. This resulted in the isolation of a second Lf binding protein of 84 kDa from isolates of *Neisseria meningitidis, Moraxella catarrhalis* and *M. bovis*, presumptively identified as LbpB (Figure 1, lanes 2 and 6, solid arrowhead)[14].

After demonstrating the isolation of a 100 kDa LbpA and a 84 kDa 'LbpB' from several strains of *M. catarrhalis*, further studies were performed to examine receptors in a larger collection of isolates. An additional band migrating just below LbpA was observed in several isolates, and this band was also isolated, albeit less efficiently, under high stringency conditions[15]. This prompted us to re-evaluate whether the presumptive identification of the 84 kDa protein as LbpB was valid. Since a previously described iron-regulated outer membrane protein, CopB[16], was of a similar molecular weight, we initiated studies to specifically determine if the 84 kDa protein was CopB. We tested the reactivity of the 84 kDa protein with an anti-CopB monoclonal antibody (mAb) and performed the low stringency affinity isolation experiment with a CopB-ve mutant. The inability to isolate the 84 kDa protein from the CopB-ve mutant and the reactivity of the 84 kDa protein with the anti-CopB mAb conclusively demonstrated that the 84 kDa Lf binding protein was in fact CopB[15] and not LbpB. The identity of the 84 kDa protein in *N. meningitidis* is not as readily apparent, as preliminary experiments with an isogenic mutant deficient in FrpB, a protein with substantial homology (71% similarity, 51% identity) to CopB, resulted in continued isolation of the 84 kDa protein. This suggests that the 84 kDa protein in *N. meningitidis* is not FrpB and thus might be LbpB or possibly another CopB homologue.

In order to determine whether the additional high molecular weight (95 kDa) Lf binding protein found in in several strains of *M. catarrhalis* was LbpB, several of its properties were examined. We chose avenues of investigation based on the precedent of certain Tbp properties. Since the ability to bind labeled transferrin after SDS-PAGE and electroblotting was a characteristic property of TbpB and not TbpA, we examined the ability of the affinity isolated proteins from a number of *M. catarrhalis* isolates to bind labeled human Lf (hLf). These experiments demonstrated that the putative 'LbpB', and not LbpA, was capable of binding HRP-hLf[15]. Furthermore, the observation that convalescent antisera reacted with TbpB and not TbpA[17], prompted us to examine sera reactivity of the proteins isolated with Lf affinity resins. The results demonstrated that there was little or no reactivity against LbpA but strong reactivity against the putative 'LbpB' (Figure 4)[15,17]. Finally, we examined the specificity of binding by performing affinity isolation experiments with immobilized hLf and bovine Lf (bLf). As with the transferrin receptor proteins (TbpA&B), LbpA & LbpB were only isolated with glycoprotein from the natural host (hLf, Figure 1, lane 1) and not with Lf from a different host species (bLf, lane 3). Taken together these results strongly suggest that the additional high molecular weight protein is LbpB and the increasing genetic evidence for a gene encoding this protein[12,18] further strengthens this conclusion.

Although 'LbpB' was evident in some strains of *M. catarrhalis,* it was not readily resolved from LbpA (Figure 1, lane 1) in many of the strains, even when attempting to use SDS-PAGE conditions designed to optimally resolve proteins of this size. However, the reactivity of this protein band with convalescent antisera provided presumptive evidence for the presence of LbpB, and in some instances, we also demonstrated binding by HRP-hLf[15,17]. In addition, antiserum that had been prepared against purified receptor from one such strain, reacted with both LbpA and LbpB thus implying that both proteins were present in the original receptor preparation[15]. This reactivity with both LbpA and LbpB was also observed with antiserum prepared against receptor isolated from a strain of *M. bovis,* suggesting that this receptor preparation also contained both receptor proteins even though only a single band was evident on SDS-PAGE gels.

In contrast to the results we obtained with LbpA, we were unable to isolate any peptides derived from LbpB after exposing intact *M. catarrhalis* cells to protease[15]. These results suggest that LbpB may be largely exposed at the cell surface and might also account for the fairly strong immune response against this protein that is observed in convalescent antiserum (Figure 4)[17]. These characteristics, also found in TbpB, have been attributed to the topology of a largely hydrophilic protein which is anchored into the outer membrane by fatty acyl chains linked to the N-terminal cysteine[5]. Although the lipoprotein nature of LbpB has not been established, the predicted protein sequence of the *lbpB* gene contains a signal peptidase II recognition sequence with an N-terminal cysteine[18] that is normally lipidated during processing.

3.3. Characterization of CopB

CopB is an iron-regulated, 81–84 kDa, surface-exposed, major outer membrane protein in *M. catarrhalis* that has been previously characterized[16]. Although no specific function was attributed to this protein, its predicted amino acid sequence was similar to TonB-dependent outer membrane proteins of other gram negative bacteria[16], implying that it may be involved in transport processes across the outer membrane. In our affinity isolation experiments using immobilized hLf under low pH and salt conditions CopB was effectively isolated (Figure 1, lane 2, solid arrowhead)[15]. Unlike LbpA and LbpB, CopB is also readily isolated when bLf is used as the affinity ligand (Figure 1, lane 4). It is interesting that a protein of a similar molecular weight is also isolated from strains of *N. meningitidis* using hLf (Figure 1, lane 6) or bLf (lane 8) under these conditions. This presumably may be a CopB homologue and one might have presumed it would be FrpB, due to its known homology with CopB[19,20]. However, our preliminary results with an isogenic *frpB* mutant suggests that the 84 kDa protein isolated by immobilized hLf is not FrpB, thus implying that there is another CopB homologue present in the meningococcus. Searches of the partially completed gonococcal genome has unfortunately not yet revealed any additional genetic loci encoding a obvious CopB homologue.

4. Lbp-LACTOFERRIN INTERACTIONS

The Lbp proteins possess the characteristic specificity for the proteins from the host that is also observed with the Tbp proteins and if a similar mechanism of iron removal is utilized by these two types of receptors, the ligand-receptor interaction would be expected to be similar. The overall conclusion from a series of studies evaluating the Tf-Tbp interaction in different bacterial species is that TbpA reacts primarily with regions on the C-lobe of Tf and that TbpB may interact with regions on both the C-lobe and N-lobe[5]. The comparable studies on the Lf-Lbp interaction concluded that regions on both lobes of Lf are involved in binding to LbpA[21] but these results will have to be reinterpreted in light of the recent identification of LbpB and particularly in view of the comigration of the two receptor proteins in strains of *M. catarrhalis*. The development of more refined tools for delineation of the receptor-ligand interaction, such as recombinant chimeric transferrin-lactoferrins[22], should greatly facilitate this analysis.

5. THE *lbp* OPERON

Prior studies reported on the cloning and sequencing of the the *lbpA* gene from *Neisseria meningitidis*[12] and *N. gonorrheae*[13]. In spite of the limited upstream sequence that was available, the presence of an *lbpB* gene was inferred by the presence of a predicted amino acid sequence that was characteristic of the known TbpB proteins. We have subsequently

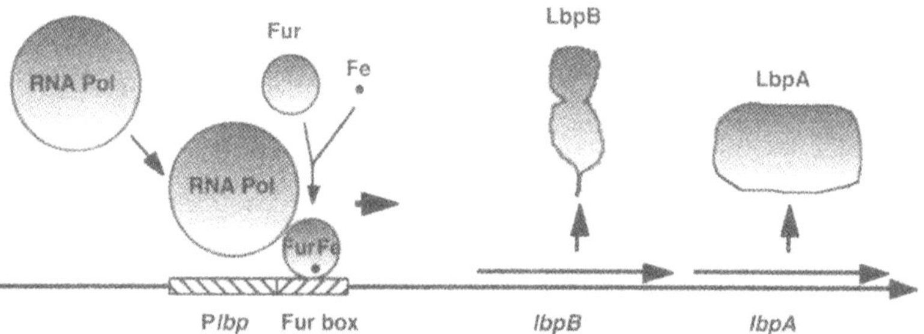

Figure 2. Proposed organization and regulation of the *lbp* operon. Schematic illustration of the organization of the *lbp* operon in *M. catarrhalis* and *N. meningitidis*. The *lbpB* gene, encoding LbpB, is immediately upstream of the *lbpA* gene encoding, LbpA. The two genes constitute an operon that is regulated by a promoter found immediately upstream of the *lbpB* gene. In the presence of increasing concentrations of Fe^{++}, a complex of Fe^{++} and the iron repressor, Fur, is formed. The repressor complex binds to the Fur box which overlaps the promoter region and inhibits binding of RNA polymerase and thus inhibits production of the mRNA required for production of LbpB and LbpA.

cloned and sequenced the upstream region and identified an open reading frame representing the entire *lbpB* gene[18]. Insertion of an antibiotic resistance cassette containing transcriptional terminators in the *lbpB* gene eliminates expression of both LbpB and LbpA, confirming that the two respective genes constitute an operon. There is a promoter region immediately upstream of the *lbpB* gene in which the −10 site is overlapped with a sequence with high identity with the Fur box consensus sequence[18]. This implies that the iron regulation of expression of these genes is mediated by the Fur repressor protein as depicted in Figure 2. However, more direct experimental evidence will be required to support this proposal.

The genes encoding LbpB and LbpA from *M. catarrhalis* have also been cloned and sequenced (Sheena Loosemore, unpublished observations) and have a similar arrangement as the meningococcal genes. As with *N. meningitidis,* expression of both LbpB and LbpA is abrogated by insertion of an antibiotic resistance cassette containing transcriptional terminators into the *lbpB* gene (data not shown). This indicates that the two genes constitute an operon and infers that an iron-regulated promoter is also present upstream of the *lbpB* gene.

6. THE IRON ACQUISITION PATHWAYS

Due to the more extensive studies on the bacterial Tf receptor-mediated iron acquisition pathway, several models for the pathway have been proposed which have similar features[5,23]. The obvious parallels between the Tf and Lf receptors lead to the obvious assumption that the organization of the Lf receptor-mediated pathway will be essentially the same as the Tf receptor-mediated pathway[2]. It is clearly advantageous to make comparisons between the two pathways but direct evidence is nevertheless required for establishing the organization of the Lf receptor-mediated pathway.

Isogenic LbpB mutants of *N. meningitidis*[18] and *M. catarrhalis* (data not shown) are not noticably impaired in growth on Lf as the sole iron source *in vitro* (Table 1). This is somewhat remniscent of the studies with isogenic TbpB-ve mutants in *N. gonorrheae*[24] and *H. influenzae*[25] which were capable of supporting growth *in vitro* with Tf as the sole iron source. In spite of a lack of direct evidence for a role in iron acquisition, the Lf binding properties and apparent surface accessibility of LbpB suggests that it participates in this process (Figure 3A), possibly by mediating initial binding of Lf *in vivo*.

Table 1. Growth of isogenic mutant strains

	Growth on indicated iron source		
Strain*	Lf	Tf	Study/source
wild-type	+++	+++	
LbpB-ve	+++	+++	18
LbpA-ve	—	+++	2,6,7
TonB-ve	—	—	31
FbpA-ve	—	—	26
CopB-ve	—	—	15,16
FrpB-ve	+++	+++	19,20

*LbpB-ve and LbpA-ve mutants of *N. meningitidis* and *M. catarrhalis* were used. TonB-ve, FbpA-ve and FrpB-ve mutants were only from *N. meningitidis*. The CopB-ve mutant was in *M. catarrhalis*.

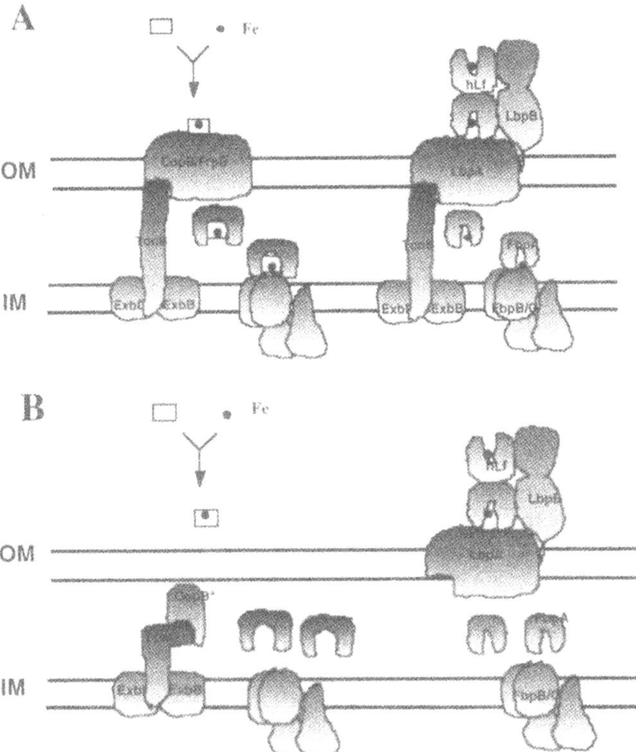

Figure 3. Organization of the iron acquisition pathways. **A.** A model for the Lf receptor-mediated iron acquisition pathway and for a postulated iron acquisition pathway involving the CopB or FrpB outer membrane proteins is outlined. The open rectangle is meant to represent a hypothetical exogenous siderophore or iron chelating molecule recognized by the CopB/FrpB proteins. **B.** The effect of production of a partial product of the CopB gene (CopB*) on the functioning of the CopB-mediated and Lbp-mediated iron acquisition pathways.

Isogenic LbpA-ve strains are specifically deficient in the use of Lf as a source of iron for growth (Table 1), indicating that LbpA plays a critical role in the iron acquisition process. The homology of LbpAs to other TonB-dependant outer membrane receptors suggests that they may serve as gated pores for transport of iron accross the outer membrane, with the TonB complex providing the energy required for this process (Figure 3A). The inability of isogenic TonB-ve strains to utilize Lf as a source of iron for growth (Table 1), supports this model.

Once iron has traversed the outer membrane it is proposed to be bound by the periplasmic iron-binding protein FbpA and then shuttled to an inner membrane transport complex composed of FrpB and FrpC (Figure 3A). A recently constructed isogenic mutant in FbpA[26] was shown to be deficient in iron acquisition from both Lf and Tf (Table 1), which is the first direct evidence to support its proposed role in the Lf receptor-mediated pathway.

Studies with isogenic mutants of FrpB in *N. meningitidis* and CopB in *M. catarrhalis* have provided contradictory and somewhat perplexing results (Table 1), and the lack of a known function for either of these proteins has limited the insights into this dilemma. The impairment of growth on either Tf or Lf in the CopB-ve mutant[16] could be interpreted to mean that it was an essential component of both iron acquisition pathways in *M. catarrhalis*. If so, it would seem logical that a homologous protein found in other bacterial species would serve a similar role. However, an isogenic mutant in FrpB, the homologue present in *N. meningitidis* , has no observable effect on iron acquisition from either Tf or Lf[19,20]. Thus one would expect that there either there is another more closely related homologue in *N. meningitidis* that has not been previously identified, or that the organization and funtion of the pathways is decidedly different in these two species. Similarly, there is no obvious homologue of CopB encoded by loci in the *Haemophilus influenzae* genome[15], suggesting that it is unlikely to serve an essential role in the iron acquisition pathway from Tf.

An alternate explanation is that the effect of the CopB deficiency on iron acquisition from Tf and Lf is indirect and due to impairment of some other component necessary for iron acquisition from these two iron sources, such as TonB (Figure 3B). In this hypothetical example, CopB and FrpB are postulated to be TonB-dependant outer membrane receptors responsible for mediating acquistion of iron by an exogenously supplied siderophore or other iron chelating molecule (Figure 3A). In the isogenic mutant generated by insertional activation of the *copB* gene, a defective CopB protein (CopB*) retains the ability to complex with TonB and thus interferes with TonB-mediated functions such as iron acquisition from Tf and Lf. Unfortunately there are presently no other TonB-dependent functions in *M. catarrahis* that can be readily evaluated to test this hypothesis. However, this model could be tested by construction of an isogenic mutant in which the entire CopB coding sequence is replaced, as it would be predicted to have little or no effect on growth dependent upon Tf or Lf.

7. THE RECEPTOR PROTEINS AS VACCINE ANTIGENS

Ever since their first identification[10,27] the bacterial transferrin and lactoferrin receptor proteins have been touted as potentially valuable vaccine antigens due to their inherent surface accessibility and their postulated critical role *in vivo*. To date most of the efforts at evaluating the potential efficacy of these surface antigens has been focussed on the transferrin receptor proteins. A variety of studies have demonstrated the potential utility of TbpB as a vaccine antigen by demonstrating its immunogenicity in the native host, its

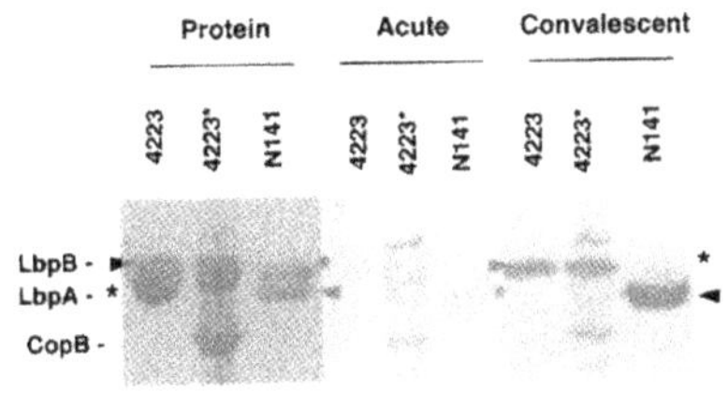

Figure 4. The human immune response against LbpB from *M. catarrhalis*. Lf binding proteins were isolated with hLf-Sepharose from the infecting strain of *M. catarrhalis* (N141) and a heterologous strain (4223) using standard, high stringency conditions, or modified (low salt, low pH) low stringency conditions (4223*). The proteins were eluted in SDS-PAGE buffer, electroblotted and stained for protein or reacted with acute or convalescent patient antisera.

ability to induce bactericidal antibody and its efficacy in passive and active immunotherapy in animal infection models[28–30]. The ability to induce an effective immune response is at least partially offset by the antigenic hetereogeneity that is observed in this protein in a number of species. In contrast, TbpA appears to be relatively nonimmunogenic in the native host and there is a paucity of data to support its potential as a vaccine antigen.

In conjunction with the evaluation of the the immune response against Tbps from *M. catarrhalis* in convalescent patient sera, we have also evaluated the response against the Lf binding proteins[17]. When the reactivity of convalescent antisera against proteins isolated with immobilized hLf is analyzed, there is relatively little reactivity against LbpA (Figure 4, *) or CopB. However, there is a strong immune response against LbpB (Figure 4, arrowhead) and this reactivity is at least partially crossreactive. These results are reminiscent of the findings with TbpB, and suggests that LbpB may also be a useful target for antibody-mediated defense mechanisms. However, it will be important to determine whether LbpB results in the induction of functional (bactericidal or opsonic) antibody and whether it is capable of inducing a protective immune response in the host. The degree of antigenic hetereogeneity is another obvious issue that will have to be evalutated if its utility as a vaccine antigen is seriously considered.

8. THE PHYSIOLOGICAL ROLE OF BACTERIAL LACTOFERRIN RECEPTORS

Although it is clear that the Lbps are capable of mediating acquisition of iron from Lf to support growth of bacteria *in vitro*, it is less clear whether this is the primary role of these proteins *in vivo*. If the Lbps were essential for acquisition of iron for growth in the mucosal environment, it is salient to ask why these receptor proteins aren't present in members of the Pasteurellaceae that share similar ecological niches and have a similar spectrum of disease. A second possibility is that the Lf receptors primary role is to maintain a low concentration of free iron in the immediate vicinity of the bacterial cell surface and thus reduce the potential toxic effects of the metal when interacting with oxygen. The corollory of this hypothesis would be to suggest that members of the Neisseriaceae, which posses Lf receptors, are more sensitive to the Fe-mediated oxygen toxicity than members of the Pasteurellaceae. Finally, it is also possible that the Lf receptors primary function is to complex Lf at the surface in a way that effectively reduces the antimicrobial effects of Lf mediated by the cationic peptide region. However, it is relevant to question whether the number of Lf receptors present on the bacterial surface would be sufficient to accomplish this task. Thus although we are rapidly gaining a greater appreciation of the structural and functional features of these fascinating surface receptor proteins, there is a great deal to be learned about the potential role that these proteins play *in vivo*.

REFERENCES

1. Brock JH. Lactoferrin Structure-Function Relationships: An Overview. In: Hutchens TW, Lonnerdal B. eds. Lactoferrin: Interactions and Biological Functions. Totowa: Humana Press, 1997; 3–23.
2. Bonnah RA, Yu R-H, Schryvers AB. Bacterial lactoferrin receptors in the *Neisseriaceae*. In: Lönnerdal B, Hutchens W. eds. Lactoferrin: Interactions and biological functions. Totowa: Humana Press, 1997; 277–301.
3. Elass-Rochard E, Roseanu A, Legrand D, Trif M, Salmon V, Motas C, Montreuil J, Spik G. Lactoferrin-lipopolysaccharide interaction: Involvement of the 28–34 loop region of human lactoferrin in the high-affinity binding to *Escherichia coli* 055B5 lipopolysaccharide. Biochem J 1995; 312: 839–845.
4. Naidu SS, Svensson U, Kishore AR, Naidu AS. Relationship between antibacterial activity and porin binding of lactoferrin in *Escherichia coli* and *Salmonella typhimurium*. Antimicrob Agents Chemother 1993; 37: 240–245.
5. Gray-Owen SD, Schryvers AB. Bacterial transferrin and lactoferrin receptors. Trends in Microbiol 1996; 4: 185–191.
6. Pettersson A, Maas A, Tommassen J. Identification of the *iroA* gene product of *Neisseria meningitidis* as a lactoferrin receptor. J Bacteriol 1994; 176: 1764–1766.
7. Quinn ML, Weyer SJ, Lewis LA, Dyer DW, Wagner PM. Insertional inactivation of the gene for the meningococcal lactoferrin binding protein. Microb Pathog 1994; 17: 227–237.
8. Dhaenens L, Szczebara F, Husson MO. Identification, characterization, and immunogenicity of the lactoferrin-binding protein from *Helicobacter pylori*. Infect Immun 1997; 65: 514–518.
9. Naidu AS, Andersson M, Forsgren A. Identification of a human lactoferrin-binding protein in *Staphylococcus aureus*. J Med Microbiol 1992; 36: 177–183.
10. Schryvers AB, Morris LJ. Identification and characterization of the human lactoferrin-binding protein from *Neisseria meningitidis*. Infect Immun 1988; 56: 1144–1149.
11. Schryvers AB, Lee BC. Comparative analysis of the transferrin and lactoferrin binding proteins in the family *Neisseriaceae*. Can J Microbiol 1989; 35: 409–415.
12. Pettersson A, Klarenbeek V, van Deurzen J, Poolman JT, Tommassen J. Molecular characterization of the structural gene for the lactoferrin receptor of the meningococcal strain H44/76. Microbial Pathogenesis 1994; 17: 395–408.
13. Biswas GD, Sparling PF. Characterization of *lbpA*, the structural gene for a lactoferrin receptor in *Neisseria gonorrhoeae*. Infect Immun 1995; 63: 2958–2967.
14. Bonnah RA, Yu R-H, Schryvers AB. Biochemical analysis of lactoferrin receptors in the Neisseriaceae: Identification of a second bacterial lactoferrin receptor protein. Microbial Pathogenesis 1995; 19: 285–297.
15. Bonnah RA, Yu R-H, Wong H, Schryvers AB. Biochemical and immunological properties of lactoferrin binding proteins from *Moraxella (Branhamella) catarrhalis*. Microbial Pathogenesis 1997; (submitted)
16. Aebi C, Stone B, Beucher M, Cope LD, Maciver I, Thomas SE, McCracken GH, Jr., Sparling PF, Hansen EJ. Expression of the CopB outer membrane protein by *Moraxella catarrhalis* is regulated by iron and affects iron acquisition from transferrin and lactoferrin. Infect Immun 1996; 64: 2024–2030.
17. Yu R-H, Bonnah RA, Ainsworth S, Myers L, Harkness RE, Schryvers AB. Analysis of the immunological responses to transferrin and lactoferrin receptor proteins from *Moraxella* (*Branhamalla*) *catarrhalis*.. Infect Immun 1997; (submitted).
18. Bonnah RA, Wong H, Schryvers AB. Cloning and characterization of the gene encoding lactoferrin binding protein B from *Neisseria meningitidis*. J Bact 1997, (submitted).
19. Pettersson A, Maas A, Van Wassenaar D, Van der Ley P, Tommassen J. Molecular characterization of FrpB, the 70-kilodalton iron-regulated outer membrane protein of *Neisseria meningitidis*. Infect Immun 1995; 63: 4181–4184.
20. Beucher M, Sparling PF. Cloning, sequencing, and characterization of the gene encoding FrpB, a major iron-regulated, outer membrane protein of *Neisseria gonorrhoeae*. J Bacteriol 1995; 177: 2041–2049.
21. Yu R-H, Schryvers AB. Regions located in both the N-lobe and C-lobe of human lactoferrin participate in the binding interaction with bacterial lactoferrin receptors. Microb Pathog 1993; 14: 343–353.
22. Wong H Schryvers AB. Construction of recombinant chimeric human lactoferrin/bovine transferrins. In: Spik G. *et al.* eds Advances in Lactoferrin Research. New York: Plenum Publishing Co. 1997.
23. Cornelissen CN, Sparling PF. Iron piracy: Acquisition of transferrin-bound iron by bacterial pathogens. Mol Microbiol 1994; 14: 843–850.
24. Anderson JA, Sparling PF, Cornelissen CN. Gonococcal transferrin-binding protein 2 facilitates but is not essential for transferrin utilization. J Bacteriol 1994; 176: 3162–3170.

25. Gray-Owen SD, Loosemore S, Schryvers AB. Identification and characterization of genes encoding the human transferrin binding proteins from *Haemophilus influenzae*. Infect Immun 1995; 63: 1201–1210.
26. Khun HH, Kirby SD, Lee BC. A *Neisseria meningitidis fbpA* mutant is incapable of using nonheme iron for growth. J Bact 1997 (submitted).
27. Schryvers AB, Morris LJ. Identification and characterization of the transferrin receptor from *Neisseria meningitidis*. Mol Microbiol 1988; 2: 281–288.
28. Watts T, Schryvers AB, Lo RYC, Potter AA. Protective capacity of *Pasteurella haemolytica* transferrin-binding proteins Tbp1 and Tbp2 against experimental disease. Infect Immun 1997; (submitted).
29. Danve B, Lissolo L, Mignon M, Dumas P, Colombani S, Schryvers AB, Quentin-Millet MJ. Transferrin-binding proteins isolated from *Neisseria meningitidis* elicit protective and bactericidal antibodies in laboratory animals. Vaccine 1993; 11: 1214–1220.
30. Rossi-Campos A, Anderson C, Gerlach G-F, Klashinsky S, Potter AA, Willson PJ. Immunization of pigs against *Actinobacillus pleuropneumoniae* with two recombinant protein preparations. Vaccine 1992; 10: 512–518.
31. Stojiljkovic I, Srinivasan N. *Neisseria meningitidis tonB*, *exbB*, and *exbD* genes: Ton-dependent utilization of protein-bound iron in neisseriae. J Bacteriol 1997; 179: 805–812.

EVIDENCE FOR THE EXISTENCE OF A SURFACE RECEPTOR(S) FOR FERRICLACTOFERRIN AND FERRICTRANSFERRIN ASSOCIATED WITH THE PLASMA MEMBRANE OF THE PROTOZOAN PARASITE *Leishmania donovani*

Bradley E. Britigan,[1,2] Troy S. Lewis,[1] Michael L. McCormick,[1,2] and Mary E. Wilson[1–4]

[1]Research Service
VA Medical Center-Iowa City
Iowa City, Iowa 52246
[2]Department of Internal Medicine
[3]Department of Microbiology
[4]Program in Immunology
University of Iowa, College of Medicine
Iowa City, Iowa 52242

1. SUMMARY

Previous work has demonstrated the ability of the promastigote form of the protozoan parasite Leishmania chagasi to utilize iron chelated to lactoferrin and transferrin for growth and metabolism. We have obtained evidence suggesting that the promastigote form of the parasite possesses specific binding sites for lactoferrin and transferrin. Lactoferrin binding appears to be: 1) independent of whether or not the protein contains iron; 2) not inhibited by transferrin; and 3) independent of whether the organism is in log or stationary phase of growth. Transferrin binding is: 1) markedly greater if the protein is iron loaded; 2) inhibited by the presence of lactoferrin; and 3) independent of whether the organism is in log or stationary growth phase. Preliminary ligand blot analyses are consistent with the presence of a protein or proteins which bind lactoferrin and/or transferrin. The relationship to these binding sites to those described in other protozoan species requires further investigation.

Advances in Lactoferrin Research, edited by Spik *et al.*
Plenum Press, New York, 1998.

2. INTRODUCTION

Species of Leishmania, a protozoan parasite, cause a spectrum of disease in humans. This ranges from disease which is limited to the skin to one in which the organism disseminates to involve multiple areas of the host reticuloendothelial cell system. The nature of the infection is dependent on the species of Leishmania with which the host becomes infected[1]. In the case of South American visceral leishmaniasis it is *L. chagasi* which is the responsible pathogen[1]. Cutaneous and visceral leishmaniasis remains a major cause of morbidity and mortality throughout the world. At present there is no vaccine proven to prevent the disease and available therapies are complicated by the significant toxicities of the drugs required[1].

All species of Leishmania exist in two distinct morphologic forms. In the proboscis of the sandfly vector, it is the promastigote form[1]. The promastigote is inoculated into the skin of the human (or other mammalian) host during ingestion of a blood meal by the fly. The promastigote is then ingested by host macrophages. A portion of the ingested promastigotes manage to evade macrophage-mediated killing mechanisms following which they undergo conversion to the intracellular amastigote form of the parasite. Within the macrophage phagolysosome the parasite grows and replicates. The life cycle is completed when the sandfly acquires an amastigote-laden macrophage or monocyte in a blood meal when it feeds on an leishmania-infected mammal. This is followed by sequential morphologic changes of the parasite which culminate in the mature promastigote in the fly proboscis.

Although there are limited data available, it seems likely that like nearly all living organisms L. chagasi is dependent on the availability of exogenous iron for growth and metabolism. Promastigote growth in vitro requires exogenous iron[2-4]. Work from our laboratory and others indicate that L. chagasi promastigotes can utilize iron in the form of hemin, ferriclactoferrin, or ferrictransferrin for in vitro growth whereas ferritin does not seem to be utilized under similar experimental conditions[3]. To our knowledge there is no information as to the type of iron sources, if any, which may be utilized by the amastigote form of the parasite as it replicates within mammalian macrophages.

The mechanism whereby Leishmania promastigotes acquire iron from the above iron chelates has been an area of interest of our laboratories and others[3,5,6]. Voyiatzaki and Soteriadou have presented evidence suggesting that other Leishmania species (L. infantum, L. major) possess a surface protein which is capable of binding apotransferrin[5,6]. However, no clear linkage between this protein and iron acquisition by these species of Leishmania has been made. In previous work, we obtained data indicating that production/excretion of siderophores or proteases is not involved in iron acquisition from lactoferrin or transferrin by L. chagasi[3]. In the work reported herein we describe data obtained by our laboratories to date which are consistent with the hypothesis that L. chagasi possess surface receptor(s) for ferriclactoferrin and ferrictransferrin.

3. METHODS

3.1. Promastigote Culture

L. chagasi promastigotes were cultured in modified minimal essential medium supplemented with 10% fetal calf serum and 8 μM hemin[3,5]. After initial seeding at 10^6/ml promastigotes were harvested at either mid-log (8–12 × 10^6/ml) or stationary (5–7 × 10^7/ml) growth phase as desired[5].

3.2. Binding Studies

Plasticware was pretreated with 1% gelatin hydrolysate in phosphate buffered saline (PBS).Promastigotes (0.5–1.6 × 10^7/ml) were suspended in Hanks balanced salt solution (HBSS) containing 0.2–2% humans serum albumin (HSA) and 2 mM HEPES at 4°C. They were then exposed for 30 min (4°C) to ferric or apo forms of [^{125}I]lactoferrin or [^{125}I]transferrin previously generated using the iodogen method (Pierce, Rockford, IL). Some experimental tubes also contained 100 fold of unlabeled protein. Following the incubation period cell-associated and free protein were separated by centrifugation of the cell suspension through an oil cushion comprised of dinonyl phthalate and dibutyl phthalate (Sigma Chemical, St. Louis, MO) at a ratio of 1:4 (vol/vol). The promastigotes and cell-bound ligand sediment through the cushion. The bottom of the tubes containing the cell pellet were then flash frozen in dry ice/acetone following which this section of the tube was cut off and the cell-associated [^{125}I]protein quantitated using a γ counter.

3.3. Ligand Blot

Promastigote samples were solubilized in Laemmli solubilizing buffer and subjected to SDS-PAGE following which they were transferred to nitrocellulose. The nitrocellulose was blocked in 4% BSA/TBS for 6–8 hours at 4°C. [^{125}I]lactoferrin or [^{125}I]transferrin were then incubated with the blots at a concentration of 100 ng/ml overnight at 4°C. The blots were then washed in 0.1% tween-20/TBS for 4 hours at room temperature, exchanging the wash every 10–15 min. Location of bound [^{125}I] was determined by autoradiography) 48–72 hour exposure of Kodak X-OMAT AR at −80°C.

4. RESULTS

As shown in Figure 1 binding studies performed with [^{125}I]lactoferrin indicate a specific (i.e. "cold ligand" inhibitable) binding of lactoferrin to L. chagasi promastigotes. Binding appeared to be unaffected by the growth phase of the organism (log vs. stationary) or whether the lactoferrin was iron-saturated or iron free (apo). In contrast to the re-

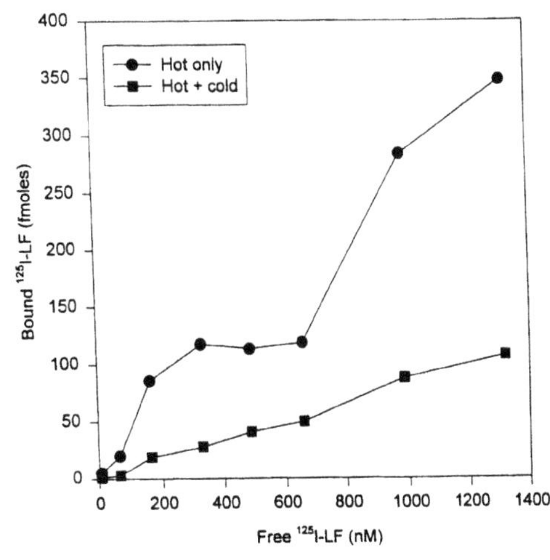

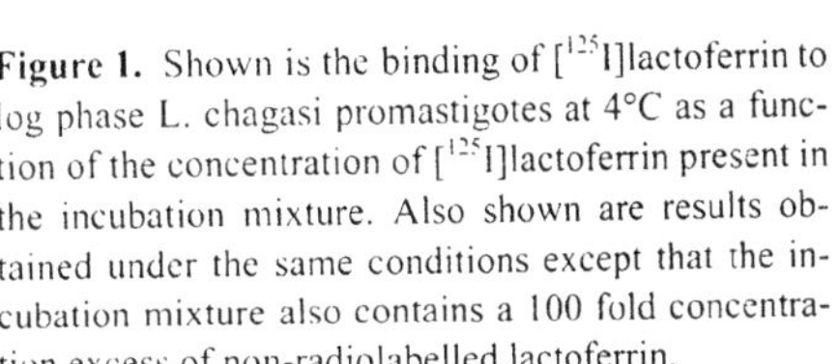
Figure 1. Shown is the binding of [^{125}I]lactoferrin to log phase L. chagasi promastigotes at 4°C as a function of the concentration of [^{125}I]lactoferrin present in the incubation mixture. Also shown are results obtained under the same conditions except that the incubation mixture also contains a 100 fold concentration excess of non-radiolabelled lactoferrin.

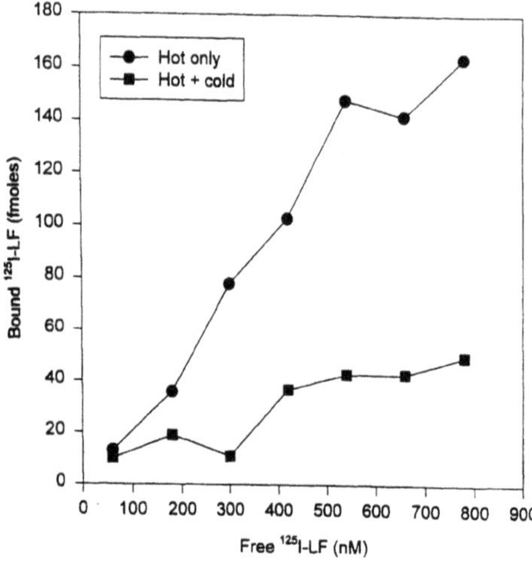

Figure 2. Shown is the binding of [^{125}I]ferrictransferrin to log phase L. chagasi promastigotes at 4°C as a function of the concentration of [^{125}I]ferrictransferrin present in the incubation mixture. Also shown are results obtained under the same conditions except that the incubation mixture also contains a 100 fold concentration excess of non-radiolabelled transferrin.

sults with "cold" lactoferrin, binding of [^{125}I]lactoferrin was not altered by the presence of 100 fold excess unlabelled ferrictransferrin. Binding studies performed over a wide concentration range of [^{125}I]lactoferrin suggest the possible presence of two populations of receptors for the protein (Figure 1). Alternatively the second phase of cell association could be caused by lactoferrin-lactoferrin interactions.

As noted above, other investigators have described the presence of a transferrin binding protein on the surface of the promastigotes of other Leishmania species[5,6]. Consistent with these findings, we were able to demonstrate saturable binding of [^{125}I]ferrictransferrin as evidenced by the "cold" transferrin inhibitable binding of [^{125}I]ferrictransferrin shown in Figure 2.

The magnitude of binding observed with transferrin was less than that seen with lactoferrin. The specific binding observed over a wide concentration range also raises the possibility of two receptor populations. In contrast to the results with lactoferrin (Figure 1) we were unable to detect binding of the apo form of transferrin to the promastigote surface. Also in contrast to results with lactoferrin we find that "cold" lactoferrin inhibits binding of [^{125}I]transferrin to the same extent to that we observe with "cold" diferric transferrin. Once again, binding of ferrictransferrin appears to be similar with log and stationary forms of L. chagasi promastigotes.

Given the above data suggesting the presence of surface receptor(s) on L. chagasi promastigotes, we have undertaken steps to characterize and eventually identify that receptor(s). Using ligand blot techniques, we have obtained preliminary data indicating the presence of a membrane-associated protein(s) capable of binding lactoferrin and transferrin. Consistent with these being specific binding sites, binding of [^{125}I]lactoferrin and [^{125}I]transferrin to the nitrocellulose blots was decreased by the presence of "cold" lactoferrin and transferrin.

5. DISCUSSION

We have previously demonstrated that L. chagasi can utilize iron chelated to lactoferrin and transferrin to meet their metabolic and growth requirements for iron. In the present work we extend our earlier observations[3,7] which suggest the presence of a surface binding molecule for lactoferrin and transferrin on the promastigote surface. In fact, in the case of lactoferrin our current data raise the possibility that two populations of lactoferrin binding sites exist. Further studies are required to confirm this possibility. Additional studies in this regard will also be needed to assess this possibility with transferrin.

Although evidence has been previously presented that some Leishmania species are capable of binding apotransferrin via a surface receptor, to our knowledge there have been no reports other than our own[3,7] which have suggested that these organisms also may be able to bind lactoferrin by this or any other mechanism. A heterodimeric receptor (ESAG-6 and ESAG-7) for transferrin has been reported in Trypanosoma brucei[8–11], another protozoan parasite. Preliminary screening studies using cDNA for ESAG-6 and ESAG-7 have as yet not revealed homologous genes in L. chagasi. A "lactoferrin receptor" has been described in Trichomonas vaginalis, a flagellated protozoan[12] but this has not been extensively characterized.

Preliminary data using ligand blot analysis has identified a potential candidate for a "lactoferrin binding protein" associated with the plasma membrane of L. chagasi promastigotes. Similar studies are underway using transferrin as the ligand as are approaches to further characterize and purify these putative receptors. It remains unclear whether lactoferrin and transferrin bind to the same receptor molecule with different affinities or whether they bind to two separate sites but that sufficient similarity exists such that lactoferrin binds to the transferrin "receptor" as well. In bacterial species which possess the ability to bind both lactoferrin and transferrin separate receptors are responsible[13,14].

In summary, we have obtained data suggesting the existence of specific surface binding sites for both lactoferrin and transferrin on the membrane surface of the promastigote form of the causative agent of visceral leishmaniasis, L. chagasi. Whether the binding sites we have detected are involved in the ability of this organisms to utilize iron chelated to lactoferrin and transferrin to meet its metabolic/growth requirements for this essential metal cation remains to be determined. Such ability to acquire iron from lactoferrin and transferrin by the promastigote form of this parasite could play a role in pathogenesis of infection with this organism as it would likely encounter these iron chelates shortly after introduction into the extracellular environment of the mammalian host, or in the case of lactoferrin within the gut and mouth of the sandfly vector. Thus, further characterization of these "receptors" in the iron metabolism of this important human parasite are indicated.

ACKNOWLEDGMENTS

This work was supported in part by awards from the Department of Veterans Affairs Research Service (BEB and MEW) and the National Institutes of Health AI-34954 and AI-32135. Drs. Britigan and Wilson are Established Investigators of the American Heart Association.

REFERENCES

1. Wilson ME. Leishmaniasis. Curr Opin Infect Dis 1993;6:331–341.
2. Chang K-P, Hendricks LD. Laboratory cultivation and maintenance of *Leishmania*. In: Chang K-P, Bray RS, eds. Leishmaniasis. New York: Elsevier, 1985:214–244.

3. Wilson ME, Vorhies RW, Andersen KA, Britigan BE. Acquisition of iron from transferrin and lactoferrin by the protozoan *Leishmania chagasi*. Infect Immun 1994;62:3262–3269.
4. Soteriadou K, Papavassiliou P, Voyiatzaki C, Boelaert J. Effect of iron chelation on the in-vitro growth of *Leishmania* promastigotes. J Antimicrob Chemother 1995;35:23–29.
5. Voyiatzaki CS, Soteriadou KP. Identification and isolation of the *Leishmania* transferrin receptor. J Biol Chem 1992;267:9112–9117.
6. Voyiatzaki CS, Soteriadou KP. Evidence of transferrin binding sites on the surface of *Leishmania* promastigotes. J Biol Chem 1990;265:22380–22385.
7. McCormick ML, Wilson ME, Lewis TS, Vorhies RW, Britigan BE. Specific binding of ferrilactoferrin and ferritransferrin in the protozoan *Leishmania chagasi*. In: Hutchens TW, Lönnerdal B, eds. Lactoferrin: Interactions and Biological Functions. Totowa, NJ: Humana Press Inc. 1997:333–342.
8. Steverding D, Stierhof YD, Fuchs H, Tauber R, Overath P. Transferrin-binding protein complex is the receptor for transferrin uptake in Trypanosoma brucei. J Cell Biol 1995;131:1173–1182.
9. Ligtenberg MJL, Bitter W, Kieft R, Steverding D, Janssen H, et al. Reconstitution of a surface transferrin binding complex in insect form *Trypanosoma brucei*. EMBO J 1994;13:2565–2573.
10. Steverding D, Stierhof Y-D, Chaudhri M, Ligtenberg M, Schell D, et al. ESAG 6 and 7 products of *Trypanosoma brucei* form a transferrin binding protein complex. Eur J Cell Biol 1994;64:78–87.
11. Maier A, Steverding D. Low affinity of *Trypanosoma brucei* transferrin receptor to apotransferrin at pH 5 explains the fate of the ligand during endocytosis. FEBS Lett 1996;396:87–89.
12. Peterson KM, Alderete JF. Iron uptake and increased intracellular enzyme activity follow host lactoferrin binding by *Trichomonas vaginalis* receptors. J Exp Med 1984;160:398–410.
13. Cornelissen CN, Sparling PF. Iron piracy: Acquisition of transferrin-bound iron by bacterial pathogens. Mol Microbiol 1994;14:843–850.
14. Schryvers AB, Lee BC. Analysis of bacterial receptors for host iron-binding proteins. J Microbiol Methods 1993;18:255–266.

17

LACTOFERRIN SECRETION INTO MOUSE MILK

Development of Secretory Activity, the Localization of Lactoferrin in the Secretory Pathway, and Interactions of Lactoferrin with Milk Iron

Margaret C. Neville,[1] Katie Chatfield,[1] Linda Hansen,[1] Andrew Lewis,[1] Jenifer Monks,[1] Jan Nuijens,[2] Michelle Ollivier-Bousquet,[3] Floyd Schanbacher,[4] Valery Sawicki,[1] and Peifang Zhang[1]

[1]Department of Physiology
University of Colorado, Health Sciences Center
Denver, Colorado 80262
[2]Pharming, B.V.
Leiden, The Netherlands
[3]INRA, Laboratoire de Biologie Cellulaire et Moleculaire
78352 Jouy-en-Josas, France
[4]Ohio Agricultural Research and Development Company
Ohio State University
Wooster, Ohio

1. INTRODUCTION

Lactoferrin, one of four major protein components of human milk, has long been thought to protect the breast and the infant against infection through its bacteriostatic and bacteriocidal activity[1]. However, it has also been reported to stimulate growth of the infant intestine[2] and to modulate the activity of immune cells[3]. Lactoferrin shows both tissue-specific and species-specific regulation. In the mouse uterus, for example, it is clear that estrogen is a major regulator of lactoferrin secretion[4]. In humans and cows lactoferrin appears to be present in milk at higher levels during the onset of lactation[5,6] and to decrease during established lactation. However, the actual rate of secretion is highly species-dependent. Lactoferrin secretion is especially low during established lactation in ruminants; correspondingly cows' milk has less than 0.1 mg/ml lactoferrin[5,7]. In both human and cow mammary secretions lactoferrin concentrations are high at involution and in milk from the post-mastitic gland, a phenomenon that has been best documented in the cow[5,8].

Advances in Lactoferrin Research, edited by Spik *et al.*
Plenum Press, New York, 1998.

Cellular regulation of lactoferrin secretion in the mammary gland has received relatively little attention although some interesting results are available from Teng and Schanbacher[9–11]. In the mouse uterus transcriptional regulation by estrogen is clear[4] whereas there is no evidence that estrogen plays a role in the regulation of mammary lactoferrin. Evidence from the Schanbacher laboratory has suggested that post-transcriptional processes also play a role in regulating the rate of lactoferrin synthesis[12,13].

Understanding the mechanisms by which lactoferrin secretion is regulated in the mammary gland is important for several reasons: lactoferrin is a potentially important molecule present at relatively high concentrations in breast milk. Because of the low concentration in cows' milk lactoferrin is a very minor component of infant formula. Understanding the regulation of lactoferrin secretion may have the practical application of allowing enhancement of secretion in the ruminant. Another, more speculative and therefore exciting, possibility is that lactoferrin may actually serve as a signaling molecule in the mammary gland. If, as data to be presented below suggest, lactoferrin is secreted in large quantities in the stressed or involuting gland it may promote immune reactions that help fight infection or even decrease tumorigenesis.

We have begun to study the regulation of lactoferrin secretion in the mammary glands of a laboratory species, the mouse. This animal has the advantages that it breeds relatively rapidly, that tools such as antibodies as well as cDNA and genomic clones are available for many molecules, and that transgenic technology allows manipulation of its genetic framework, providing a means for the testing of hypotheses about the molecular mechanisms by which lactoferrin secretion is regulated. The present studies, some of which are necessarily descriptive given the current state of our knowledge, provide a physiologic framework within which the *in vivo* regulation of lactoferrin secretion can be understood. In this article the literature on lactoferrin regulation and secretion will be summarized and recent results from our laboratories described.

2. DEVELOPMENTAL REGULATION OF LACTOFERRIN SECRETION

2.1. Mammogenesis and Lactogenesis

The mammary gland progresses through distinct developmental stages in the adult animal. In most mammalian species the non-pregnant, nonlactating female has a limited amount of mammary tissue that consists primarily of epithelium-lined ducts coursing through a copious adipose stroma. With the onset of pregnancy there is marked proliferation of alveolar and adipose cells, the formation of true alveoli and differentiation of alveolar cells that are competent to secrete milk. However milk secretion is held in check by the high levels of circulating progesterone[14]. Scant amounts of a secretion product containing up to 10 mg/ml lactoferrin, depending on the species, are produced during this time. When progesterone falls at the termination of pregnancy, the differentiated mammary gland undergoes a state change, called *lactogenesis*. At this time the secretion of true milk components is increased several-fold and, importantly, the secretion of certain colostral components including lactoferrin and immunoglobulins is transiently up and then down-regulated at least in humans (Figure 1). In cows and other ruminants the down-regulation is severe resulting in very low levels of milk lactoferrin during lactation as illustrated in Figure 2[15].

In the experiment depicted lactation was followed in a heifer through pregnancy and lactation. Once lactogenesis occurred the concentration of lactoferrin was depressed and

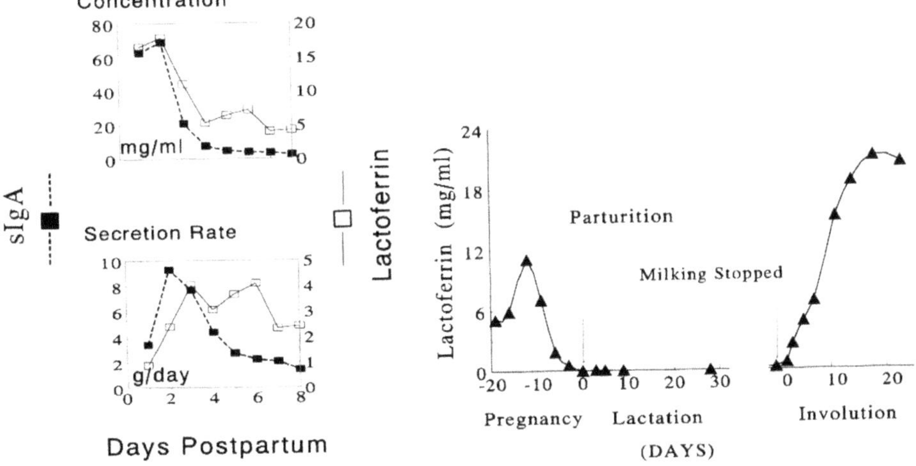

Figure 1. Lactoferrin concentration and secretion rate in human milk. Figure from ref. 38 based on data from 6 and 39. Used by permission.

Figure 2. Lactoferrin in the secretion product of the udder of a single dairy cow progressing through late pregnancy, early lactation and involution. Redrawn from reference 15.

remained low as long as the gland was milked, then rose dramatically with cessation of milk removal. Casein secretion changed in a reciprocal manner (data not shown) illustrating the independence of the regulation of lactoferrin secretion from that of other milk proteins. A similar pattern was also reported in guinea pigs[7].

In the mouse, Teng[16] found that lactoferrin mRNA increased on day one postpartum, fell slightly on day ten and rose to higher levels at weaning on day 21. In earlier work in rodents only milk was analyzed and the developmental sequence is not clear. Interpretation of the data in rodents is made more difficult by the fact that some rodents have transferrin in their milk, either in place of lactoferrin as in the rat[17] and rabbit[7] or in addition to lactoferrin as in the mouse[18,19]. Coomassie staining of mouse skim milk proteins on two dimensional gels suggested that mature mouse milk has substantial quantities of both lactoferrin and transferrin[20]. In outstanding early experiments using mammary explants from lactating mice Green and Pastewka[21] positively identified mouse lactoferrin but not transferrin in the medium after treatment with prolactin.

We have used an antibody specific to lactoferrin to quantitate the protein in mouse milk throughout lactation (Figure 3). Levels varied widely from 0.05 to 0.8 mg/ml in the milk of two strains (CD-1 and B6CBA). There was no systematic variation with duration of lactation. The mean lactoferrin concentration was about 0.14 mg/ml. One prepartum sample had a lactoferrin concentration about 0.3 mg/ml, suggesting that, unlike human milk, the colostral phase in the mouse is not associated with a greatly increased lactoferrin secretion. This conclusion must, however, be treated with caution, since sampling of prepartum and early post-partum milk in the mouse is a serious challenge. The data on the mouse subjected to multiple milkings (closed circles, Figure 3 is discussed below.

2.2. Involution and Mastitis

The lactoferrin content of the mammary secretion has been found to be increased during involution as well as in the mastitic gland. Figure 2 illustrates involutional changes

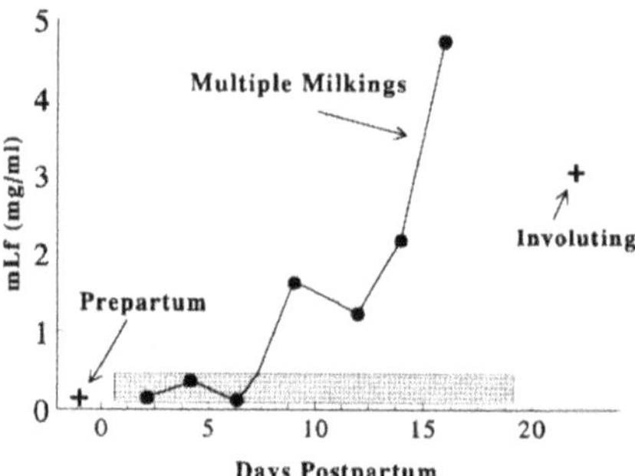

Figure 3. Lactoferrin in mouse milk. The figure shows results from CD1 (Charles River Labs) and B6CBA (Jackson labs) mice measured by ELISA on whole milk diluted 16,000 to 64,000 from lactating mice. All data fell within the stipled box. Single samples taken four hours prepartum or from a 22 day postpartum mouse are shown by +. One CD-1 mice was milked serially every 2 to 3 days. After the third milking the lactoferrin in the milk began to rise.

in bovine milk and Figure 4 illustrates lactoferrin concentrations in the secretion product of one woman after abrupt voluntary weaning.

The lactoferrin concentration in the small amount of secretion product that could be expressed from the breast rose about tenfold during the month after termination of breastfeeding[22]. Figure 5 shows changes in whey protein concentrations in the secretion product of a quarter of a bovine gland after experimental infection with *E. coli*[8].

Acute increases in the IgG and albumin concentration in the mammary secretion were thought to signal a breakdown of the mammary epithelium. These symptoms subsided after about three days. However, the lactoferrin concentration in the mammary secretion increased gradually to nearly 5 mg/ml, a 50-fold increase over the concentration in the milk from quarters that had not been infected. This increase was observed at a time when milk secretion, which had ceased at the height of the infection, was gradually returning to normal. High rates of lactoferrin secretion were maintained for four weeks, indicating that the post-mastitic state continues for quite some time after resumption of secretion of normal milk components.

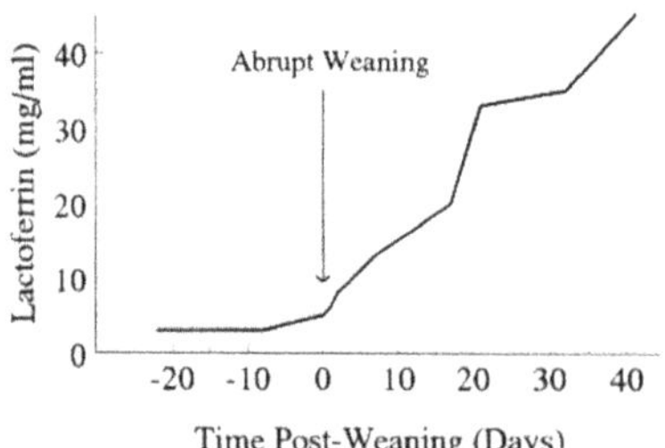

Figure 4. Effect of abrupt termination of breastfeeding on the concentration of lactoferrin in the mammary secretion of one woman. Modified from reference 22.

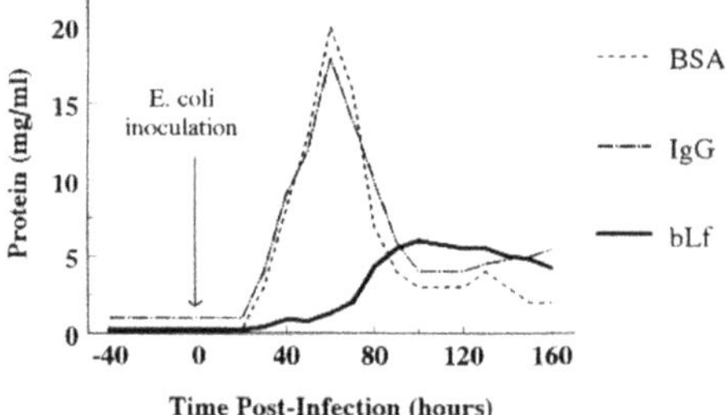

Figure 5. Experimental mastitis in a cow. Within 24 hours of inoculation of *E.coli* in one quarter of a mastitis-free cow the concentrations of serum albumin (BSA) and immunoglobulin (IgG) increased many fold reaching a peak at about 60 hours. Milk secretion declined to nearly zero at the same time (not shown). As the udder began to recover the concentrations of BSA and IgG fell, but lactoferrin began to rise leveling off at about 5 mg/ml. Milk secretion began to increase at the same time and lactoferrin secretion from this quarter continued at the rate of about 2 g/day for more than 30 days. Replotted from reference 8.

We were able to obtain one sample of mammary secretion after termination of lactation in a postpartum day 22 mouse. In this sample the lactoferrin had risen about tenfold to 3.5 mg/ml suggesting that lactoferrin secretion is enhanced in the involuting mouse gland as in the cow and human. Interestingly, we milked one mouse every two or three days throughout lactation. After the third milking the lactoferrin level rose about tenfold (Figure 3). We surmise that the procedure for milking a mouse, which involves a fair amount of suction, is not completely benign, and induces some damage to the gland bringing about an inflammatory condition that results in the mouse, as in the cow, in an enhancement of lactoferrin secretion.

2.3. Immunocytochemical Analysis

Using immunocytochemistry with specific antibodies to mouse lactoferrin (the kind gifts of C. Teng from NIEHS and J. Nuijens from Pharming, B.V.) we were able to examine the mammary distribution of murine lactoferrin through the entire developmental cycle. In the virgin gland we observed significant staining in the ductal system (Figure 6A).

With pregnancy lactoferrin continued to be evident throughout the developing epithelium with particularly intense staining in the lumina (data not shown). In the lactating gland stain was detected in variable amounts throughout the epithelium lumina. At higher magnification localization of bright granules near the apical membrane of the epithelial cells could be seen (Figure 6B). Notably, there was often a line of stain near the basal surface of the cell. This observation raises the possibility that some of the lactoferrin is secreted in the basal direction as has been noted for casein and growth hormone in transgenic mice expressing human growth hormone under the regulation of a mammary specific promoter[23]. Cytoplasmic staining varied significantly and occasional cells showing much brighter stain than their neighbors were observed (Figure 6C). These cells often showed morphological abnormalities. In general we did not see the alternating alveoli, some of which stained and some of which did not as reported by Molenaar *et al.*[24] and Hurley[25] in sheep and cattle, respectively, although occasional alveoli had large numbers of brightly staining cells. We suspect that especially bright staining is associated with cell deterioration, possibly due to partial involution or as a response to localized inflammation. In the ruminant gland, where lactoferrin secretion is repressed during lactation, staining might be expected only in lumina where cells are subjected to some sort of stress. In the involuting gland, lactoferrin staining was significantly enhanced in both cells and lumina. Lactoferrin continued to be seen in the cells and lumen of the parenchyma, long after natural involution.

It is important to note that lactoferrin is not confined to the luminal cells in the mammary gland. We observed occasional interstitial cells with large granules staining very brightly for lactoferrin (Figure 6D). These cells have the conformation of macrophages. The observation that they take up fluoresceinated serum albumin is consistent with this identity.

2.4. Messenger RNA for Murine Lactoferrin

Lactoferrin mRNA was low in the mammary gland of the virgin animal, reflecting the paucity of epithelial tissue. The level increased during pregnancy, remained approximately constant during lactation and then increased several fold during late lactation and involution. Such data on lactoferrin mRNA as exist[11,12,15] suggest that protein production is regulated by the amount of mRNA present. However, no mRNA measurements appear to

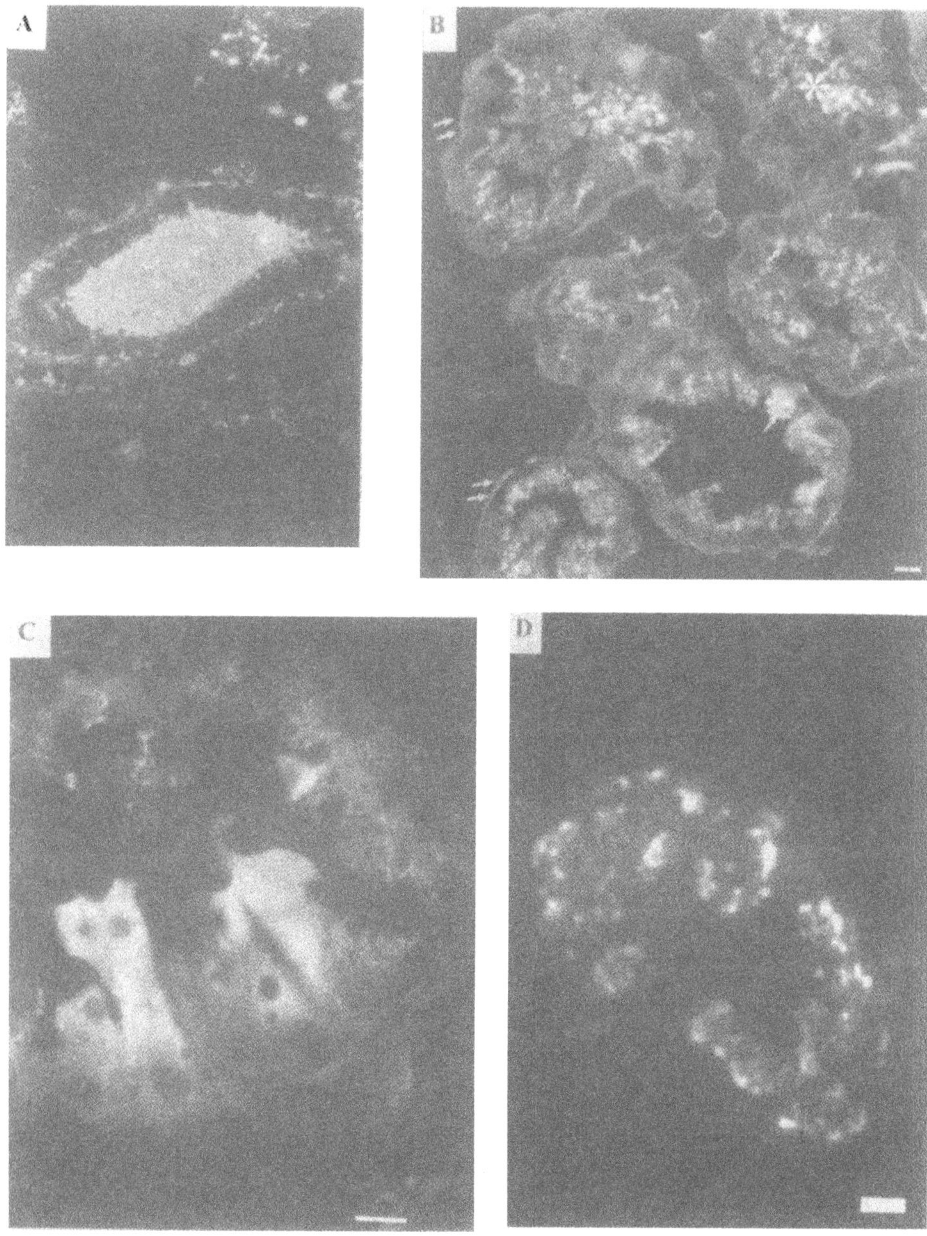

Figure 6. Immunocytochemical views of the mouse mammary gland. A highly specific antibody for mouse lactoferrin was used to stain sections of the mouse mammary gland that had been fixed with paraformaldehyde and embedded in polymethacrylate medium. The second antibody was goat anti rabbit IgG, labeled with rhodamine or FITC. **A.** Staining for lactoferrin in a mammary duct in a virgin mouse. **B.** Staining in a section of mammary gland from a 12 day lactating mouse showing apical staining of vesicular structures (single arrows) and a diffuse line of basal staining (double arrows). **C.** A few alveolar cells containing bright, diffuse stain in a mammary gland from a 2 day lactating mouse. **D.** An interstitial cell containing very bright staining granules thought to be a macrophage. Bar 10 microns, except in D where it is 2 microns.

be available for the mastitic gland. A major question raised by these and previous findings during involution is the mechanism by which lactoferrin mRNA becomes elevated at a time when the mRNAs' for other milk proteins are diminishing and the epithelial cells are beginning to undergo apoptosis.

3. *IN VITRO* STUDIES OF LACTOFERRIN SECRETION

Primary mouse mammary epithelial cultures and mouse cell lines have been used to compare secretion of transferrin and lactoferrin with their respective mRNA levels in cells cultured with and without lactogenic hormones on various types of substratum. In the earliest of these studies Lee *et al.*[20] used two dimensional electrophoresis to identify transferrin in both mouse milk and the medium of primary epithelial cells. Although not mentioned by the authors, lactoferrin is clearly visible in most of the gels as an 80 kD extended band with slower anodal migration, as expected from its higher isoelectric point. Autoradiograms of ^{35}S-methionine labeled proteins secreted into the medium of cells grown on plastic showed only bands corresponding to transferrin whereas the medium of attached and floating collagen gels clearly showed both proteins to be equally and heavily labeled. In later studies Lee *et al.* interpreted their data to mean that transferrin secretion from primary cultures of mouse mammary gland is much less sensitive to the influence of lactogenic hormones and substratum than is the secretion of ß-casein[19,26]. Using Comma 1D cells[27], Parry and his coworkers showed[28] that a protein corresponding in molecular weight to lactoferrin (80 Kd) was secreted from cultures growing on both attached matrices, floating collagen gels and nitrocellulose filters in Millicell chambers. Casein was secreted only when the cultures on collagen gels were released or when the cells were cultured on filters. In a more recent study Barcellos-Hoff *et al.*[29] showed that primary mouse cells growing on a reconstituted basement membrane matrix secreted both lactoferrin and transferrin, in both the apical and basal directions. In primary cultures of bovine mammary epithelial cells lactoferrin appeared to be up regulated along with casein[10]. This type of regulation is not consistent with the *in vivo* depression of lactoferrin secretion in the lactating bovine mammary gland and suggests that mammary cultures may not entirely mimic the *in vivo* situation.

The evidence for basal lactoferrin staining described above brings up the question of directionality of lactoferrin secretion. Although caseins are thought to be secreted almost exclusively in the apical direction in full lactation and in most *in vitro* models of mammary secretion[29,30], other proteins such as components of the extracellular matrix[31] as well as proteinases[32] are secreted basally. There are a number of reports suggesting that lactoferrin may be secreted in both the apical and basolateral directions under certain conditions *in vivo* as well as in *in vitro* models of the mammary gland. Thus, Hurley[25] observed basal staining of mammary ducts in sections of involuting bovine mammary gland. The lactating human mammary gland showed only intraluminal staining for lactoferrin by immunocytochemistry, whereas epithelial cells in the inactive gland showed clear staining on both apical and baso-lateral surfaces[33].

In Comma 1D cells grown on filters an 80 kD protein was secreted both apically and basolaterally whereas casein was secreted exclusively in the apical direction[28]. In the above-mentioned studies[29] of primary mouse epithelial cells grown on Matrigel^{R} basolaterally secreted proteins could be identified by their appearance in the medium. Luminally secreted proteins were released when medium calcium was chelated with EGTA, opening the junctional complexes between the cells. In these experiments both transferrin and lactoferrin were identified immunocytochemically and found to be secreted equally in both directions whereas 70 to 90% of the caseins were secreted apically.

Recently we have used an accurate ELISA to quantitate mLf in the medium of a cell line, CIT_3 cells, a derivative of the Comma 1D line[27]. This cell line can be grown on membrane filters where it forms a tight epithelium[34] so that apical and basal secretion can be distinguished. After four days of culture the concentration of murine lactoferrin in the apical medium was 9 mg/ml and that in the basal medium about 1.4 mg/ml (Table 1) indicating preferential apical secretion.

These numbers reflect an apical secretion rate of 0.8 mg/day/cm^2 and a basal secretion rate of 0.2 mg/day/cm^2. These cultures did not respond to lactogenic hormones with an increase in lactoferrin secretion. It is currently not clear whether lactoferrin secretion in this cell line has lost its ability to respond to prolactin or whether the lactoferrin responds differently from other cell components. These results, like those of the Immunocytochemical analyses described above, suggest that lactoferrin is a constitutive secretion product of the normal mouse mammary epithelium and that it can be secreted both apically and basally.

The questions we are currently unable to answer are: What is the proportion of basal secretion during the various developmental stages of the mammary gland? Is the protein secreted primarily into the lumen where it could serve as a powerful anti-infective agent or is it also secreted basally under certain circumstances where it might serve a regulatory function?

4. THE LACTOFERRIN SECRETION PATHWAY

Because lactoferrin appears to be regulated differently from other milk proteins such as casein and because its secretion pathway appeared to differ somewhat from that of casein, we used the same anti murine lactoferrin antibodies to examine lactoferrin localization in the mouse mammary gland by immuno-electron microscopy (Michelle Ollivier-Bousquet, unpublished data). In the lactating mouse mammary gland lactoferrin was clearly localized in secretory vesicles also containing casein micelles. It was also seen over areas containing rough endoplasmic reticulum and Golgi complexes providing strong

Table 1. Lactoferrin secretion by cultures of murine mammary cells grown on membrane filters

	Murine lactoferrin	
Medium	Concentration (mg/ml ± S.E.M.)	Secretion rate (mg/24 hr/cm^2 ± S.E.M)
Apical		
Growth medium	9.0 ± 1.4	0.77 ± 0.10
Secretion medium	9.0 ± 0.5	0.77 ± 0.01
Basal		
Growth medium	1.4 ± 0.3	0.20 ± 0.04
Secretion medium	1.1 ± 0.1	0.16 ± 0.05

CIT_3 cells were grown on membrane filters for 9 days as described[34]. After the cultures acheived confluence at 5 days, the medium was changed to fresh growth medium containing insulin and EGF or secretion medium containing insulin, cortisol and prolactin and incubation was continued another 4 days. The concentration of murine lactoferrin was determined by ELISA on medium samples from triplicate wells.

evidence that this protein follows the normal apical secretory pathway in the lactating mammary gland.

Lactoferrin was also found in the interstitial space at the basal surface of the mammary epithelium and associated with connective tissue and other interstitial structures. While this finding is consistent with basally directed lactoferrin secretion, it is important to point out that lactoferrin is present in the bloodstream so that the source of the interstitial lactoferrin cannot be determined simply from immunocytochemical localization.

5. LACTOFERRIN AND MILK IRON

The mechanism of secretion of iron into milk and the mechanisms by which milk iron is regulated are currently unknown. It is clear that the concentration of calcium in milk is related to the concentrations of calcium binding ligands such as casein and citrate[35]. One question is whether iron is subject to similar regulation by the concentration of iron binding proteins in the secretory vesicles. A second question is why the lactoferrin in human milk is only about 5% saturated with iron[36] whereas that in mouse milk appears to be completely iron saturated (Jan Nuijens, unpublished data). Data from transgenic mice over-expressing human lactoferrin in their mammary glands under the control of a mammary specific promoter shed some light on both these questions.

The iron concentration was measured in milk from CD1 and B6CBA strains as well as in milk from transgenic B6CBA lines expressing about 4 and 14 mg/ml human lactoferrin in their milks (These mice were the kind gift of Pharming, B.V.). In all strains, milk iron varied linearly with the concentration of iron in the serum (P. Zhang and M.C. Neville, unpublished). Moreover, the milk of the transgenic mice fell on the same regression line as that of the controls providing conclusive evidence that the concentration of lactoferrin, and presumably iron-binding proteins in general, does not alter the concentration of iron in milk.

The milk from the high expressing line of transgenic mice had a definite pink cast suggesting that a significant fraction of the human lactoferrin in this milk was iron-saturated. To test this hypothesis we utilized a technique developed by J. Nuijens, Pharming, B.V. After defatting and centrifugation to remove casein the milk sample was diluted in SDS sample buffer and immediately run on a polyacrylamide gel, or boiled with β-mercaptoethanol to denature the lactoferrin prior to gel electrophoresis. Lactoferrin was identified by Western blot with a highly specific anti-human lactoferrin antibody. Human hololactoferrin ran more rapidly than apolactoferrin (Figure 7, middle lanes).

In human milk, boiling had little effect on the migration rate (Figure 7, left hand lanes), implying that the majority of human lactoferrin in milk is present as apolactoferrin as known from previous work[36]. The human lactoferrin present in the milk of the transgenic mice ran more rapidly prior to boiling (Figure 7, right hand lanes 5 and 6) indicating that this lactoferrin was at least partially saturated with iron. These data show that human lactoferrin is capable of being saturated with iron in secreted milk. Why then is it present mainly as apolactoferrin in human milk?

The answer to this question is clear from the data in Table 2. Human milk has very low concentrations of iron (less than 7 mM) at least one-third of which is bound to xanthine oxidase in the milk fat globule membrane. Lactoferrin is present in human milk at a concentration of about 2 mg/ml or 25 mM . Since each molecule can bind 2 atoms of iron the total iron-binding capacity is 50 mM. Based on these figures the maximum possible saturation is about 14%. Because not all iron is available to be bound to lactoferrin, an ac-

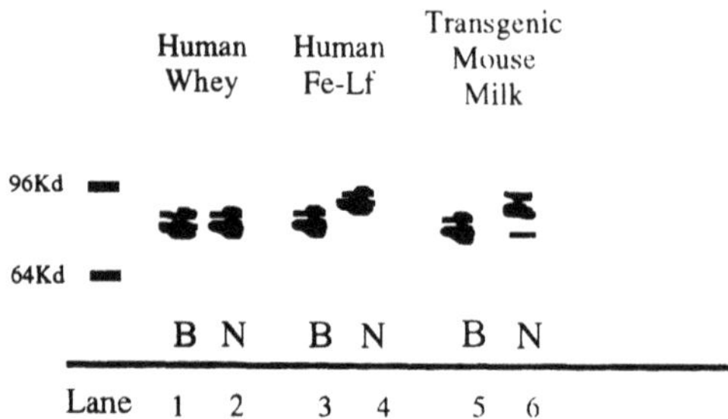

Figure 7. An immunoblot of human whey, iron saturated purified human lactoferrin and whey from the milk of a transgenic mouse secreting about 15 mg/ml human lactoferrin in its milk. N, sample was not boiled prior to loading on SDS poly-acrylamide gel; B, sample was boiled prior to loading. Iron-saturated lactoferrin runs more rapidly than apolactoferrin under the conditions used (J. Nuijens, unpublished). Boiling in SDS-b-mercaptoethanol sample buffer denatures the lactoferrin releasing the iron. These results show that the human lactoferrin secreted into the milk of transgenic mice is at least partially saturated with iron.

tual iron saturation of 5% is reasonable. In the mouse, on the other hand, iron is present at a concentration nearly 200 mM, whereas the iron binding capacity of the lactoferrin in mouse milk is only about 7 mM based on data from this laboratory. In the milk of the transgenic animals there was sufficient iron to saturate about 50% of the human lactoferrin (Table 2). These data provide evidence that human milk lactoferrin is likely capable of being saturated with iron in milk when the iron levels are sufficiently high. They provide support for the contention that no special mechanisms, other than low milk iron, are responsible for the low iron saturation of the lactoferrin from human milk.

6. SUMMARY AND PERSPECTIVE

The findings from both the literature and the limited studies presented here in mice suggest the following hypothesis about the secretion of lactoferrin from the mammary gland:

> Lactoferrin is a constitutive secretion product of the normal mammary epithelium and is present in the epithelial compartment of virgin, pregnant and post-lactational mammary glands. In ruminants secretion of the protein is down-regulated during lactation to produce very low concentrations in milk. Lactoferrin is up-regulated by inflammation and at involution and under certain conditions, particularly in the post-mastitic gland and in some tissue culture cells, this up-regulation appears to be permanent, sometimes in cells that continue to secrete other

Table 2. Iron binding by lactoferrin in human and mouse milks

	Human milk	Mouse milk	Transgenic mouse milk (1590)
Iron content (mg/ml)	0.3–0.9*	11.4 ± 0.6 (N = 9)	~10
Lactoferrin concentration (mg/ml)	2	0.3	14
Iron content (mM)	6.9	197	172
Lactoferrin iron binding capacity (mM)	50	7.5	350
Maximum % saturation of lactoferrin	14%	100%	49%

*ref. 37.

milk components. Lactoferrin can be secreted basally as well as apically by mammary epithelial cells. Physiologic changes in the lactoferrin concentration in the mammary secretion product appear to be the result of changes in the concentration of its mRNA.

The first part of this hypothesis is supported by the finding that lactoferrin is present in the ducts of the virgin and postinvolutional gland in the mouse, as well as in a normal mammary cell line. It would be of considerable interest to determine when lactoferrin synthesis is actually manifest in the embryo or prepubertal gland. Whereever it has been examined, with the exception of the mouse where only one milk sample is available, the lactoferrin content of the secretion product that can be extracted from the prepartum gland is in the 5 to 10 mg/ml range. After involution the concentration has been observed to go as high as 40 mg/ml in the human and 22 mg/ml in the cow. These high levels suggest that the protein may serve a protective function during periods of milk stasis. Since the combination of lactoferrin and lysozyme has a potent antibacterial effect, it would be of interest to know whether the lysozyme content of the non-lactational secretion product is equally high.

Immunocytochemical analysis of the lactoferrin distribution in the mouse mammary gland suggests that during involution cells that are getting ready to undergo apoptosis actually upregulate the synthesis of lactoferrin and that some of this lactoferrin is diffusely localized in the cytoplasm rather than in secretory vesicles. This observation raises two questions: Is lactoferrin actually localized to apoptosing cells? Do the high concentrations of lactoferrin secreted during involution have any regulatory effects on, for example, the immune system cells that enter the gland at this time?

One question raised by earlier observations in ruminant mammary glands is whether lactoferrin is secreted by the same cells that secrete casein, transferrin and other milk proteins. Thus Molinaar *et al.*[24] observed groups of alveoli in the involuting sheep gland that were involved in casein secretion and others involved in lactoferrin secretion. Our observations that most if not all alveoli in the mouse mammary gland stain for lactoferrin and that lactoferrin coincides with casein in the secretory pathway show that lactoferrin is secreted into milk by the normal apical secretion pathway and, in mice at least, by the same cells that secrete other milk components. A possible interpretation of the observations in ruminants is that in normal ruminant alveoli lactoferrin synthesis is downregulated. If localized inflammation is present, certain alveoli may have a diminished capacity for milk secretion, but may up-regulate lactoferrin secretion. Our observations of occasional very bright staining cells in the mammary epithelium of the lactating mouse are in concordance with this interpretation.

These studies provide a framework for future studies of the mechanisms of lactoferrin up-regulation during involution and mastitis and down-regulation during lactation in ruminants. Since changes in mRNA levels appear to underlie changes in lactoferrin synthesis in all these situations, the first question that should be answered is to what extent mRNA transcription is upregulated and to what extent changes in mRNA stability play a role. Run-on experiments using nuclei isolated from mammary glands in various reproductive and inflammatory stages are necessary to answer this question. If as we suspect, mRNA synthesis is upregulated *in vivo*, it will be of great importance to find an *in vitro* system in which lactoferrin production is also regulated and in which analysis of relevant promoter elements is possible.

In conclusion, this is an exciting time for lactoferrin researchers because of the availability of a large variety of tools that have allowed us to describe the regulation of lactoferrin in the mammary gland and that should, in the future, allow us to determine the mechanisms underlying these observations.

REFERENCES

1. Brock JH: Lactoferrin in human milk: its role in iron absorption and protection against enteric infection in the newborn infant. Arch. Dis. Child. 55:417–421, 1980.
2. Nichols BL, McKee KS, Henry JF, Putman M: Human lactoferrin stimulates thymidine incorporation into DNA of rat crypt cells. Pediatr. Res. 21:563–567, 1987.
3. Kijlstra A: The role of lactoferrin in the nonspecific immune response on the ocular surface. Reg. Immunol. 3:193–197, 1991.
4. Teng CT, Liu Y, Yang N, Walmer D, Panella T: Differential molecular mechanism of the estrogen action that regulates lactoferrin gene in human and mouse. Mol. Endocrinol. 6:1969–1981, 1992.
5. Sanchez L, Aranda P, Pérez MD, Calvo M: Concentration of lactoferrin and transferrin throughout lactation in cows' colostrum and milk. Bioc. Hoppe-Seyler 369:1005–1008, 1988.
6. Lewis-Jones DI, Lewis-Jones MS, Connolly RC, Lloyd DC, West CR: Sequential changes in the antimicrobial protein concentrations in human milk during lactation and its relevance to banked human milk. Pediatr. Res. 19:561–565, 1985.
7. Masson P: La Lactoferrine: Protéine des sécrétions externes et des leucocytes neutrophiles. Paris: Librairie Maloine S.A. 1970.
8. Harmon RJ, Schanbacher FL, Ferguson L, Smith KL: Changes in lactoferrin, immunoglobulin G, bovine serum albumin, and a-lactalbumin during acute experimental and natural coliform mastitis in cows. Infect. Immun. 13:533–542, 1976.
9. Liu Y, Teng CT: Characterization of estrogen-responsive mouse lactoferrin promoter. J. Biol. Chem. 32:21880–21885, 1991.
10. Talhouk RS, Neiswander RL, Schanbacher FL: In vitro culture of cryopreserved bovine mammary cells on collage gels: synthesis and secretion of casein and lactoferrin. Tissue Cell 22:583–599, 1990.
11. Goodman RE, Schanbacher FL: Bovine lactoferrin mRNA: sequence, analysis, and expression in the mammary gland. Biochem. Biophys. Res. Comm. 180:75–84, 1991.
12. Schanbacher FL, Pattanajitvilai S, Neville MC: Post-transcriptional regulation of bovine and human lactoferrin. In Hutchens TW, Lonnerdal B: (eds): Lactoferrin: Interactions and Biological Functions. Ottowa, N.J. Humana Press, 1997, 81–95.
13. Nuijens JH, Van Berkel PHC, Schanbacher FL: Structure and biological actions of lactoferrin. J. Mammary Gland Biol. Neoplasia 1:285–296, 1996.
14. Kuhn NJ: The biochemistry of lactogenesis. In Mepham TB: (ed): Biochemistry of Lactation. Amsterdam: Elsevier, 1983, 351–380.
15. Schanbacher FL, Goodman RE, Talhouk RS: Bovine mammary lactoferrin: implications from messenger ribonucleic acid (mRNA) sequence and regulation contrary to other milk proteins. J. Dairy Sci. 76:3812–3831, 1993.
16. Teng CT, Pentecost BT, Chen YH, Newbold RR, Eddy EM, McLachlan JA: Lactotransferrin gene expression in the mouse uterus and mammary gland. Endocrinology 124:992–999, 1989.
17. Grigor MR, Carne A, Geurson A, Flint DJ: Effect of extended lactation and diet on transferrin concentrations in rat milk. J. Nutr. 118:669–674, 1988.
18. Leclercq YJ, Sawatzki G, Wieruszeski J-M, Montreuil J, Spik G: Primary structure of the glycans from mouse serum and milk transferrins. Biochem. J. 247:571–578, 1987.
19. Lee EY-H, Barcellos-Hoff MH, Chen L-H, Parry G, Bissell MJ: Transferrin is a major mouse milk protein and is synthesized by mammary epithelial cells. In Vitro Cell. Develop. Biol. 23:221–226, 1987.
20. Lee EY-H, Parry G, Bissell MJ: Modulation of secreted proteins of mouse mammary-epithelial cells by the collagenous substrata. J. Cell Biol. 98:146–155, 1984.
21. Green MR, Pastewka JV: Lactoferrin is a marker for prolactin response in mouse mammary explants. Endocrinology 103:1510–1513, 1978.
22. Hartmann P, Kulski JK: Changes in the composition of the mammary secretion of women after abrupt termination of breast feeding. J. Physiol. 275:1–11, 1978.
23. Devinoy E, Stinnakre MG, Lavialle F, Thépot D, Ollivier-Bousquet M: Intracellular routing and release of caseins and growth hormone produced into milk from transgenic mice. Exp. Cell Res. 221:272–280, 1995.
24. Molenaar AJ, Davis SR, Wilkins RJ: Expression of α-lactalbumin, α-S1-casein and lactoferrin genes is heterogeneous in sheep and cattle mammary tissue. J. Histochem. Cytochem. 40:611–618, 1992.
25. Hurley WL, Rejman JJ: Bovine lactoferrin in involuting mammary tissue. Cell Biol. Int. Rep. 17:283–289, 1993.

26. Lee EY-H, Lee W-H, Kaetzel CS, Parry G, Bissell MJ: Interaction of mouse mammary epithelial cells with collagen substrata: Regulation of casein gene expression and secretion. Proc. Natl. Acad. Sci. ,USA 82:1419–1423, 1985.
27. Danielson KG, Oborn CJ, Durban EM, Butel JS, Medina D: Epithelial mouse mammary cell line exhibiting normal morphogenesis in vivo and functional differentiation in vitro. Proc. Natl. Acad. Sci. ,USA 81:3756–3760, 1984.
28. Parry G, Cullen B, Kaetzel CS, Kramer R, Moss L: Regulation of differentiation and polarized secretion in mammary epithelial cells maintained in culture: extracellular matrix and membrane polarity influences. J. Cell Biol. 105:2043–2051, 1987.
29. Barcellos-Hoff MH, Aggeler J, Ram TG, Bissell MJ: Functional differentiation and alveolar morphogenesis of primary mammary epithelial cells cultured on a reconstituted basement membrane. Development 105:223–235, 1989.
30. Neville MC, Stahl L, Brozo LA, Lowe-Lieber J: Morphogenesis and secretory activity of mouse mammary cultures on EHS biomatrix. Protoplasma 160:110–123, 1991.
31. Streuli CH, Bissell MJ: Expression of extracellular matrix components is regulated by substratum. J. Cell Biol. 110:1405–1415, 1990.
32. Talhouk RS, Chin JR, Unemori EN, Werb Z, Bissell MJ: Proteinases of the mammary gland: developmental regulation in vivo and vectorial secretion in culture. Development 112:439–449, 1991.
33. Campbell T, Skilton RA, Coombes RC, Shousha S, Granham MD, Luqmani YA: Isolation of a lactoferrin cDNA clone and its expression in human breast cancer. Br. J. Cancer 65:19–26, 1992.
34. Toddywalla VS, Kari FW, Neville MC: Active Transport of nitrofurantoin across a mouse mammary epithelial monolayer. J. Pharmacol. Exp. Therap. 280:669–676, 1997.
35. Neville MC, Keller RP, Casey CE, Allen JC: Calcium partitioning in human and bovine milk. J. Dairy Sci. 77:1964–1975, 1994.
36. Fransson G-B, Lonnerdal B: Iron in human milk. J. Pediatr. 96:380–384, 1980.
37. Casey CE, Smith A, Zhang P: Macrominerals in human and animal milks. In Jensen RG: (ed): Handbook of Milk Composition. San Diego: Academic Press, 1995, 622–674.
38. Neville MC: Lactogenesis in women: A cascade of events revealed by milk composition. In Jensen RD: (ed): The Composition of Milks. San Diego: Academic Press, 1995, 87–98.
39. Neville MC, Allen JC, Archer P, Seacat J, Casey C, Lutes V, Rasbach J, Neifert M: Studies in Human Lactation: Milk volume and nutrient composition during weaning and lactogenesis. Am. J. Clin. Nutr. 54:81–93, 1991.

LACTOFERRIN IN DUODENAL ASPIRATES DURING CHILDHOOD

G. Sawatzki,[1] G. Georgi,[1] Th. Richter,[2] G. Moro,[3] and G. Boehm[1]

[1]Milupa Research
Milupa GmbH & Co KG, Friedrichsdorf
[2]Department of Pediatrics
University of Leipzig, Germany
[3]Center for Infant Nutrition
Milan, Italy

1. INTRODUCTION

Lactoferrin is a major protein in human milk. Since its concentration is higher in human colostrum than in mature human milk, it is assumed that this iron-binding protein may be part of the protective system of human milk like the immunoglobulin A[1]. Lactoferrin is also found in secretory fluids like saliva and pancreatic juice[2,3]. Industrial produced infant formulas do not contain significant amounts of bovine milk derived lactoferrin, neither they contain the species-specific human lactoferrin. Therefore, it remains an open question, whether human milk is the only significant source of lactoferrin reaching the gastrointestinal (GI) tract of the newborn human infant. To clarify this, a study was implemented to quantify the lactoferrin synthesised by one specific site of the GI tract, namely the pancreas. Testing was performed by measuring the lactoferrin in samples of duodenal aspirates. To get age-related information, these samples were taken from infants and children starting very early after birth until the age of about 8 years.

2. PATIENTS AND METHODS

This prospective study was based on 12 healthy preterm infants (mean gestational age: 29.4 weeks; range: 26–32 weeks) at a mean postnatal age of 17.6 days (range: 8–29) and 20 children (mean age: 46.5 months; range: 8–103 months).

The preterm infants were admitted to the Department of Perinatal Pathology, Macedonio Melloni Hospital, Milan, Italy. The gestational age was determined by menstrual

Advances in Lactoferrin Research, edited by Spik *et al.*
Plenum Press, New York, 1998.

history and clinical assessment according to Dubowitz et al.[4] The criteria for inclusion in the study were:

- predominantly enteral feeding from the first week of life (e.g. during the first week of life the enteral feeding > 50% of total fluid intake)
- birth weight appropriate for gestational age according to intrauterine growth chart of Lubchenco et al.[5]
- mechanical ventilation or continuos positive airway pressure treatment no longer than 24 hours and no requirements for supplementary oxygen after the first week of life or at time of study
- no intrauterine or postnatally acquired infection
- no major malformations

In the preterm infants, the examination of the duodenal aspirates were performed as part of a metabolic monitoring for individualisation of the nutritional management[6]. The indication for the measurements was the introduction of a cow's milk preterm formula due to a lack of mother's milk. Until the time of the study, all infants were fed with pasteurised mother's milk. Up to the 36th week of their postmenstrual age[7] the infants received the mother's milk individually fortified with a bovine human milk fortifier adjusted to give a protein content of 2 g/dl and an energy density of 75 kcal/dl[8]. The techniques for placement of the duodenal tube and for aspiration of the duodenal content have been described previously[9]. 0.5 ml of duodenal content was preprandially (i.e. 3 to 4 hours after the last meal) aspirated. Immediately before aspiration of duodenal juice, 1.0 ml of gastric aspirate was collected via a second tube. The confirmation of the correct tube placement was done comparing pH value and colour of the duodenal aspirate with those of simultaneously aspirated gastric content and by ultrasonography as described by Greenberg et al.[10]

The children were submitted to Department of Gastroenterology, Paediatric Hospital, University of Leipzig, Germany. Due to different clinical reasons the children were subject of a pancreozymin-secretin test according to the guidelines for diagnosis of exocrine pancreas insufficiency[11]. Children who's test results proved a normal exocrine pancreatic function related to the respective age were enrolled in this study. The placement of the tube and the control of accuracy of placement was performed by measuring pH in the aspirate and by ultrasound. After an eight hour starvation period a first duodenal juice sample was obtained. A second sample was obtained 20 minutes after stimulation with pancreozymin (15 ng/kg intravenously as 15 minutes infusion) and a third sample after stimulation with secretin (2,9 µg/kg intravenously as bolus). After aspiration, the duodenal samples were frozen and stored at –70°C until analysis.

For quantification of lactoferrin an ELISA assay kit (Calbiochem, Cat. No.: 427275) was used. The wells of the microplate have been coated with a primary monoclonal antibody to lactoferrin. 100 µl of pancreatic juice or lactoferrin standard solutions (1.6 ng/ml to 100 ng/ml) were pipetted into the wells and incubated at 37°C for 1 hour. Each well was washed 5 times with washing buffer (50 mM Tris-HCl buffer, pH 7.8, containing 150 mM NaCl, 0.1% Tween 20). Then, 100 µl of the biotinylated anti-lactoferrin (diluted in 20 mM phosphate buffer, pH 7.4, with 150 mM NaCl, 20 mg/ml bovine serum albumin and 0.1% Tween 20) was added. The microplate was incubated for 1 hour at 37°C. After washing, 100 µl avidin-horseradish peroxidase (diluted in 20 mM phosphate buffer, pH 7.4, with 150 mM NaCl, 20 mg/ml bovine serum albumin and 0.1% Tween 20) was added and incubated for 15 minutes at 37°C. Finally, 100 µl of the substrate o-phenylenediamine (OPD) were pipetted into the washed wells. After an incubation time of 5–10 minutes (until the absorbance of the highest concentrated standard (100 ng/ml) reaches approximately

2.5) the reaction was stopped by 50µl stop solution (1 M H_2SO_4). The absorption was read by the EASY READER 340 AT at 420 nm. For calculation of the lactoferrin concentration the four parameter equation was used.

The activity of lipase was measured nephelometrically using a modified Boehringer test kit for colipase-dependent lipase with triolein as substrate at pH 7.4 within 30 min after aspiration. The high pH was used to prevent simultaneous measurements of pancreatic and gastric lipase because the substrate is not specific for pancreatic lipase. The activity of trypsin was determined according to the method of Hummel et al. as described previously with N-tosyl-L-arginine methyl ester as substrate[12].

The study was approved by the Ethical Committees of both hospitals, and written informed parental consent was obtained prior to enrolment of each infant in the study.

Mean values and standard deviation were calculated for each of the variables. Regression analysis after logarithmic transformation of the lactoferrin concentrations were used for statistical evaluation. A $p < 0.05$ was taken as level of significance.

3. RESULTS

During the first week of life the mean duodenal lactoferrin concentration was with 1.38 ± 0.93 ng/ml extremely low and near the detection limit of the method. There was a significant ($p < 0.001$) increase of duodenal lactoferrin with age (Fig. 1) resulting in lactoferrin levels of about 50 ng/ml at an age of about 8 years.

Stimulation with pancreozymin or secretin resulted in higher lactoferrin concentrations but the increase was less pronounced as for the activities of pancreatic enzymes (Fig. 2).

4. DISCUSSION

This study demonstrates that some lactoferrin is synthesised even in the GI tract of the newborn human infant. The amount of this protein in duodenal aspirates gathered directly after birth (less than 2 ng/ml) is far beyond the concentration witch has been found later in childhood (about 50 ng/ml) or in adults, where levels of about 2000 ng/ml were measured[13]. Taking into account, that the level of lactoferrin in human milk is greater than 2 mg/ml[14] it is obvious that especially for the infant the only important source of species-specific lactoferrin is the ingested human milk. Even infant formulas do not supply con-

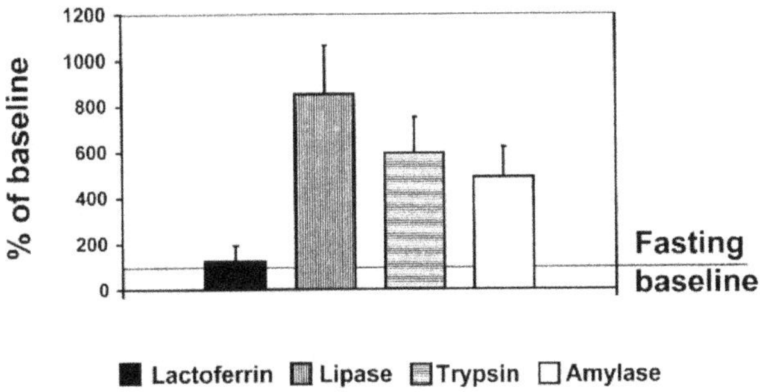

Figure 1. Influence of pancreatic stimulation on Lf concentration and activity of pancreatic enzymes in duodenal aspirates.

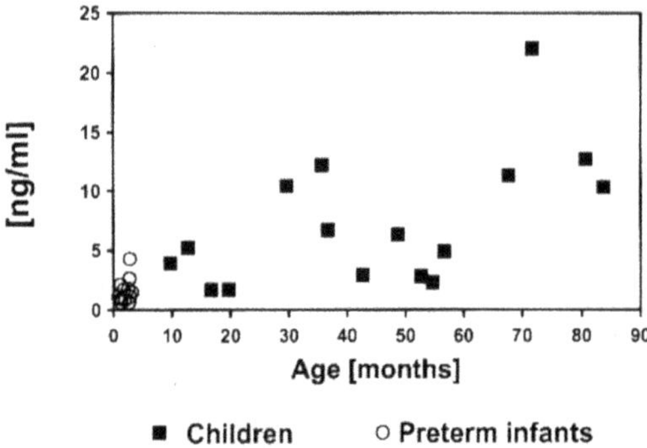

Figure 2. Lactoferrin concentration in duodenal aspirates.

siderable amounts of bovine lactoferrin (non species-specific!) since cows milk[14] is very low in lactoferrin (about 0.1 mg/ml) and this amount may also be lower due to heat treatment of the milk during the production of infant formulas. The role of lactoferrin is discussed as being part of the non-specific immune system or in iron absorption[15,16]. There seems to be a lot of unresolved questions whether lactoferrin has any role in iron absorption especially during the first three months of life[17]. The interesting finding of this study is that whereas the lactoferrin concentration in human milk is decreasing during lactation, the synthesis in the gut is increasing, but at a relative low level. If there is any protective role of lactoferrin in infancy this increase of the lactoferrin level is in certain congruence to the immunoglobulin synthesis. Nevertheless, it must be pointed out, that the supply of lactoferrin by human milk to the infant is about 6 magnitudes higher than the amount which is synthesised by the infant. If lactoferrin has any important role for the infant, there is no source more important than the own mothers milk. This might have consequences for the infant formulas to be produced in the future, but in addition the question whether bovine or human lactoferrin is the right molecule has also to be clarified.

REFERENCES

1. Prentice A, Ewing G, Roberts SB, Lucas A, MacCarthy A, Jarjou LMA, Whitehead RG: The nutritional role of breast-milk IgA and lactoferrin. Acta Paediatr Scand, 1987; 76: 592–598
2. Masson PL, Heremans JF, Dive C: An iron-binding protein common to many external secretions. Clin chim Acta, 1966; 14: 735–739
3. Estevanon JP, Sarles H, Figarella C: Lactoferrin in the duodenal juice of patients with chronic calcifying pancreatitis. Scand J Gastroenterol, 1975; 10:327–330
4. Dubowitz LM, Dubowitz V, Goldberg C: Clinical assessment for gestational age in the newborn infant. J. Pediatr. 1970, 77:1–10
5. Lubchenco LO, Hausmann C, Dressler M: Intrauterine growth as estimated from life-born weight data at 24 - 42 week's gestation. Pediatrics 1983, 32:793–807
6. Boehm G: Stoffwechsel und Ernährung von untergewichtigen Neugeborenen in der Adaptationsphase und der weiteren Entwicklung. Zschr. klin. Med. 1990, 45:1531–1535
7. Boehm G, Räihä NCR: Postmenstrual age correlates to indices of protein metabolism in very low birth weight infants. J. Pediatr. Gastroenterol. Nutr. 1993, 16:306–310
8. Moro G, Minoli I, Ostrom M, Jacobs JR, Picone TA, Räihä NCR, Ziegler E: Fortification of human milk: evaluation of a novel fortification scheme and of a new fortifier. J. Pediatr. Gastroenterol. Nutr. 1995; 20:162–172

9. Minoli I, Moro G, Ovadia MF: Nasoduodenal feeding in high risk newborns. Acta Paediatr. Scand. 1978, 67:161–168
10. Greenberg M, Bejar R, Asser S: Confirmation of transpyloric feeding tube placements by ultrasonography. J. Pediatr. 1993, 122:413–415
11. Kolacek S, Puntis JWL, Lloyd DR, Brown GA, Booth IW: Ontogeny of pancreatic exocrine function. Arch. Dis. Childh. 1990; 65:178–181
12. Boehm G, Bierbach U, DelSanto A, Moro G, Minoli I: Activity of trypsin and lipase in duodenal aspirates of healthy preterm infants. Effect of gestational and postnatal age Biol. Neonate,1995; 67: 248–253
13. Hayakawa T, Kondo T, Shibata T, Murase T, Harada H, Ochi K, Tanaka J: Secretory component and lactoferrin in pure pancreatic juice in chronic pancreatitis. Digestive Disease Sciences, 1993; 38:7–11
14. Masson PL, Heremans JF: Lactoferrin in milk from different species. Comp Biochem Physiol, 1971; 39B:119–129
15. Spik G, Montreuil J: Role de la lactotransferrine dans les mecanismes moleculaires de la defense antibacterienne. Bull Eur Physiopathol Respir 1983; 19:123–130
16. Brock JH: Lactoferrin in human milk: its role in iron absorption and protection against enteric infection in the newborn infant, 1980; 55:417–421
17. Sawatzki G: Lactoferrin in infant formulas; how and why?, 1997; in: Lactoferrin: interactions and biological functions (Eds: Hutchens TW, Lönnerdal B) Humana Press Inc., Totowa, NJ: p 389–397

19

IRON IN SYNOVIAL FLUID

Removal by Lactoferrin and Relationship to Iron Regulatory Protein (IRP) Activity

C. Guillen, I. B. McInnes, and J. H. Brock

Department of Immunology
Wetern Infirmary
Glasgow, United Kingdom

1. INTRODUCTION

Rheumatoid arthritis (RA) is a common human autoimmune disease with an incidence of about 1%. It is a chronic inflammatory polyarthritis characterised by infiltration of the synovial membrane by macrophages, T cells, B cells and activated fibroblasts[1], leading in most cases to progressive destruction of cartilage and bone.

While there has been progress in defining the etiology and pathogenesis of RA, these are still incompletely understood. However, it has been suggested that iron may play a role. Iron accumulates in the synovium, particularly in macrophages, and there are high levels in the synovial fluid during a chronic inflammatory response[2]. Synovial fluid iron may exceed the binding capacity of proteins such as transferrin, lactoferrin and ferritin and may then catalyse the production of hydroxyl radicals (OH•) from superoxide (O_2^-) and hydrogen peroxide (H_2O_2). OH• may be responsible for membrane damage, hyaluronic acid degradation, inactivation of proteinase inhibitors and destruction of several antioxidants within the synovial membrane[3]. If levels of synovial iron-binding proteins could be increased, "free" iron, and hence tissue damage, might be reduced. Lactoferrin, with its high affinity for iron even at moderately acid pH, would be particularly effective in this respect.

The reason why iron accumulates in the synovium is not fully understood. Control over cellular iron homeostasis is normally exerted by a system that operates post-transcriptionally in the cytoplasm. A *cis*-acting RNA motif, the iron-responsive element (IRE), and *trans*-acting cytoplasmic iron regulatory proteins (IRP's) control iron storage and up-

Advances in Lactoferrin Research, edited by Spik *et al.*
Plenum Press, New York, 1998.

take through expression of ferritin and the transferrin receptor. Iron deprivation stimulates IRE-binding by IRP, thereby repressing ferritin mRNA translation but protecting transferrin receptor mRNA from degradation[4,5]. However, in inflammatory conditions where iron accumulates the function of IRP may be altered, possibly through the action of inflammatory mediators such as nitric oxide and H_2O_2 which are known to modulate IRP activity[6].

The objectives of this work were therefore to determine levels of "free" iron in synovial fluid from RA patients and relate them to lactoferrin levels, study the ability of exogenous lactoferrin to bind "free" iron in synovial fluid, and relate IRP activity in synovial cells to synovial fluid iron concentration and the patients's clinical parameters.

2. MATERIALS AND METHODS

Synovial fluids were obtained by knee aspiration from 30 RA patients. The samples were centrifuged and the supernatants used to determine "free" iron by the bleomycin assay[7]. Patients were routinely monitored for clinical parameters.

The cells were used to assay IRP activity by mobility shift assay, using a ^{32}P-labelled oligoribonucleotide probe containing the ferritin IRE sequence[8]. Total and active IRP was determined by incubating samples with or without 2 mercaptoethanol (2%) for 10 minutes prior to assay, and the relative intensities of bands corresponding to IRE-IRP complexes quantified by densitometry.

To examine the binding of synovial fluid iron by lactoferrin and transferrin, fluids known to contain "free" iron were incubated overnight at 4°C with various concentrations of human or mouse lactoferrin or transferrin, and "free" iron then determined by the bleomycin assay.

Lactoferrin in synovial fluid was determined by a sandwich ELISA.

Data were analysed using Pearson's correlation or Students's t-test as appropriate. A p value of < 0.05 was considered to indicate a significant difference between groups.

3. RESULTS

3.1. "Free" Iron in Synovial Fluids of RA Patients

When the bleomycin assay was performed at pH 7.4, none of the fluids demonstrated unbound iron. In contrast, when the assay was performed at pH 5.3, comparable to that found in the microenviroment of cells such as activated macrophages which produce reactive oxygen compounds, 9 of 30 samples analysed contained "free" iron, the maximum concentration being 10 μM. All but two of the remainder gave "negative" values, i.e. they demonstrated a net iron-binding capacity of up to 10 μM Fe. There was no significant correlation between "free" iron/iron-binding capacity and any of the clinical parameters.

3.2. Correlation between Lactoferrin and "Free" Iron in Synovial Fluid

Lactoferrin levels were significantly lower ($p < 0.01$) in synovial fluids containing "free" iron than in those which did not (Fig. 1).

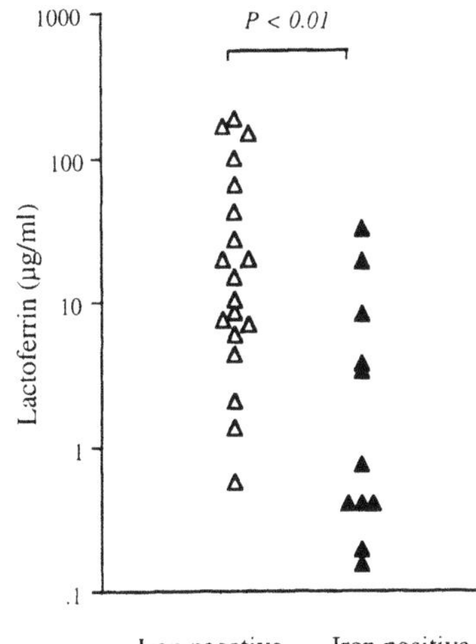

Figure 1. Lactoferrin in synovial fluid from RA patients. Each symbol represents the result from an individual synovial fluid sample.

3.3. Binding of "Free" Iron in Synovial Fluid by Lactoferrin and Transferrin

To study whether exogenous lactoferrin and transferrin could bind "free" iron in synovial fluid, aliquots of fluids known to contain "free" iron were incubated overnight with lactoferrin or transferrin. In general, human lactoferrin reduced the amount of "free" iron in synovial fluid, while transferrin had no effect, or sometimes actually increased it. Mouse lactoferrin, unlike human lactoferrin, was unable to reduce free iron level. A typical result is shown in Fig. 2.

3.4. IRP Activity in Synovial Fluid Cells

Total and active IRP was determined in cells isolated from the synovial fluid of 14 RA patients, and the ratio of active to total IRP calculated. There was no correlation between IRP activity and the amount of "free" iron in synovial fluid. However, there was a significant positive correlation ($p < 0.02$) with serum C-reactive protein (CRP) (Fig. 3).

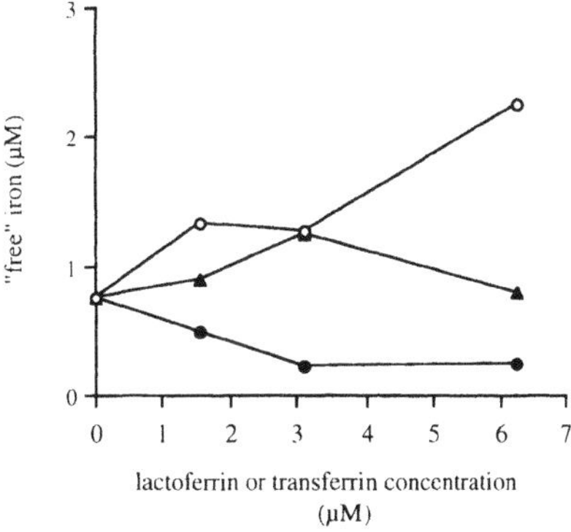

Figure 2. Binding of "free" iron in synovial fluid by exogenous lactoferrin and transferrin. Samples of synovial fluid were incubated overnight with different concentrations of human lactoferrin (●), human transferrin (▲) or mouse lactoferrin (○), and "free" iron determined by the bleomycin assay.

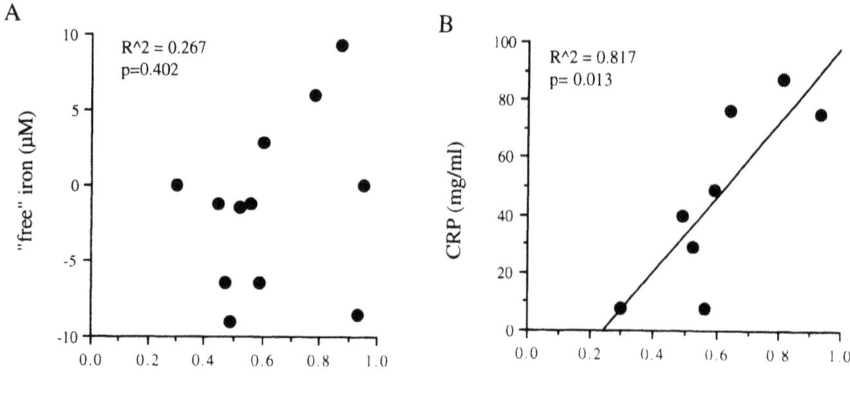

Figure 3. Iron regulatory protein (IRP) activity. IRP activity was determined by mobility shift assay, and related to "free" iron concentration in synovial fluid (A) and serum CRP (B).

4. DISCUSSION

In this study it was found that about one third of the synovial fluids examined contained "free" iron. It should be noted that such iron was only detectable when the assay was performed at pH 5.3; at pH 7.4 no "free" iron was detectable in any sample. However, the lower pH is comparable to that found in the microenviroment of an activated macrophage[9], where concentrations of hydrogen peroxide and superoxide are likely to be highest. At this pH transferrin is an ineffective iron binder, but lactoferrin is still effective[10] and could be a key molecule in determining whether "free" iron is present in the synovium. Our results provide two pieces of evidence to support this. Firstly, lactoferrin levels were significantly higher in synovial fluids in which "free" iron was present than in those where it was not detected. Secondly, addition of lactoferrin, but not transferrin, to fluids containing "free" iron reduced the "free" iron concentration. Interestingly, mouse lactoferrin, was also ineffective in decreasing "free" iron in synovial fluid.

This work has also addressed possible underlying mechanisms for the accumulation of iron in the synovium. We have found that in cells from synovial fluid the proportion of active IRP, which normally control iron homeostasis, did not correlate with synovial fluid "free" iron levels, but did correlate with serum CRP, a key marker of disease activity. This suggests that the effect of inflammatory mediators such as H_2O_2 or NO on IRP activity is more important than that of iron itself.

In conclusion, this work has shown that lactoferrin is a key molecule in reducing the amount of potentially toxic "free" iron in the synovium, and that such iron may accumulate, at least in part, because inflammatory cells in the synovium override the normal iron-mediated homeostatic effect of IRP on iron uptake and storage.

REFERENCES

1. Cush JJ and Lipsky PE (1988). Phenotypic analysis of synovial tissue and peripheral blood lymphocytes isolated from patients with rheumatoid arthritis. Arthritis Rheum. 31: 1230–1238.

2. Blake DR, Gallagher PJ, Potter AR et al. (1984). The effect of synovial iron on the progression of rheumatoid disease. Arthritis Rheum. 27: 495–501.
3. Burkhardt H, Schwingel M (1986). Oxygen radicals as effectors of cartilage destruction. Arthritis Rheum. 29: 379–387.
4. Melefors Ö, Hentze MW (1993). Iron regulatory factor - the conductor of cellular iron regulation. Blood Rev. 7: 251.
5. Klausner RD, Rouault TA, Harford JB (1993). Regulating the fate of mRNA: the control of cellular iron metabolism. Cell. 72: 19.
6. Pantopoulos K, Weiss G, Hentze MW (1996). Nitric oxide and oxidative stress (H_2O_2) control mammalian iron metabolism by different pathways. Mol. Cell. Biol. 16: 3781–8.
7. Gutteridge JMC, Rowley DA, Halliwell B (1981). Superoxide-dependent formation of hydroxyl radicals in the presence of iron salts. Biochem. J. 199: 263–265.
8. Gray NK, Quick S, Goossen B, Constable A, Hirling H, Kühn LC, Hentze MW (1993). Recombinant iron-regulatory factor functions as an iron-responsive-element-binding protein, a translational repressor and an aconitase. A functional assay for translational repression and direct demonstration of the iron switch. Eur. J. Biochem. 218: 657–667.
9. Etherington DJ, Pugh G, Silver IA (1981). Collagen degradation in an experimental inflammatory lesion: studies on the role of the macrophage. Acta Biol. Med. Ger. 40: 1625–1636.
10. Aisen P, Leibman A (1972). Lactoferrin and transferrin: a comparative study. Biochim. Biophys. Acta. 257: 314–323.

20

THE GUT

A Key Metabolic Organ Protected by Lactoferrin during Experimental Systemic Inflammation in Mice

Marian L. Kruzel, Yael Harari, Chung-Ying Chen, and Gilbert A. Castro

Department of Integrative Biology, Pharmacology and Physiology
University of Texas Medical School
Houston, Texas

1. SUMMARY

The gastrointestinal tract may be viewed as an ecologic system in which a balance between the host and bacterial flora exists. Two major host components appear to be involved in maintaining this balance. The first is a non-specific structural barrier provided by the epithelial layer of the gastrointestinal mucosae. The second component involves functional immunological elements found in the mucosal and submucosal compartments, e.g., gut associated lymphoid tissue. When gut integrity is disrupted by invasive pathogens or by trauma, a myriad of pro-inflammatory mediators are released from cells in the gut wall that exert actions in the tissue or gut lumen[1]. One of these mediators is lactoferrin, an iron binding protein found in high concentration in most human exocrine secretions. Despite controversies on its physiological role, evidence is emerging that lactoferrin plays an important role in host defense against toxic metabolites and antigenic components of potential pathogens[2–4]. This manuscript is intended to provide an overview of work related to lactoferrin's modulatory roles in inflammation, and to present observations from experimental studies on the preservation of intestinal structure and function by lactoferrin during intestinal inflammation. The possibility that lactoferrin limits the autodestructive inflammatory responses presents a new alternative for the future management of systemic inflammation.

2. INTRODUCTION

The significance of lactoferrin in health and disease has been the subject of several reviews[5–6]. Lactoferrin is thought to be an important component of the defense system, active at mucosal surfaces. It is a multifunctional protein expressed in a variety of cell

Advances in Lactoferrin Research, edited by Spik *et al.*
Plenum Press, New York, 1998.

types under different mechanisms of control[7]. For example, its expression in neutrophils is part of the maturation process, whereas in uterus and mammary gland the expression of lactoferrin is regulated by hormones (e.g. estrogen and prolactin). Lactoferrin has been implicated in modulation of many biological effects, such as, down regulation of myelopoiesis *in vitro*[8–10] and *in vivo*[11], mediation of monocyte/macrophage cytotoxic activity[12–13], and inhibition of *in vitro* antibody synthesis[14]. Recently, Zimecki et al. reported that lactoferrin down regulates the proliferative responses of peripheral blood mononuclear cells (PBMC) and up regulates the tumor necrosis factor in trauma patients, or patients in post-operative and septic shock[15].

Because of the high concentration of lactoferrin in colostrum, the importance of lactoferrin has been studied mostly in host defense responses in infants[16–18]. A working premise is that lactoferrin in human milk is not for the benefit of the mother, but for her offsprings, protecting against pathogens during adaptation to non-uterine life. Several studies of human milk have focused on lactoferrin's role in rendering breast-fed infants more resistant to the development of microbe-induced gastroenteritis when compared to formula-fed babies[17]. It has been suggested that lactoferrin protects the infant's gut by competing with bacteria for environmental iron[19–20]. However, supplementation of infant formula with lactoferrin has no influence on gut microflora, suggesting that mechanisms other than iron chelation might be responsible for protection against bacteria-induced injury[20]. Lactoferrin can kill a wide variety of Gram-negative and Gram-positive bacteria by direct interaction with the cell surface, a mode of action that is not dependent on iron[21–22]. In addition, the protective effect of lactoferrin in septicemia has been reported to be iron independent[23].

Physiological differences between species, the differences in the experimental designs and methods employed to measure some biological activities of lactoferrin, all play a part in the heterogeneity of results regarding its biological effects. The purpose of this report is to focus on information indicating that one of the physiological roles of lactoferrin is the modulation of the inflammatory responses.

3. LACTOFERRIN AND INFLAMMATION

Inflammation is fundamentally a protective response to cell injury. When excessive in magnitude or duration, however, the otherwise beneficial effects of inflammation may be deleterious, impacting negatively on the recovery or healing of the host. As illustrated in Fig. 1, acute inflammation may develop into chronic conditions or to systemic effects that lead to multiple organ disfunction, also known as multiple organ failure (MOF).

The roles of lactoferrin in modulating inflammation is not well understood. There are several reports in the literature that an increase in plasma lactoferrin is paralleled by an increase in the neutrophil count, or visa versa[4,24]. This correlation is not surprising, because lactoferrin is a constituent of secondary granules in neutrophils and can be released from these cells by exogenous stimuli such as bacteria. Neutrophils are the first phagocytic cells to arrive at the site of injury to ingest bacteria, dead cells, and cellular debris. A high level of lactoferrin in plasma has been suggested to be a predictive indicator of sepsis-related morbidity and mortality (reviewed in ref. 4). When tissue damage occurs, particularly if it is induced by infection during trauma, the vascular effects of the repair mechanisms are immediate. The tissue becomes inflamed at the site of injury, with the tissue spaces and the lymphatics blocked by fibrin clots. The fluid barely flows through the inflamed tissue, therefore the spread of bacteria and/or their toxic products is delayed. Un-

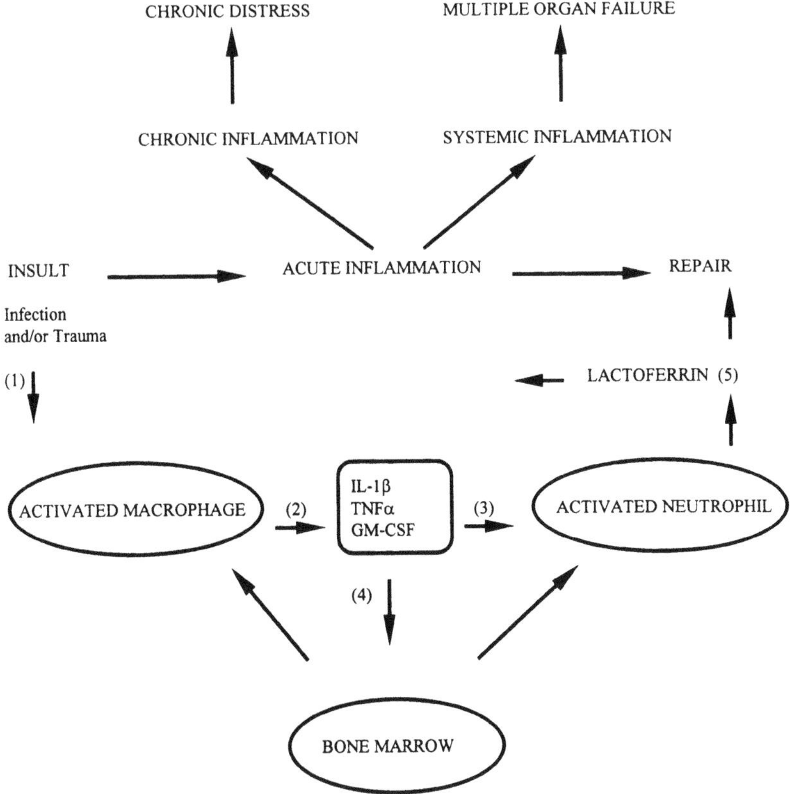

Figure 1. Outcomes of acute inflammation and the modulatory effect of lactoferrin. Insult (1) defined by infection, toxic mediators (LPS) or trauma leads to activation of monocyte/macrophage system and stimulates the production of IL-1β, TNFα and GM-CSF (2), which in turn activates circulating neutrophils (3) and stimulates the production of fresh neutrophils and monocyte/macrophages from the bone marrow (4). Activated neutrophils degranulate at the site of injury and release lactoferrin (5). By binding to the specific receptor on monocytes, lactoferrin reduces the production of IL-1β and GM-CSF, which decreases the production of fresh monocyte/macrophages and neutrophils. In addition, lactoferrin released from neutrophils can bind LPS, reducing its ability to oprime neutrophils.

like the immune responses, which may take days to develop, the vascular effects of inflammation occur in seconds and coincide with the burst of inflammatory cytokines, such as tumor necrosis factor (TNFα) and interleukin-1 (IL-1β), from activated monocytes/macrophages. Subsequent to the release of cytokines is an acute increase in neutrophils in the blood. Within an hour after inflammation begins, the bone marrow may increase the production of fresh neutrophils up to fivefold from a normal level. Large numbers of neutrophils begin to invade the tissues that attract these cells. The feedback control of inflammation begins with degranulation of neutrophils and massive release of lactoferrin, which in turn inhibits the production and release of IL-1β and granulocyte/macrophage colony stimulating factor (GM-CSF) (Fig. 1).

In addition, lactoferrin released at the site of the injury sequesters iron and binds to LPS, blocking its ability to prime fresh neutrophils. Although generally beneficial to the host, inflammatory processes are intrinsically destructive to the surrounding tissues and potentially can result in major tissue injury. Furthermore, the inflammatory response can spread from the local environment and induce generalized systemic response which may become self-perpetuating by overproduction of proinflammatory cytokines[1,25]. By blocking deleterious effects of those cytokines, lactoferrin has the potential to control the development of systemic inflammation.

It is becoming increasingly evident that the gastrointestinal tract is a portal of entry for bacteria, which cause life threatening infection in immunocompromised patients[26]. Therefore, it is of great importance to preserve the structure and function of the gastrointestinal tract during acute infection or trauma to avoid the systemic translocation of the enteric bacteria, or their metabolic products.

In studies that we performed, lactoferrin's ability to maintain gut structure and function during local and systemic inflammation in mice were examined. Effects of human lactoferrin were studied in two experimental settings: a) *Trichinella spiralis*-induced gut inflammation[27]; b) LPS-induced endotoxemia (unpublished results). It was evident from our studies that lactoferrin protects intestinal epithelium from damage during these deleterious conditions. In some respects, it is not surprising that lactoferrin attenuates the impairment caused by inflammation of the gut, since this is the primary protective function of lactoferrin in the process of colonization of neonate intestine[18]. During parturition the neonate is exposed to a variety of bacteria from mother's skin flora as well as from the environment. Many factors play roles in determining how the gut is colonized. Since colostrum is produced by the mother for a few days after parturition, and lactoferrin constitutes up to 30% of its total protein, it is suggested that lactoferrin plays a primary role in regulating the colonization of the neonate gut. After the first week of lactation, the level of lactoferrin in mothers milk decreases up to fivefold. By that time the gastrointestinal tract has transformed to cope with the stress associated with bacterial colonization.

The protective function of lactoferrin for intestinal epithelium discussed in this paper can also be explained in view of Wang *et al.*, who demonstrated that inactivation of LPS by lactoferrin may lead to the reduction of tissue-damaging oxygen radicals by blocking the ability of LPS to prime fresh neutrophils[28]. Moreover, iron-free lactoferrin has a similar ability to inhibit iron-catalyzed hydroxyl radical formation in the presence or absence of high concentrations of LPS[29]. Lactoferrin-LPS interaction may also be relevant to the role of lactoferrin in infection. Miyazawa *et al.* suggested that a dynamic interaction between LPS and lactoferrin during acute infection can modulate the biologic effects of each molecule. Both the antiproliferative effects of lactoferrin on myelopoiesis and the ability of LPS to prime neutrophils may be decreased[30]. Recently, two LPS binding sites on the lactoferrin molecule have been described[31]. A high affinity site was found in the N-terminal fragment whereas a low affinity binding site is located at the C-terminus.

4. CLINICAL IMPLICATIONS

The gastrointestinal tract appears to be the primary source of systemic bacterial infection because most terminal infections associated with MOF are caused by enteric bacteria[32]. However, as many as 70% of patients who die of MOF following trauma do so without evidence of bacterial sepsis or infectious focus[25]. One explanation of these contradictory results is that the inflammatory response induced by injury may elicit the production of cytokines in the gut even in the absence of detectable portal bacteremia[33]. Cytokines are the most likely common mediators in the pathophysiology of MOF. Once released via the portal vein into the systemic circulation, they can stimulate other inflammatory cells, notably neutrophils. Thus, the gut has been described as the " motor" of

MOF following trauma, while neutrophils serve as the effector cells of distant organ injury[34]. Therefore, the control of cytokine production in the gut is of great importance for patients who may develop MOF.

In our recent studies on LPS-induced endotoxemia in mice (unpublished results), we found that lactoferrin attenuates the release of TNFα and IL-1β in plasma. Miyazawa *et al.* demonstrated that lactoferrin interacts with specific receptors on mononuclear cells and suppress the release of IL-1β, an effect mediated by the interaction with LPS, as discussed earlier. Also, it has been demonstrated *in vitro*[10] and *in vivo*[35] that lactoferrin suppresses the release of IL-6 and TNFα from lymphocytes. This may explain the inhibition of granulopoiesis by lactoferrin, which is secondary to the suppression of TNFα and IL-1β required for the production of GM-CSF[36]. In addition, because both TNFα and IL-1β enhance leukocyte adhesion to endothelial cells, suppression of TNFα and IL-1β by lactoferrin at the inflammatory site provides a feedback mechanism to prevent excessive neutrophil recruitment and activation. These effects suggest that lactoferrin regulates nonspecific and immune-mediated inflammatory responses.

The clinical benefit of lactoferrin given orally has been reported by Teraguchi *et al.*, who demonstrated that bovine lactoferrin decreased bacterial translocation in mice[37]. He further suggested that this is due to lactoferrin's ability to suppress bacterial overgrowth. The bacteriostatic effect was dependent on the concentration of lactoferrin, the duration of treatment and the bacterial species, with *Clostridium ramosum* C1 being the most profoundly suppressed[38]. It is likely that in parallel to the suppression of bacterial overgrowth, lactoferrin is acting through the cytokine network to suppress the production of TNFα, IL-1β and IL-6 and reduce bacterial translocation from the gut. By virtue of its multifunctional property, lactoferrin appears to activate different levels of the host defense system, therefore control the development of systemic inflammation.

5. CONCLUSION

In addition to carrying out its digestive, absorptive and secretory functions, the gastrointestinal tract serves as a major barrier to bacteria and/or bacterial products, such as endotoxin, preventing their contact with systemic organs and tissue. Disruption or impairment of the intestinal defense system by invasive bacteria or trauma promotes bacterial translocation or increases the local production of cytokines[25]. Moreover, in many clinical situation, notably trauma, patients may become immunocompromised with their intestinal barrier function lost. Because the gut contains the body's largest collection of lymphoid cells, it is likely that injury to the gut barrier, secondary to an inflammatory insult, would result in the activation of immunological elements and compartments of the mucosal immune system. We postulate that lactoferrin plays an important role in regulating the mucosal immune system and, thereby blocks the generalized autodestructive systemic inflammatory response.

Available evidence suggests that human lactoferrin given orally is capable of attenuing the intestinal injury caused by inflammation. Although the mechanism of lactoferrin action is not clear, it is likely to be a combined effect of different activities, namely its ability to sequester iron, bind endotoxin, inhibit the release of cytokines and to reduce the production of hydroxyl radicals. In the search of treatment for the systemic inflammatory response syndrome, lactoferrin may be important. However many questions remain to be answered. Further studies will be required to address two clinically important issues: 1)

Whether the beneficial effects of lactoferrin are species specific, and 2) Whether the parenteral administration of lactoferrin is as beneficial as the oral administration.

Based on our knowledge of the biology of lactoferrin and mechanisms by which lactoferrin blocks specific mediators in septic responses, it may be possible in the near future to design a treatment for systemic inflammation where multiple agents with different actions are combined to produce the desired therapeutic effect.

REFERENCES

1. Deitch E.A. 1992 Multiple organ failure: Pathophysiology and potential future therapy. Ann. Surg. 216, 117–134.
2. Spik G., Montreuil J., 1983 The role of lactotransferrin in the molecular mechanisms of antibacterial defense. Bull. Eur. Physiopathol. Resp. 19 (2), 123–130.
3. Brock J. 1995 Lactoferrin: a multifunctional immunoregulatory protein? Immunology Today 16, 417–419.
4. Bayens R.D., Bezwoda W.R. 1994 Lactoferrin and the inflammatory response. In: Lactoferrin: Structure and Function, eds. T.W.Hutchens *et al.*, Plenum Press, pp.133–141.
5. Lonnerdal B., Iyer S. 1995 Lactoferrin: molecular structure and biological function. Annu. Rev. Nutr. 15, 93–110.
6. Sanches L., Calvo M., Brock J.H. 1992 Biological role of lactoferrin. Arch. Dis. Child. 67, 657–661.
7. Teng C.T., Liu Y., Yang N., Walmer D., Panella T. 1992 Differential Molecular Mechanism of the Estrogen Action That Regulates Lactoferrin Gene in Human and Mouse. Mol. Endo. 6, 1969–1981.
8. Broxmeyer H.E., Smithyman A., Eger R.R., Meyers P.A., de Sousa M. 1978 Identification of lactoferrin as the granulocyte-derived inhibitor of colony stimulating activity (CSA)-production. J. Exp. Med. 148, 1052–1067.
9. Broxmeyer H.E., de Sousa M., Smithyman A., Ralph P., Hamilton J., Kurland J.I., Bognacki J. 1980 Specificity and modulation of the action of lactoferrin, a negative regulator of myelopoiesis. Blood 55 (2), 324–333.
10. Crouch S.P.M., Slater K.J., Fletcher J. 1992 Regulation of cytokine release from mononuclear cells by the iron-binding protein lactoferrin. Blood 80, 235–240.
11. Broxmeyer H.E., Williams D.E., Hangoc G., Cooper S., Gentile P., Shen R-N., Ralph P., Gillis S.. Bicknell. 1987 The oposing action *in vivo* on murine myelopoieis of purified preparations of lactoferrin and the colony stimulating factor. Blood Cells 13 (1–2), 31–48.
12. McCormick J.A., Markey G.M., Morris T.M.C. 1991 Lactoferrin-inducible monocyte cytotoxicity for K562 cells and decay of natural killer lymphocyte cytotoxicity. Clin. Exp. Immunol. 83, 154–156.
13. Nishiya K., Horwitz D.A. 1982 Contrasting effects of lactoferrin on human lymphocyte and monocyte natural killer activity and antibody-dependent cell-mediated cytotoxicity. J. Immunol. 129, 219–2523.
14. Duncan R.L., McArthur W.P. 1981 Lactoferrin-mediated modulation of mononuclear cell activities. Suppression of the murine *in vitro* primary antibody response. Cell Immunol. 63, 308–320.
15. Zimecki M., Adamik B., Wlaszczyk A., Kubler A., Zagulski T. 1997 Immunoregulatory activity of lactoferrin in healthy individuals and in patients: Clinical studies *in vitro*.Third International Conference on Lactoferrin, Le Touquet, France.
16. Howie P.W., Forsyth J.S., Ogston S.A.. Clark A., du V.Florey C. 1990 Protective effect of breast feeding against infection. B. M. J. 300, 11–16.
17. Brock J.H. 1980 Lactoferrin in human milk: its role in iron absorption and protection against enteric infection in the newborn infants. Arch. Dis. Child. 55, 417–421.
18. Roberts A.K. 1986 Prospects for further approximation of infant formulas to human milk. *Human Nutrition:Applied Nutrition* 40A, Suppl.1, 27–37.
19. Roberts A.K., Chierici R., Sawatzki G., Hill M.J., Volpato S. & Vigi V. 1992 Supplementation of an adapted formula with bovine lactoferrin. I. Effect on the infant feacal flora. Acta Paediatr. Scand. 81, 119–124.
20. Balmer S.E., Scott P.H. , Wharton B.A. 1989 Diet and fecal flora in the newborn: lactoferrin. Arch. Dis. Child. 64, 1685–1690.
21. Bellamy W., Takase M., Wakabayashi H., Kawase K., Tomita M. 1992 Antimicrobial spectrum of lactoferricin B, a potent bactericidal peptide derived from the N-terminal region of bovine lactoferrin. J. Appl. Bacteriol. 73, 472–479.

22. Wakabayashi H., Bellamy W., Takase M., Tomita M. 1992 Inactivation of *Listeria monocytogenes* by lactoferricin, a potent antimicrobial peptide derived from cow's milk. J. Food Protect. 55, 238–240.
23. Zagulski T., Lipinski P., Zagulska A., Broniek S., Jarzabek Z. 1989 Lactoferrin can protect mice against a lethal dose of *Escherichia coli* in experimental infection in vivo. Br. J. Exp. Pathol. 70, 697–704.
24. Baynes R.D., Bezwoda W.R., Khan Q., Mansoor N. 1986 Relationship of plasma lactoferrin content to neutrophil regeneration and bone marrow infusion. Scand. J. Haematol. 36, 79–84.
25. Goris R. J.A., Boekhorst T.P.A., Nuytinck J.K.S., Gimbrere J.S.F. 1985 Multiple organ failure. Arch. Surg. 120, 1109–1115.
26. Deith E.A., Xu D., Qi L., Berg R.D. 1991 Bacterial translocation from the gut impairs systemic immunity. Surgery 109, 269–276.
27. Kruzel M.L., Harari Y., Chen C.Y., Castro G.A. 1997 Preservation of Mucosal Structure and Function by Lactoferrin during *Trichinella spiralis*-Induced Enteritis in Mice. Manuscript submitted to the Int. J. Exp. Pathol.
28. Wang D., Pabst K.M., Aida Y., Pabst M.J. 1995 Lipopolysaccharide-inactivating activity of neutrophils is due to lactoferrin. J. Leukoc. Biol. 57, 865–874.
29. Cohen M.S., Mao J., Rasmussen G.T., Serody J.S., Britigan B.E. 1992 Interaction of Lactoferrin and Lipopolysaccharide (LPS): Effects on the Antioxidant Property of Lactoferrin and the Ability of LPS to Prime Human Neutrophils for Enhanced Superoxide Formation. J.Infect. Dis.166, 1375–1378.
30. Miyazawa K., Mantel C., Lu L., Morrison D.C., Broxmeyer H.E. 1991 Lactoferrin-lipopolysaccharide interactions. Effect on lactoferrin binding to monocyte/macrophage-differentiated HL-60 cells. J. Immunol. 145, 723–729.
31. Elass-Rochard E., Roseanu A., Legrand D., Trif M., Salmon V., Motas C., Montreuil J., Spik G. 1995 Lactoferrin-lipopolysaccharide interaction: involvment of the 28–34 loop region of human lactoferrin in the high-affinity binding to *Escherichia coli* 055B5 lipopolysaccharide. Biochem. J. 312, 839–845.
32. Marshall J.C., Christou N.V., Horn R., Meakins J.L. 1988 The microbiology of multiple organ failure. Arch. Surg. 123, 309–315.
33. Mainous M.R., Ertel W., Chaudry I.H., Deith E.A. 1995 The gut: a cytokine-generating organ in systemic inflammation? Shock 4 (3),193–199.
34. Botha A.J., Moore F.A., Moore E.E., Sauaia A., Banerjee A., Peterson V.M. 1995 Early neutrophil sequestration after injury: a pathogenic mechanism for multiple organ failure. J. Trauma 39, 411–417.
35. Machnicki M., Zimecki M., Zagulski T. 1993 Lactoferrin regulates the release of tumor necrosis factor alpha and interleukin 6 *in vivo*. Int. J. Exp. Pathol. 74, 433–439.
36. Bagby G.C., McCall E., Layman D.L. 1983 Regulation of colony-stimulating activity production: Interactions of fibroblasts, mononuclear phagocytes, and lactoferrin. J. Clin. Invest. 71, 340–344.
37. Teraguchi S., Shin K., Ogata T., Kingaku M., Kaino A., Miyauchi H., Fukawatari Y., Shimamura S. 1995 Orally administered bovine lactoferrin inhibits bacterial translocation in mice fed bovine lactoferrin. Appl. Environ. Microbiol. 61 (11), 4131–4134.
38. Teraguchi S., Shin K., Ozawa K., Nakamura S., Fukuwatari Y., Tsuyuki S., Namihira H., Shimamura S. 1995 Bacteriostatic Effect of Orally Administered Bovine Lactoferrin on Proliferation of Clostridium Species in the Gut of Mice fed Bovine Lactoferrin. Appl. Environ. Microbiol. 61 (4), 501–506.

THE ANTIBACTERIAL ACTIVITY OF LACTOFERRIN AND NEONATAL *E. coli* INFECTIONS

A Selective and Critical Review

Bruno Reiter[1] and Jean-Paul Perraudin[2]

[1]Child Health Institute
University of Bristol
United Kingdom
[2]Laboratoires Biopole
75 Bvd Ed Machtens
1080 Bruxelles, Belgium

It aint necessarily so
The things that your preacher
Is liable to teach yer
It aint necessarily so.

1. INTRODUCTION

In 1960, Montreuil *et al.*[1] among others described a method for isolating and purifying an iron-chelating protein from human milk now known as lactoferrin (Lf). In their discussion they made a suggestion which we are still trying to substantiate: "une activité antibiotique vis-à-vis de certains germes pathogènes pour le Nourrisson". The suggestion that lactoferrin is antibacterial was based on the seminal paper by Schade & Caroline in 1944[2]. They used raw hen egg white to stabilize a bacteriophage during lyophilization and observed that its host, *Salmonella dysenteriae*, was inhibited by the egg white; this inhibition could be reversed by the addition of iron. After a suitable 'incubation time' of about 20 years, this discovery, by serendipity, became the beginning for all aspects of iron and bacterial infection and immunity, as we now know. In 1961, Hanson[3] discovered secretory immunoglobulin A (sIgA) which eventually led to the concept of Mucosal Associated Lymphoid Tissue (MALT)—the gut and lung, mammary, salivary and lacrymal glands, and the genital tract. In addition to sIgA, secretions can also contain other non-antibody

Advances in Lactoferrin Research, edited by Spik *et al.*
Plenum Press, New York, 1998.

protective proteins such as lactoferrin (Lf), lysozyme (LZ), lactoperoxidase (LP) and xanthine oxidase (XO). LP is the enzyme which catalyses the bacterial activity of the lactoperoxidase-thiocyanate-hydrogen peroxide system (LP-system)[4].

There is evidence that these systems have a long evolutionary history preceding the immune systems of the vertebrates. A transferrin-like iron chelator in spiders and crabs is a forerunner of Lf. LZ has its forerunner in bacteriophage lysin; other small molecular weight proteins (cecropins, attacins, etc) occur in insects. Superoxide dismutase, catalase and peroxidases make aerobic life possible by protecting against the toxic effects of oxygen-free radicals. XO can generate H_2O_2 to activate for instance the LP-system. The antibacterial role of Lf should therefore not be considered in isolation but as one of the non-antibody protective proteins which exist *a priori* without antigenic reactions. The real and deciding impetus for research on Lf came from the classic paper by Bullen *et al.* in 1972[5]. As it appeared in a medical journal, the results with human milk and extrapolation from guinea piglets to infants was readily accepted and is invariably cited, but it receives little critical discussion now.

2. BACTERIOSTASIS AND IRON CHELATION

Masson *et al.*[6] demonstrated the presence of Lf in all bronchial glands. On solid agar Lf, isolated from human milk, was shown to inhibit *Staphylococcus albus* but not *Staph. aureus*, reversible by Fe. Independently, Oram and Reiter[7] purified Lf from bovine milk and colostrum and showed that in a synthetic medium, Lf inhibited the growth of *Bacillus stearothermophilus*. Considering that Fe is essential for all forms of life, it was accepted that the capacity of Lf to chelate Fe is inhibitory to bacteria.

The proof that bacteriostasis does not solely depend on Fe chelation was further demonstrated in a paper by Stuart *et al.*[8] They established that when all the Fe from the medium had been chelated, raising the level of Lf further increased the inhibition. Brock suggests that *Escherichia coli* could acquire Fe from lactoferrin by synthesizing increasing amounts of siderophores, thus providing Fe (pers. commun.). The nutrient composition of the medium ensured good growth for the control. Lf delayed growth for only a few hours, eventually reaching the same density of growth as the control. Prior to these experiments, the same serotype of *Escherichia coli* (0111 B_4) was shown to be more inhibited—the growth curves differed greatly because the medium contained ammonia instead of the amino acids. Even *in vitro* it is necessary to recognize that the degree of inhibition always depends on the rate of growth. There is one recorded instance that Fe is not essential for growth: *Lactobacillus plantarum* has no requirement for Fe. Unfortunately, the lack of Fe requirement is often extended to all lactic acid bacteria. This is not true because Group N streptococci, for instance, need iron, albeit at only 1/100 that of *Escherichia coli*.

3. BINDING OF Lf TO BACTERIA

Based on the experience with bacteriophage, 'phage lysin and LZ, all of which need to attach to surfaces, it was obvious that Lf ought to exert its effect at least in part by attachment. This was indirectly proven when an enteropathogenic strain of *Escherichia coli* was suspended in bovine colostrum. Some results[9] suggest that inhibition by an Fe chelator is not simply due to removal of Fe from the medium but involves an interaction between the protein and the surface of the organism. Valenti *et al.*[9] found that Lf was absorbed to different degrees by different bacteria but an attempt to correlate absorption of

Lf with the type and amount of siderophores failed. The degree of Lf absorption was extended to *Escherichia coli* isolates from human intestinal infections by Naidu *et al.*[10] Enterotoxigenic (ETEC) strains were found to bind a significantly higher amount of human Lf than enteropathogenic (EPEC), enteroinvasive (EIEC) and enterohemorrhagic (EHEC) strains. The lowest binding was observed with *Escherichia coli* isolated from a 'normal' flora (does this explain the commensal strains as defined by Spik *et al.*[11]? It was also confirmed that there was no difference in the ability of the strains to produce siderophores. The binding was inversely related to the change in metabolic rate of the bacteria by lactoferrin. The degree of Lf binding may be related to the serotypes and is not blocked by fucose, mannose or other carbohydrates. It is now well established that the binding proteins on the bacterial surface are the porins - OmpF and OmpC - which can be shielded by the polysaccharide chains as in mutants of a wild strain of *Salmonella typhimurium*. Such mutants are less readily inhibited. Incidentally it is interesting to note that both the lipopolysaccharides of *Escherichia coli* and the porins have been known for a long time to be involved in the attachment of T4 coli phage. Naidu *et al.* demonstrated[10] the high binding capacity of staphylococci, depending on the source of isolation and the growth conditions. They observed that staphylococci cultured in carbohydrate-salt-rich medium failed to bind Lf because the cocci produced slime or capsules.

4. LACTOFERRIN AND PERMEABILITY OF THE OUTER MEMBRANE OF GRAM-NEGATIVE BACTERIA TO LYSOZYME

Ellison *et al.*[12] were also not satisfied that bacteriostasis of Gram-negative bacteria depends solely on iron chelation or that antibodies to siderophores suppress bacterial Fe metabolism. They based their work on the effects of EDTA on the outer membrane. By an imaginative leap, they considered whether Lf could have the same role as EDTA, notwithstanding that Lf, unlike EDTA, is known not to chelate Ca and Mg. Against expectations, Lf released the LPS at nearly the same rate as EDTA. Considering the similar distribution of LZ and Lf, they investigated whether LZ could gain access and degrade the underlying peptidoglycan after increasing the permeability by prior exposure to Lf. Indeed, this synergistic effect of Lf/LZ kills and lyses not only *Escherichia coli* but also *Vibrio cholerae* and *Salmonella typhimurium* (it is well known for instance that the complement-activated bacteriocidal activity of specific antibodies is further enhanced by LZ). They also observed that *Escherichia coli* would be either inhibited on reaching the same level as the control after 24 h or it would be reduced to very low counts depending on the media in which the tests were performed. It became evident that killing progressively decreased as the media osmomolarity increased. No killing occurred at the osmolarity approaching 100 osm. This is an important finding because it explains why inhibition by Lf depends not only on the size of the inoculum, the nutrient composition of the medium, the metabolic state of the organisms—whether they are in the lag, stationary, or logarithmic phase—but also and importantly on the osmolarity.

Now it is possible to explain the results of Arnold *et al*[13] obtained from 1977 onwards, when bacteria were suspended in distilled water or unphysiological salt solution: the bacteria which are killed by Lf in distilled water were possibly on the way to becoming 'ballooned' spheroplasts as observed by Ellison in media of low osmolarity. Nevertheless, in 1981 Arnold *et al.*[14] published some important findings that can be re-interpreted.

In the presence of apolactoferrin, the uptake of glucose by *Streptococcus mutans* was inhibited, as was the incorporation of uracil, thymidine and lysine. This metabolic inhibition by *Streptococcus mutans* in the presence of Lf can be compared with the effect of the LP-system which influences the metabolism of both Gram-negative and Gram-positive microorganisms[15]. It would therefore be of great interest if the Lf experiments in distilled water were tested for leakage, which is almost certain to be the case. Lassiter *et al.*[16] made the important observation that under anaerobic conditions *Streptococcus mutans* was not killed by Lf in distilled water. Their suggestion that hydroxy radicals may be involved was based on the erroneous conclusion that Lf can produce free radicals. This was disproved by Halliwell and Guttteridge in 1985[17]. *Streptococcus mutans* is killed under aerobic conditions which can be reversed by catalase and to a lesser degree by peroxidase. The requirement for oxygen points to H_2O_2 produced by *Streptococcus mutans* which may become bacteriocidal in deionized water at pH 5. In the case of the LP system, bacteria are killed when plated out under aerobic conditions but can recover 1–2 h after exposure to the system when incubated anaerobically. This power of recovery of damaged organisms is now well documented, and was first observed with *Staphylococcus aureus*. When heated at sublethal temperatures, the 'heat-shocked' cells recovered after prolonged incubation (extended lag phase) when plated on non-selective agar media. Recovery on selective media (e.g. egg yolk glycine pyruvate tellurite), diagnostic for staphylococci, was much lower because of the inherent toxicity of the medium. To avoid this complication, it may therefore give a truer estimate of bacterial content in samples such as fresh feces to assess the ATP activity instead of plating out (see later, *in vivo*).

According to Bullen *et al.*[5] (and see later), the bacteriostatic activity of Lf for *Escherichia coli* 0111 can be appreciably increased by specific antibodies. They used the high titer antiserum (1:20000), probably containing predominantly IgG, a strongly agglutinating class of immunoglobulins. The increased inhibition may therefore be due to a reduction in colony-forming units because of 'clumping' of the bacteria. Although this may be an artefact *in vitro*, it may be an advantage *in vivo*. Agglutinated bacteria are less likely to attach to the intestinal epithelium and may therefore be rapidly removed by peristalsis. It was known for a long time that many Lf preparations were contaminated with sIgA and it is now known that Lf and sIgA form complexes. Lf isolated from human milk is predominantly contaminated with LZ, and from bovine milk with LP. Although the concentration of LZ in bovine milk is very low, it was difficult to exclude it during purification. Unfortunately, in the past too many Lf preparations purchased from biochemical suppliers were found to be heavily contaminated, up to 70%. Clumping was also observed by the combined effect of Lf and LZ. After exposure of *Micrococcus luteus*[18] to LZ, the protoplasts were heavily aggregated by Lf. It was calculated that each cell has 1.4×10^6 binding sites per cell. Modifying the Lf by succinylation (4/39 lysines) abolished clumping without impairing the Fe chelating activity. A similar effect was obtained with *Escherichia coli* 0111[19] in the combination of LZ with apo-Lf and partially Fe saturated Lf. Since Fe-Lf is less susceptible to digestion the authors concluded that LZ/Lf may have a significant effect in protecting breast fed infants against intestinal infections. When sIgA became the subject of intensive research after 1960, immunologists tended to complain that their preparations of sIgA were contaminated with Lf.

5. ATTACHMENT OF BACTERIA BY ADHESINS, PILI, OR FIMBRIAE

At birth, the intestine is sterile, but becomes rapidly colonized with microphilic and anaerobic bacteria. Commensal or non-pathogenic strains of *Escherichia coli*, lactobacilli and other microphilic bacteria are a minor part of the flora. Exposure to pathogenic *Escherichia coli, Shigella, Campylobacter, Yersinia* etc would change the otherwise remarkably stable intestinal flora. As far back as 1971, Sojka[20] observed that certain serotypes of *Escherichia coli* were frequently isolated from the feces of newborn piglets (and later from calves and lambs) with diarrhea. One of the *Escherichia coli* serotypes was K_{88} antigen, now generally known as an adhesin (pili or fimbriae). They enable the organism to attach to the intestine and overcome mechanical clearing by peristalsis as was discovered by Jones and Rutter[21]. Enterotoxin producing strains of *Escherichia coli* (ETEC) are mainly associated with serotypes K_{88}, K_{89}, 947P etc in animals and CFI/CFII etc (colonizing factors) in man. ETEC strains cause diarrhea not by colonization *per se* but by the production of enterotoxins, depending on plasmid DNA. Enterotoxins affect electrolyte transport and cause loss of fluid, hence loose watery feces and eventual dehydration. Adhesion is therefore a specific phenomenon requiring interaction with receptors which can be prevented by antisera (e.g. against K_{88}). There exist also phenotypes of pigs which do not possess the specific receptors for K_{88}—they are naturally immune to *Escherichia coli*. It must be emphasized also that initial attachment happens in the upper section of the small intestine, but only several weeks after birth, as adult pigs cannot be infected by the same strain of *Escherichia coli* as neonates. Camara *et al.*[22] very recently investigated some EPEC strains of specific serotypes (0111 and 0119) which are the main cause of acute diarrhea in infants in Brazil. *In vivo* they caused localized destruction of brush border microvilli and *in vitro* they attached to HeLa, Hep-2 cells. Whole human colostrum and specifically each sIgA prevented attachment. It appeared that sIgA recognizes an EPEC surface protein (94 kDa).

The significance of the capacity of bacteria to adhere to the intestinal epithelium (brush border cells, enterocytes) cannot be overemphasized. *In vitro* studies on Lf eventually have to be followed up in animals and man. Small laboratory animals are killed at intervals - monogastric calves can be cannulated instead but this is a luxury rarely available - and differential bacterial counts made on the luminal contents of the different sections of the gut; the fecal flora largely represents only the content of the lower large intestine and the colon. So far, however, attachment to the epithelium has been neglected in Lf research. The colonization of the microvilli can easily be seen under the microscope, and assessed by counting the number of bacteria attaching per cell *in vitro*[23]. It was also shown that exposure of *Escherichia coli* to the LP-system prevents attachment. However, the test is cumbersome and time consuming, and the first barrier, the intestinal mucin, is largely lost during the isolation of the cells.

As an alternative, milk-fat globule (FG) or the fat globule membrane (FGM) were considered[23]. The fat in milk is 'packaged', the lipid core is surrounded by a true membrane derived from the apical plasma membrane of the mammary epithelial cell. It was found that FG is agglutinated by cold agglutinins in milk and from blood. Agglutination of erythrocytes (RBC) by *Escherichia coli* possessing $K_{88,}$ K_{89} was prevented by FG or FGM. *Escherichia coli* was found to attach directly to the FGM. The receptor-destroying enzyme of *V. cholerae*, which is known to destroy the receptor on the RBC, was also found to destroy the receptor on the FG and FGM for *Escherichia coli*. It was then considered that the true cell membrane on the FG may have the same or similar capacity to bind bacteria as the brush border cells. Indeed, enteropathogenic bovine strains of *Escherichia coli* attached only to bovine FG and FGM, while human strains of *E. coli* possessing the colonization factor CFI and CFII attached only to the FG of human milk, but there was also some low attachment rate to porcine FG, while the porcine strains attached specifically only to porcine FG. The bovine and porcine strains of *Escherichia coli* failed to attach to their specific FG after exposure to the LP-system. The human strains remained unaffected by the LP-system; they are mannose-sensitive (fail to attach to RBC in the presence of mannose), and were also inhibited by mannose from attaching to FG.

It was questioned whether the attachment to FG and FGM could be so strong that it could hinder attachment to the intestinal wall *in vivo*. It has been known for a long time that FG are surrounded by a glycocalyx consisting of mucin-like glycoproteins and it is therefore feasible that the FG test represents the *in vivo* conditions better than either the brush border or hemagglutination tests. The tests can be simplified by suspending bacteria in their specific milk, and leaving them standing at 37°C for 30 min; after centrifugation, from the bacterial count in the skim-milk portion the degree of attachment can be calculated[24]. The concept that FG or FGM may hinder attachment *in vivo* was supported by Atroshi *et al.*[25] They even succeeded in distinguishing between K_{88} sensitive and resistant pig phenotypes using FGM instead of brush border cells. They also discussed a scenario whereby FG transport *Escherichia coli* from the sensitive posterior part of the small intestine to the proximal part of the intestine. They base this on the visual observation of intact FG in the lumen through the first days of life when lipase, proteinase and pancreatic secretions are very low. FGM have actually been recovered from the feces of breast fed infants.

Searching the literature, L.Å. Hanson[26] in a symposium on human milk discussed the attachment of *Escherichia coli* to spermatozoa which are carried up into the female genital tract. He compared this with the binding of the same bacteria to FG, but questioned whether the binding could be a good or bad thing. The conclusion was that "this should be further investigated". Investigations with infants on the effect of Lf normally depend on analysis of the fecal flora. We need therefore a method of distinguishing between commensal and pathogenic strains of *Escherichia coli* isolated from infants' feces. In the same symposium, Guggenbichler[27] reported that he isolated *Escherichia coli* from the upper small intestine of 30 infants with long-standing diarrhea. Of 10^5 organisms more than 80% (sic) had some adhering capacity on isolated human intestinal cells. In vain did he try to block adhesion by breast milk, antibody concentrate from milk, whole colostrum or by a carbohydrate fraction of human milk. He succeeded however in blocking adhesion by using an oligosaccharide fraction isolated from carrots (also produced commercially). Apparently this is a traditional European remedy for acute diarrhea. In the same discussion Hanson admits that he has not yet studied the effect of human milk on adherence. While the necessity for such a distinction of the intestinal flora in the case of clinical diarrhea is evident, we will have to attempt to distinguish between *Escherichia coli* with and without adhesins in the feces of apparently healthy infants. It is not surprising that sIgA was shown to prevent the hemagglutination of *Escherichia coli* with CFI adhesin, but it is surprising that uncomplexed secretory component also inhibits attachment. Since it is known that Lf combines directly with IgA and secretory component, many of the previous results having shown that sIgA increases this inhibition by Lf *in vitro*, may now have a new explanation: see papers by Spik *et al.* on the milk-sensitive and -resistant strains[11]. There seems to be only one reference to the invasiveness of *Escherichia coli* (EIEC): it was shown that Lf prevented the invasion of a genetically engineered strain of *Escherichia coli* into HELA cells.

In summary, we need to establish whether Lf bound to bacteria promotes or prevents attachment like sIgA to the intestinal epithelium. We know now about the existence of specific Lf receptors in the intestine—does that mean that Lf facilitates attachment, or does Lf bound to the bacterial surface prevent attachment? Simple experiments, unfortunately not yet even considered, could answer these vital questions. Perhaps even more importantly, we are not certain whether Lf interferes with enterotoxin production of Gram-negative pathogens—like the LP-system (see later).

6. *IN VIVO* ROLE OF LACTOFERRIN

6.1. Trials in Animals

The seminal paper by Bullen *et al.*[5] contains some controversial points which have been discussed personally in the past and do not reduce its value; they are comprehensively dealt with in the book on *Iron and Infection*[28]. In the original paper it was claimed that bovine colostrum and milk only inhibit *Escherichia coli* 0111 B_4 H_2 after adjustment of the pH to above 6.95. Subsequently it was established that colostrum is bacteriostatic below pH 6.95, but only after dialysis. In the dialysate citrate was identified and shown to prevent bacteriostasis unless counteracted by bicarbonate. High concentrations of citrate favour the release of Fe from Lf, making it available for growth.

This dependency of bacteriostasis on citrate also occurs *in vivo*[29]. When a low dose of a highly virulent strain of *Escherichia coli* (250–300 cfu) was infused into lactating or nonlactating (involuted) udders, 13 out of 17 lactating quarters became infected but none of the involuted quarters. Before calving (4–6 weeks) milking is discontinued and components of the milk are reabsorbed, including citrate. The resistance of involuted quarters is due to the high concentration of Lf (up to 70 mg/ml) and the absence of citrate. Two days before calving, secretion of colostrum begins, which includes citrate. The reappearance of citrate has been regarded as the 'harbinger of lactogenesis'. At that time, 2 out of 2 quarters became infected after infusion of *Escherichia coli*. Indeed, it is known that the incidence of mastitis can be high after calving, with dormant infections in the nonlactating period flaring up.

This dependency of bacteriostasis on the proportion of bicarbonate to citrate may be of some relevance in the formulation of infant foods. Sometimes citrate is added to stabilize the reconstituted milk powder. This may be inadvisable as long as we do not know how quickly citrate is reabsorbed in the intestine and how much bicarbonate is contributed by saliva and pancreatic secretion.

To investigate the role of Lf, Bullen *et al*[5] selected guinea pigs because their milk is rich in Lf and transferrin (TF) in about equal amounts, 0.2–2 mg/ml; the piglets are born so mature that they can be removed from the dam immediately after birth and fed artificially, even with pellets. The piglets were dosed orally with *Escherichia coli* 0111 (1×10^6) and either kept sucking or fed a guinea pig milk substitute. At 6 days the piglets were killed, their intestines homogenized and plated out for *Escherichia coli* and lactobacilli. At that time attachment had not yet been discovered; now it would be desirable to distinguish between attached bacteria and those that are free in the lumen. The identity of *Escherichia coli* was ascertained by slide agglutination tests with a high titer antiserum (see above). This is important because it distinguished commensal from pathogenic strains of *Escherichia coli*. Within 6 days, in the sucking piglets lactobacilli became dominant, 10^5/g, and *Escherichia coli* less than 10^2/g in both the small and the large intestine. In the artificially fed piglets the coli count remained high, peaking between 2 and 4 days to 10^8/g in the small and 10^{10}/g in the large intestine. Lactobacilli failed to establish themselves. Some animals died with symptoms of diarrhea and emaciation; many suffered from bacteremia, which is surprising, because *Escherichia coli* 0111 is recognized to be an ETEC strain and therefore not invasive (immature intestine?).

If Lf and TF in the milk are solely responsible for the suppression of *E. coli*, we have to assume that milk does not contain specific antibodies claimed to promote inhibition, as shown *in vitro* (and see later on cross antigenicity). To prove that an Fe compound can abolish the bacteriostatic activity, they dosed sucking piglets with hematin; after 3 days, compared with the controls the number of *Escherichia coli* increased by >4 log cycles in the small intestine and >2 log cycles in the large intestine. Hematin is of course an alternative source of Fe and the results only emphasize the beneficial effects of natural feeding. The effect of guinea pig milk is not necessarily restricted to its Lf and TF contents because it has been shown, soon after the publication of this paper, that it also contains LP[30]; it is also known to contain excessive amounts of ascorbic acid, and possibly LZ.

Antonini *et al.*[31] investigated the protective role of ovotransferrin, also referred to as conalbumin (OTR, CA) in lieu of Lf, which was then not available commercially. They dosed newborn guinea piglets with 2×10^8 cfu *Escherichia coli* 0111. Some sucked their dam and some were fed bovine milk or milk containing 5 mg/ml OTR. In each group some piglets died, but their total number was too small for statistical analysis. At 8 days the small and large intestines of the surviving guinea pigs contained respectively about 10^4 or 10^5 cfu *Escherichia coli*/g, whether suckled or fed milk supplemented with OTR. The count in the piglets fed milk was about 3 log cycles higher. These results represent the first direct proof that OTR, and by implication Lf, is protective against *Escherichia coli in vivo*.

The choice of guinea pigs as experimental animal appeared to be attractive because of the maturity of the newborn piglet. However, it is difficult to understand why *Escherichia coli* 0111 B_4H_2, originally isolated from infants with diarrhea, should affect so strongly another species, producing bacteremia and death[5]. Widdowson *et al.*[32] discovered that the intestines of suckled rabbits, dogs and piglets are markedly heavier than those of animals of the same litter killed at birth, fed water or synthetic milk. Germfree animals are also well known to have thinner or less substantial intestinal walls. When conventional calves were dosed with 10^6–10^8 cfu *Escherichia coli* 0111, the organism proved innocuous, while the organs and the spine of germfree calves became colonized (unpublished), indicating the underdeveloped nature of the intestine? Since then, Kruzel *et al.* have stated that Lf not only promotes development of the intestine in mice, but can also attenuate intestinal inflammatory responses, and may therefore have a potential clinical application.

Teraguchi *et al.*[33] investigated the inhibitory activity of bovine Lf and Lf hydrolysate (LfH) for several species of clostridia *in vitro* and in weaned mice. The minimal inhibitory concentration of Lf was rather high—mostly >16 mg/ml, compared with LfH. When mice were dosed orally with *Clostridium ramosum* and fed pellets, the organisms were not recovered from the feces after 2 days. Mice fed bovine milk supported the proliferation of clostridia to 10^{11} cfu/g feces. Supplementing milk with Lf or LfH reduced about equally the number by 6 log cycles (sic). After 14 days, Lf had little effect on the level of bifidobacteria but reduced significantly the Enterobacteriaceae by about 2 log cycles. These results indicate that ingested Lf becomes hydrolysed and therefore more bacteriostatic; only 3% of the Lf was recovered in the feces. The ability of clostridia to colonize the colon appears to be host dependent, and further studies with pathogenic clostridia will be necessary.

Calves are prone to diarrhea (scour) during the first weeks of life. Morbidity and mortality vary according to husbandry, intensity of rearing, hygiene and climate. The ultimate aim of the use of a protective system was to replace the undesirable practice of indiscriminately supplementing formula feeds with antibiotics, thus reducing the danger of transferring multiple antibody-resistant Gram-negative pathogens to man. Large scale trials with calves over several years established that the LP-system significantly reduced the incidence and duration of scouring, and increased live weight gains. It was not possible to decide whether it also reduced mortality because the incidence was too low and was often caused by viral infections[34]. At that time it was already established that Lf by itself failed to prevent the death of small laboratory animals. Therefore a combination of Lf and LP was tried in infected calves[35]. Colostrum fed calves up to the age of 38 hours were infected with 10^{11} cfu *Escherichia coli*: serotype $0101{:}K_{30}K_{99}F_5F_{44}$. This strain attaches in the intestine, produces a heat-stable enterotoxin and induces diarrhea within 24 hours in calves. A combination of Lf and LP prevented scouring when fed two days prior to infection and 'cured' after scouring was established. All the infected calves scoured but severe scouring for 2 or more days was reduced in the treated animals (8.3%) compared with the untreated animals (50%). Duration was much reduced and none of the treated calves suffered hypothermia or clinical depression, compared with the untreated calves. In the feces, *Escherichia coli* was identified by direct immunofluorescence and confirmed by agglutination and enzyme-linked immunosorbent assay (ELISA), using a specific antibody. The surprising observation was made that treated and cured calves shed practically the same numbers of *Escherichia coli* in their feces.

6.2. Trials in Infants

It is now generally recognized that breast fed infants resist enteric infections better than when artificially fed. Perhaps the most dramatic power of breast feeding was researched by Mata *et al.*, summarized in [36]. They studied infection among mothers and their infants living under the most appalling lack of hygiene in a Mayan Indian village in Guatemala. In spite of the high infectious morbidity among pregnant women and those with malnutrition there was a high resilience in infants before weaning. Mata attributed this resistance to the establishment of bifidoflora and the presence in milk of complements, immune cells, sIgA, LZ and Lf and possibly LP could be added. Milk samples obtained from women living in The Gambia, West Africa (obtained by courtesy of the Dunn Institute of Nutrition, Cambridge, UK) showed an average of 230 units peroxidase activity up to 9 months of lactation, much higher than in the milk of women in the West.

The first attempt to supplement a formula feed with Igs was made by Hilpert *et al.*[37] They designed a 'cocktail' of *Escherichia coli* serotypes most commonly isolated from the feces of infants with enteritis and immunized cows in calf. Five days after calving they isolated and partially purified Ig-containing fractions from the colostral whey (CWP) which contained about 40% Igs. Infants with enteritis were treated with CWP at 1 g/ kg body weight daily at 2 different hospitals. Of a total of 156 cases, nearly 49% were cured, according to both clinical and bacteriological tests: in the controls, only 43 cases, nearly 28%, recovered without any treatment. Complete failure was registered in 23% of treated cases but in 67% of controls. The rest of the cases were only either a clinical or bacteriological 'success'. The much researched serotype 0111 B_4 proved to be especially refractory in the trials. Seven years later Hilpert[38] reported an improved isolation process for milk up to 30 days after calving. Although the analysis again distinguished about 40% Igs—IgG_1, IgG_2, IgA, IgM—and 5% minor proteins, no attempt was made to identify whether these proteins were Lf, LZ or LP, which must have been present in both of the preparations. In spite of the considerable clinical success any further research appears to have been abandoned.

A different approach was made by Spik *et al.*[39] They fractionated human milk by a simple method using a combined gradient of ammonium sulphate and pH. Testing different fractions, some contained an Lf/LZ complex (see before) or an Lf/sIgA complex. Both were more inhibitory for *Escherichia coli* 0111 than pure Lf. Some fractions also promoted the growth of certain bifidobacteria freshly isolated from the feces of breast fed infants. A very limited trial with 5 infants with acute diarrhea had limited success in reducing the duodenal bacterial flora. This promising start was not taken further, although a combination of inhibitors and growth promoters ought to be rewarding.

The largest trial treating infants suffering acute enteritis was undertaken with OTR, which had been shown to be as inhibitory as Lf[9]. It was found, in several hospitals, that infants given a formula feed based on bovine milk plus OTR improved in clinical status and normalized their bowel movements sooner than did the controls ($P < 0.01$).

In summary, Lf (and OTR) have some proven therapeutic effect for infants suffering enteritis by *Escherichia coli*. Infections by *Campylobacter* spp. (an increasingly common cause of infections and incidentally of can lead to Reiter's syndrome) and any other Gram-negative organisms such as *Salmonella, Shigella*, etc, known to be inhibited *in vitro* have not been investigated *in vivo*.

So far, a prophylactic role for Lf has not been investigated. Such trials would have to be on a large scale, their success depending on cases of enteritis emerging at random, unless such trials are conducted, regrettably, in developing countries. The present tendency is to try formula feeds plus Lf (or OTR) with the restricted aim of promoting the establishment and dominance of bifidobacteria in the feces as in breast fed infants. Indeed, Lf has been observed to stimulate the growth of bifidobacteria *in vitro* and *in vivo*, the latter at least in the gut of germfree mice, notwithstanding that the germfree state can be misleading in comparison with conventional animals.

In 1978 it was still valid to declare, "feeding cow's milk to infants appears to be the largest uncontrolled nutritional experiment undertaken." This comment was based on the difficulties arising with calves reared on powdered cows' milk because of the absence of any immune factor, and with piglets, of which a high percentage die or suffer nutritional diarrhea. Since then great advances have been made in humanizing artificial feeds based on cows' milk. Bullen[40] contributed greatly towards an understanding of the relationship between the composition of the feed and the bacterial flora in the intestine, the pH value, the buffering capacity and the metabolic components of fecal material and their effect on the viable count of Gram-negative bacteria in the intestine. She was under no illusion that the fecal flora represents at best only the large intestine, in particular the colon, and not the flora of the whole gut. She analyzed the feces by gas chromatography, which could be extended with present technology to an analysis of the composition of the flora without elaborate bacterial identification. After all, it is well known that the feces of breast fed infants have a not unpleasant odor of butter, slightly cheesy, and not as offensively repugnant as those of artificially fed infants. She emphasized that the low pH and the presence of acetate buffer supresses Gram-negative flora, thus promoting the flora of lactobacilli. Since this state is not achieved until about 8 days after birth, she suggested that colostral Lf and Igs play an important role in the initial protection of the infant against multiplication of Gram-negative organisms. With the decline in the level of Lf and Igs in the breast milk, a lactic acid flora becomes established in the large intestine; the protective role is taken over by the buffered acidic environment shown to be quite bactericidal against *Escherichia coli*.

At about the same time, Hambræus *et al.*[41] reduced the accepted figure for the protein content of human milk by 0.3–0.4% according to amino acid analysis; the non-protein nitrogen content was relatively high, 0.4%, of which about half is urea. This raises the question whether urea or its metabolic product, ammonia, may provide an ecological niche for bifidobacteria, as they are the only lactic acid bacteria which can utilize ammonia or urea as a nitrogen source. In addition, the high content of lactose and oligosaccharides promotes the growth of lactobacilli, but only in the large intestine, while infections in the newborn with the exception of clostridia occur in the small intestine. Wharton *et al.*[42] compared the buffering capacity of breast milk with that of cows' milk, the latter being 3 times as high, while current formula feeds were only twice as high, the lowest having demineralized whey as its basis. After 3 trials with healthy infants, they failed to change basically the pattern of intestinal or rather fecal flora with the best formula feed with or without Lf, in comparison with breast fed infants. There was no increase in bifidobacteria and lactobacilli, and the increase in *Escherichia coli* and *Streptococcus faecalis* may not be significant. They concluded that *in vitro* bacteriostasis is not reflected *in vivo*, at least as judged by the fecal flora and that future attention ought to be directed not only to trials with Lf but also to antibodies, LZ, acid/base, bicarbonate, citrate etc. Roberts *et al.*[43] investigated the effects of OTR and Lf in 2 separate trials. They found that clostridia and bacteroides were more common in feces of formula fed infants than in breast fed (Lf has never been shown to inhibit clostridia or bacteroides but LZ does inhibit clostridia). Lf at a concentration of 1 mg/ml was found to help in the establishment of bifidoflora in half the babies, but only after 3 months.

7. COMMENTS AND PROSPECTS

Based on results with human and guinea pigs, it was expected that Lf may play an important part in resistance to infantile enteritis of *E. coli*[7]. Indeed, when infants hospitalized with acute enteritis were given OTR before LF had become commercially available, some therapeutic effect was observed[44]. Another trial combining Lf with LZ led to significant weight increase in infants, and their feces contained a higher number of lactobacilli and fewer enterobacteriaceae compared with the control group. Most interestingly, none of the children given LZ/Lf developed diarrhea (or became anemic)[45]. In retrospect, these results indicate that the *in vitro* effect of LZ/Lf[30] may occur *in vivo*. Jollès and Jollès suggested that LZ may act indirectly as an adjuvant, through the hydrolytic products of peptidoglycans. This hypothesis seems to have been unknowlingly substantiated by results in earlier trials in which it was observed that feeding LZ to infants increased the secretory IgA content in their feces to the level in breast fed infant[45].

An other combination, Lf with LP-system produced prophylactic and therapeutic effects in calves infected orally with a pathogenic strain of *Escherichia coli.*[46] In calves, cannulated in the abomasum and/or anterior duodenum, sufficient oxygen seems to be present to kill *Escherichia coli* when H_2O_2 is generated by feeding glucose/glucose oxydase; they are also killed by H_2O_2 producing lactobacilli colonizing the intestine of older calves. The presence of oxygen can also be deducted from the effect that *Campylobacter* spp. grow because they are microaerophilic, i.e grow only under reduced oxygen tension (7–10%); *Campylobacter*, which has become an important cause of enteritis in man, had been regularly found in the intestines of domestic animals. There was little difference in the *Escherichia coli* count between the diarrheic and protected calves, which seemed to contradict the results with cannulated calves. In the case of *Escherichia coli,* which is aerobic and facultative anaerobic, it is known that the organism can recover after exposure to the LP-system anaerobically. It is therefore possible that damaged organisms can recover during passage through the intestine, which becomes increasingly anaerobic. Hence the number of pathogens could be similar in both groups of calves.

The pathogens may however be excreted in the feces and recover after being plated out. It is possible to investigate their viability by an other method. Recently, it has been shown that the LP-system reduces the ATP content of *Streptococcus sanguis* although the viable count is the same as for untreated cocci[47] although Gram-positive organisms like streptococci are only inhibited, unlike Gram-negative organisms such as *Escherichia coli* which are killed. Both show leakage and damage to the inner membrane, and it would be interesting to determine the ATP content in fresh feces and compare it with the total and differencial bacterial count.

An other reason why *Escherichia coli* infected calves did not become diarrheic became clear from results obtained by M. Rosen and R. Soderling (Agriculture University, Uppsala, Sweden, pers.comm.). They exposed an enterotoxigenic and adhesive strain of *Escherichia coli* to the LP-system and observed that no fluid was diffused into ligated sections of pig intestine. This is not unexpected if we consider that the LP-system and by implication Lf inhibits RNA, DNA and protein synthesis; it would therefore also inhibit the synthesis of polypeptides like enterotoxin. Instead of a protective role of LF, it is now expected only that the fecal flora of formula fed infants can be transformed by the addition of Lf to that of breast fed infants, alas with little success. Wharton in his summary[42] of four independent trials raises four different expectations, trying to explain why bovine Lf has little effect on the fecal flora of infants. Indeed, he expresses his doubts about the relevance between the *in vitro* effect of Lf to *in vivo* results. This is hardly surprising considering that the *in vitro* experiments are invariably performed with pure cultures. From birth, the intestinal flora is being evolved until a stable ecological system is established. This depends on nutrition, the surroundings, type of birth, antagonism between bacteria (but rarely do bacteriologists work with mixed cultures) containing bacteriocidins, bacteriophages, growth promoters and the immune system.

So far Lf has not been comprehensively reviewed. Ellison wrote a most illuminating chapter for the first workshop on Lf[48]. However, it was focused on the synergic effect of LZ and Lf on Gram-negative pathogens. This important finding, know now for 7 years[12] has yet been considered of sufficient importance to be tested *in vivo*. Since 1972, *Escherichia coli* 0111 dominated the Lf research, largely neglecting other serotypes and the topical question of the lack of antibiotics which can inhibit multiple antibiotic resistant strains. While the LP-system has been proven to be antibactericidalto a range of Gram-negative pathogens and antibiotic resistant pathogens[49], so far Lf research needs to be expanded.

The purpose of this review is to draw attention to some of the neglected aspects in Lf research: attachment in the presence of Lf, both *in vitro* and *in vivo*, enterotoxin production, metabolic state of the pathogens *in vivo* and *in vitro* (ATP determination), synergic effect between the different non-antibody protective proteins, not forgetting the role if Igs which has been largely avoided in this review because of the conflicting results in the literature.

In conclusion, it should be realized that research on the biological role of Lf ought to begin now in earnest.

ACKNOWLEDGMENTS

We would like to acknowledge contribution of Dr Jeremy Brock for his critical discussion and the editorial and linguistic help of Mr Brian Bone, the Librairain of the former National Institute for Research in Dairying.

REFERENCES

1. Montreuil, J, Tonnelat, J, Mullet, S (1960) Préparation et propriétés de la sidérophiline (lactotransferrine) du lait de femme. Biochim. Biophys. Acta 45: 413–421.
2. Schade, AL, Caroline, L (1944) Raw hen egg white and the role of iron in growth inhibition of *Shigella dysenteriae*, *Staphylococcus aureus*, *Escherichia coli* and Saccharomyces cerevisiae. Science 100: 14–15.

3. Hanson, LÅ (1961) Comparative immunological studies of the immune globulines of human milk and of blood serum. Int. Arch. Allergy Appl. Immunol. 18: 241–267.
4. Reiter, B, Oram, JD (1967) Bacterial inhibitors in milk and other secretions. Nature 216: 328–330.
5. Bullen, JJ, Rogers, HJ, Leigh, L (1972) Iron binding proteins in milk and resistance to *Escherichia coli* infections in infants. Br. Med. J. 1: 69–75.
6. Masson, PL, Heremans, JF, Prignot, J, Wauters, G (1966) Immunohistochemical localization and bacteriostatic properties of an iron-binding protein from bronchial mucus. Thorax 21: 538–544.
7. Oram, JD, Reiter, B (1966) Inhibitory substances present in milk and the secretion of the dry udder. Rep. Natl Inst. Res. Dairying, England p. 93.
8. Stuart, JS, Norell, S, Harrington, JP (1984) Kinetic effect of human lactoferrin on the growth of *Escherichia coli* 0111. Int. J. Biochem. 16: 1043–1047.
9. Valenti, P, Antonini, G, Fanelli, R, Orsi, N, Antonini, E (1982) Antibacterial activity of matrix-bound ovotransferrin. Antimicrob. Agents Chemother. 21: 840–841.
10. Naidu, SS, Svensson, U, Kishore, AR, Naidu, AS (1993) Relationship between antibacterial activity and porin binding of lactoferrin in *Escherichia coli* and *Salmonella typhimurium*. Antimicrob. Agents Chemother. 37: 240–245.
11. Spik, G, Cheron, A, Montreuil, J, Dolby, JM (1978) Bacteriostasis of a milk-sensitive strain of *Escherichia coli* by immunoglobulins and iron-binding proteins in association. Immunology 35: 663–671.
12. Ellison III, RT, Giehl, TJ, LaForce, FM (1988) Damage to outer membrane of enteric Gram-negative bacteria by lactoferrin and transferrin. Infect. Immun. 56: 274–281.
13. Arnold, RR, Cole, MF, McGhee, JR (1977) A bacteriocidal effect of human lactoferrin. Science 197: 263–265.
14. Arnold, RR, Russel, JE, Champion, WJ, Brewer, M, Gauthier, JJ (1982) Bactericidal activity of human lactoferrin. Differentiation from the stasis of iron deprivation. Infect. Immun. 35: 792–799.
15. Reiter, B (1978) Review of nonspecific antimicrobial factors in colostrum. Ann. Rech. Vét. 3: 205–224.
16. Lassiter, MO, Newsome, AL, Sams, LD, Arnold, RR (1987) Characterization of lactoferrin interaction with *Streptococcus mutans*. J. Dental Res. 66 480–485.
17. Halliwell, B, Gutteridge, JMC (1985) The importance of free radicals and catalytic metal ions in human diseases. Molec. Aspects Med. 8: 89–193.
18. Perraudin, J-P, Prieels, J-P (1982) Lactoferrin binding to lysozyme-treated *Micrococcus luteus*. Biochim. Biophys. Acta 718: 42–48.
19. Susuki, T, Yamauchi, K, Kawase, K, Tomita, M, Kiyosawa, I, Okonogi, S (1989) Collaborative bacteriostatic activity of bovine lactoferrin with lysozyme against *Escherichia coli* 0111. Agric. Biol. Chem. 53: 1705–1706.
20. Sojka, WJ (1965) *Escherichia coli* in domestic animals and poultry. Farnham Royal: Commonwealth Agricultural Bureaux.
21. Jones, GW, Rutter, JM (1972) Role of the K_{88} antigen in the pathogenesis of neonatal diarrhea caused by *Escherichia coli* in piglets. Infect. Immun. 6: 918–927.
22. Camara, LM, Carbonare, SB, Scaletsky, ICA, da Silva, MLM, Carneiro-Sampaio, MMS (1995) Inhibition of enteropathogenic *Escherichia coli* (EPEC) adherence to HeLa cells by human colostrum. Detection of specific sIgA related to EPEC outer-membrane proteins. In: Mestecky, J et al. eds Advances in Mucosal Immunology. New York: Plenum Press, 673–676.
23. Reiter, B (1981) The contribution of milk to resistance to intestinal infection in the newborn. In: Lambert, HP, Wood, CBS, eds Immunological Aspects of Infection in the Fetus and Newborn. New York: Academic Press (Beecham Colloquium) 155–195.
24. Reiter, R, Brown, T (1976) Inhibition of haemagglutination of red blood cells by K_{88} and K_{99} adhesion using milk fat globule membrane. Proc. Soc. Gen. Microbiol. 3: 109.
25. Atroshi, F, Alaviuhkola, T, Schild, R, Sandholm, M (1983) Fat globule membrane of sow milk as a target for adhesion of K_{88} positive *Escherichia coli*. Comp. Immunol. Microbiol. Infect. Dis. 6: 235–245.
26. Hanson, LÅ (1988) Discussion. In: Hanson, LÅ, ed. Biology of Human Milk (Nestlé Nutrition Workshop Series no. 15). New York: Raven Press, p.45.
27. Guggenbichler, J (1988) Discussion. ibidem, p.157.
28. Bullen, JJ, Griffiths, E (1987) Iron and Infection, Molecular, Physiological and Clinical Aspects. John Wiley and Sons, Chichester.
29. Reiter, B, Bramley, A (1975) Defense mechanisms of the udder and their relevance to mastitis control. Int. Dairy Fed. Annu. Bull. no. 85, 210–215.
30. Stephens, S, Harkness, RA, Cockle, SM (1979) Lactoperoxidase activity in guinea-pig milk and saliva: correlation in milk of lactoperoxidase with bactericidal activity against *Escherichia coli*. Br. J. Pathol. 60: 252–258.

31. Antonini, E, Orsi, N, Valenti, P (1977) [Effect of transferrin on the pathogenicity of Enterobacteriaceae.] Giornale di Malattie Infettive e Parassitarie 29: 481–489.
32. Widdowson, EM, Colombo, VE, Artavanis, CA (1976) Changes in the intestinal tract organs of pigs in response to feeding for the first 24 hours after birth. 2. The digestive tract. Biol. Neonate 28: 272–281.
33. Teraguchi, S, Shin, K, Ozawa, K. Nakamura, S, Fukuwatari, Y, Tsuyuki, S, Namihira, H, Shimamura, S (1995) Bacteriostatic effect of orally administered bovine lactoferrin on proliferation of *Clostridium* species in the gut of mice fed bovine milk. Appl. Environ. Microbiol. 61: 501–506.
34. Reiter, B, Fulford, RJ, Marshall, VM, Yarrow, N, Ducker, MJ (1981) An evaluation of the growth promoting effect of the lactoperoxidase system in newborn calves. Anim. Prod. 32: 297–306.
35. Prieels, J-P, Delahaut, P, Jacquemin E, Kaeckenbeeck, A, (1989) . Application du système lactopéroxydase au traitement de la diarrhée colibacillaire chez les veaux. Ann. Med. Vet, 133: 143–150
36. Mata, LJ, Urrutia, JJ (1977) Infections and infectious diseases in a malnourished population: a long-term prospective field study. In: Hambræus, L, Hanson, LÅ and McFarlane, H., eds. Food and Immunology. Almqvist & Wiksell International, 42–58.
37. Hilpert, H, Gerber, H, Amster, H, Pahud, JJ, Ballabriga, A, Arcalis, L, Farriaux, F, de Peyer, E, Nussle, D (1977) Bovine milk immunoglobulins (Ig), their possible utilization in industrially prepared infants' milk formulae. In: Hambræus, L, Hanson, LÅ & McFarlane, H. eds. Food and Immunology. Almqvist & Wiksell,
38. Hilpert, H (1984) Preparation of a milk immunoglobulin concentrate from cow's milk. In: Williams, AF, Baum, JD, eds. Human Milk Banking (Nestlé Nutrition Workshop Series vol. 5). Raven Press, 17–29.
39. Spik, G, Jorieux, S, Mazurier, J, Navarro, J, Romond, C, Montreuil, J (1984) Ibid. 133–145.
40. Bullen, CL (1977) The role of pH and buffering capacity of faeces in the control of the Gram-negative intestinal flora. In: Hambræus, L, Hanson, LÅ & McFarlane, H. Food and Immunology. Almqvist & Wiksell, 42–58.
41. Hambræus, L, Forsum, E, Lönnerdal, B (1977) Nutritional aspects of breast milk versus cow's milk formula. Ibid.
42. Wharton, BA, Balmer, SE, Scott, PH (1994) Faecal flora in the newborn: effect of lactoferrin and related nutrients. In: Hutchens, TW, Rumball, SV & Lönnerdal, B, eds. Lactoferrin Structure and Function (Advances in Experimental Medicine & Biology vol. 357). Plenum Press, 91–99.
43. Roberts, AK, van Biervliet, JP, Harzer, G (1985) Supplementation of an adapted formula with bovine lactoferrin. Effect on the infant faecal flora. In: Schaub, J, ed. Composition and Physiological Properties of Human Milk. Elsevier Science Publishers, 259–271.
44. Corda, R, Biddan, P, Corrias, A, Puxeddu, E (1983) The conalbumin in the therapeutic treatment of acute enteritis in the infant. J. Tissue React v 117–121.
45. Diadal, NV, (1986) Ontwikkeling van een babyvoeding, ungerjkt met immunfactoren. Rapport, EEG Verordening 1150/86 (Kontrakt: 1150–39.2).
46. Reiter, B, Perraudin, J-P (1990) Lactoperoxidase: biological functions. In: J Everse, EK Everse, MB Grisham, eds Peroxidases in Chemistry and Biology, vol.1. CRC Press, Boca Raton, FL, 143–180
47. Courtois, Ph, Abbeele, V, Amrani, N, Pourtois, M (1995) *Streptococcus sanguis* survival rates in the presence of lactoperoxidase-produced $OSCN^-$ and OI^- Med. Sci. Res. 23: 195–197.
48. Ellison, RT III (1994) The effects of lactoferrin on Gram-negative bacteria. In TW Hutchens, S V Rumball, B Lönnerdal, eds. Lactoferrin Structure and Function (Advances in Experimental Medicine & Biology vol. 357), Plenum Press, 71–90.
49. Reiter, B, Marshall, V,, Björck, L, Rosen, C. (1976) Nonspecific bactericidal activity of the lactoperoxidase-thiocyanate-hydrogen peroxide system of milk against *Escherichia coli* and some Gram-negative pathogens. Infect Immun. 13: 800–807

22

HOST DEFENSIVE EFFECTS OF ORALLY ADMINISTERED BOVINE LACTOFERRIN

Mamoru Tomita, Koji Yamauchi, Susumu Teraguchi, and Hirotoshi Hayasawa

Nutritional Science Laboratory
Morinaga Milk Industry Co., Ltd.
Zama-City, Kanagawa 228, Japan

1. INTRODUCTION

Lactoferrin, an iron-binding glycoprotein of the transferrin family, is present in milk, saliva, mucous secretions and many other body fluids[1]. Lactoferrin is a major component of the secondary granules in neutrophils and this protein is released from activated neutrophils in the inflammatory response[2]. Many biological functions have been attributed to this protein, including roles in bacteriostasis in milk and at mucosal surfaces, immunomodulation, regulation of cell proliferation and intestinal iron transport[3]. Since the majority of the proposed functions of lactoferrin are based on the results of in vitro experiments, it seems clear that more in vivo studies are required for better understanding of the biological role of this protein. Considering the high similarities among the physicochemical properties of lactoferrins from several species with respect to molecular weight, primary structures[4,5], three-dimensional structures[6,7], and isoelectric point[8], it seems possible that administered lactoferrin isolated from bovine milk may exert physiological activities in other animals and humans. In this paper, we outline research results demonstrating host defensive effects of orally administered bovine lactoferrin observed in several in vivo studies performed by our research group and some of our collaborators.

2. BACTERIOSTATIC EFFECT OF ORALLY ADMINISTERED LACTOFERRIN

Lactoferrin had been shown to exhibit antimicrobial activity against a wide variety of microorganisms and is thought to play an important role in the host defense system of mammals. The antimicrobial activity of lactoferrin is known to be dependent on its iron-free state and is commonly attributed to its ability to bind and sequester iron, producing an iron-deficient environment that limits microbial growth[9,10]. Evidence from several studies

Advances in Lactoferrin Research, edited by Spik *et al.*
Plenum Press, New York, 1998.

suggests that lactoferrin has bactericidal activity and kills sensitive bacteria by a mechanism distinct from sequestering iron that involves its direct interaction with the bacterial cell surface[11–13]. Recently, we have found that an enzymatic hydrolysate of lactoferrin, prepared using porcine gastric pepsin, displays increased antimicrobial activity[14]. We have also isolated a potent antimicrobial peptide named lactoferricin® from the pepsin hydrolysate of lactoferrin[15]. Other host defensive effects attributable to the glycan moiety of lactoferrin, including inhibition of bacterial attachment to animal cells, which is considered to be the initial event in bacterial colonization and infection, have been reported[16,17]. In addition to contributing directly to the destruction of invading microbial pathogens, the role of lactoferrin in the host defense appears to involve various immunomodulatory functions. Our most recent findings suggest that orally administered lactoferrin may also function to stimulate carbohydrate absorption and thereby suppress bacterial overgrowth in the intestine. Table 1 summarizes the proposed mechanisms of the antimicrobial action of lactoferrin in vivo.

It is believed that lactoferrin in milk contributes to the protection of the neonate with an immature immune system in the defense against enteric pathogens. Most ingested lactoferrin is degraded during passage through the gastrointestinal tract to low molecular mass forms[18]. In the gastrointestinal tract of newborns, especially within four months after delivery, the amount of gastric juice secreted is small, and the acidity in the stomach is not optimal for enzymatic cleavage of lactoferrin by pepsin. In fact, a clinical study of breast-fed human infants indicates that less than 17% of the lactoferrin ingested is excreted intact in the feces[19]. Taking these observations into consideration, there is a possibility that ingested lactoferrin or a fragment derived from it (e.g., lactoferricin®) remains in the gut and influences the composition of the intestinal microflora of neonates as an antimicrobial agent. The intestinal flora of human infants is formed rapidly during the first week after birth. *Escherichia coli* is the most predominant intestinal bacterium followed by streptococci and clostridia, in the phase of primary infection before the establishment of a bifidobacteria predominant flora. During this term, breast-fed infants ingest a substantial amount of lactoferrin from colostrum. Several previous studies have demonstrated the bacteriostatic effect of lactoferrin in the diet in laboratory animals and human infants. Orally administered lactoferrin has been shown to promote the establishment of a bifidobacteria predominant flora in human flora associated mice[20] and human full-term infants[21], while other studies have reported that lactoferrin had little effect in human infants[22].

There is also a possibility that ingested lactoferrin has some regulatory functions in newborn infants. Several studies have demonstrated the presence of receptors for lactoferrin on the surface of various types of cell lines including lymphocytes, enterocytes, and fibroblasts[23]. Although the metabolic pathway for ingested lactoferrin has not been completely elucidated, the finding that functionally intact lactoferrin has been detected in the urine of premature infants[24] might indicate that ingested lactoferrin is absorbed by neonates through the intestinal epithelium.

Table 1. Mechanism of antimicrobial action of lactoferrin

1. Inactivation of bacteria by iron chelation
2. Bactericidal action by binding to bacterial cells
3. Bactericidal action due to the active peptide, lactoferricin®
4. Inhibition of bacterial attachment to animal cells
5. Stimulation of nutrient absorption, which leads to suppression of bacterial growth in the intestine
6. Defensive functions by regulating the host immune system

2.1. Bovine Lactoferrin

In recent years, lactoferrin isolated on an industrial scale from bovine milk has become commercially available. Lactoferrin was first isolated from both human and bovine milk by the method of column chromatography involving adsorption of this basic protein on an ion-exchange resin[25,26]. The process for large-scale preparation of bovine lactoferrin is principally the same as the early purification method and also contains a purification step involving cation-exchange chromatography. The purified bovine lactoferrin is a powder with light red color. The purity of this product is not less than 95% as determined by HPLC analysis.

2.2. Milk-Fed Mice

Our first attempt to elucidate the influence of ingested lactoferrin on the intestinal flora was a study performed using milk-fed specific-pathogen-free (SPF) mice as an experimental model. Feeding milk to mice resulted in a great increase in numbers of *Enterobacteriaceae* in the feces of mice and also in the gastrointestinal tract. Orally administered bovine lactoferrin added to the milk at a concentration of 2% (w/v) suppressed the proliferation of intestinal *Enterobacteriaceae* and also streptococci in the milk-fed mice[27]. This bacteriostatic effect of lactoferrin was dependent on the dose of lactoferrin and the duration of administration. The pepsin hydrolysate of bovine lactoferrin showed the same bacteriostatic effects as undigested lactoferrin. These results support the view that lactoferrin exerts a bacteriostatic effect against bacteria in the gut even after it has been digested to some extent. Similarly, lactoferrin was found to suppress the proliferation of clostridia in milk-fed mice orally inoculated with these bacteria[28]. On the other hand, lactoferrin showed little effect on intestinal bifidobacteria[28,29].

In research aimed to investigate the mechanism of bacteriostasis of lactoferrin in vivo, we found evidence suggesting that lactoferrin influences carbohydrate absorption in the intestine of germ-free mice fed sterile milk[29]. Although the milk contained lactose alone, glucose and galactose as well as lactose were detected in the cecal contents of these mice after 7 days of feeding. The concentrations of these three carbohydrates in the cecal contents were significantly lower in the group fed milk containing 2% bovine lactoferrin than in the group fed milk only. This result strongly suggests that lactoferrin acts to stimulate the absorption of these carbohydrates in the intestine. In vitro experiments using everted sacs of small intestine of SPF mice also suggested that lactoferrin exerts a stimulatory effect on intestinal absorption of carbohydrates. Considering that unabsorbed carbohydrates may be used to support bacterial proliferation in the intestine, the growth of intestinal bacteria would be limited when carbohydrate absorption is stimulated by the administration of lactoferrin.

The translocation of intestinal bacteria to other organs is assumed to cause serious problems of infectious disease in premature infants and immunocompromised hosts[30]. We have examined the inhibitory effect of lactoferrin on bacterial translocation. In the case of mice fed only milk, bacterial translocation to the mesenteric lymph nodes was observed in all mice examined[31]. In contrast, when lactoferrin or its pepsin hydrolysate was added to the milk at a concentration of 2%, the incidence of bacterial translocation and the number of translocated bacteria were significantly reduced. Bacterial translocation has been reported to be evident in various animal models under the following pathophysiologic conditions: (i) disruption of the ecological balance of the indigenous microflora resulting in bacterial overgrowth with Gram-negative enteric bacilli, (ii) impairment host immune defenses, and (iii) physical injury to the mucosal barrier[30]. In the milk-fed mice, there was no apparent mucosal atrophy and no significant decrease in systemic immune response was observed during the 7-day period of feeding. These results suggest that ingested lactoferrin acts to inhibit bacterial translocation through its bacteriostatic effects in the intestine.

2.3. Human Infants

A clinical study of infant formula supplemented with lactoferrin was performed in cooperation with Showa University[32]. Nine low birth weight infants were the subjects examined. The feeding of lactoferrin-enriched formula (bovine lactoferrin 100 mg/100 ml) was started at the age of 14 to 36 days and continued for two weeks. The feeding of lactoferrin-enriched formula was found to influence the intestinal flora of the subject infants. The composition ratio of bifidobacteria in the fecal microflora was apparently increased in the case of infants given the lactoferrin-enriched formula, while the ratios of *Enterobacteriaceae*, streptococci and clostridia showed a tendency to decrease. These results suggest that ingested lactoferrin promotes the establishment of a bifidobacteria predominant flora and suppresses the proliferation of opportunistic pathogens in infants. In this study, some fecal parameters were evaluated also. Feeding of the lactoferrin-enriched formula led to a tendency of decreased fecal pH, enhanced organic acid production, increased lysozyme activity, and increased levels of immunoglobulin A in the feces after two weeks of feeding. Further studies are required to ascertain whether lactoferrin exerts an immunomodulatory effect in human infants.

3. PROTECTIVE EFFECT OF LACTOFERRIN THROUGH ITS INFLUENCE ON THE IMMUNE SYSTEM

Evidence from several in vivo studies has indicated that lactoferrin exerts a protective effect in various animals. Both human and bovine lactoferrin have been shown to protect mice infected with lethal dose of *E. coli*[33]. In this model of bacterial sepsis, the rapid lethal effect of injected *E. coli* is known to result from the acute inflammatory response of the animal to endotoxin present on the bacterial surface. Lactoferrin has been shown to suppress the systemic effects triggered by bacterial endotoxin by influencing the levels of potent inflammatory mediators[34]. Several studies have reported that lactoferrin displays preventive effects against tumor development and progression in animal models. Human lactoferrin has been shown to improve the survival of mice infected with a cancer virus[35]. Also, human lactoferrin has been shown to inhibit tumor growth and metastasis in mice through the activation of natural killer cells[36]. Recently, similar inhibitory effects of bovine lactoferrin and lactoferricin® on tumor metastasis, attributed to the inhibition of tumor-induced angiogenesis, have been observed in mice[37]. In many of these studies, lactoferrin was administered intravenously or intraperitoneally. On the contrary, the following studies focus on the therapeutic effects of bovine lactoferrin administered orally to small animals and cultured fish as demonstrated by some of our collaborators. From studies designed to elucidate the mechanisms of action, it has been revealed that orally administered lactoferrin exerts such protective effects through its influence on the host immune system.

3.1. Intractable Stomatitis in Cats

Intractable stomatitis often causes severe pain leading to loss of appetite with serious consequences in cats. It is not easy to improve the intractable stomatitis sufficiently by chemical drug treatment. Sato *et al.* have reported that oral administration of bovine lactoferrin (40 mg/kg/day) improved the symptoms of intractable stomatitis (e.g., inflammation of the tongue or lips) in feline immunodeficiency virus (FIV)-positive and FIV-negative cats[38], and this improvement coincided with an enhancement of the phagocytic activity of blood neutrophils. Healthy cats as well as ill cats were used as subjects to elucidate the therapeutic mechanism of action. The phagocytic activity of blood neutrophils was significantly increased by administration of lactoferrin (40 mg/kg/day) in healthy cats as well as ill cats. On the other hand, the adherence activity of blood neutrophils was suppressed by administration of lactoferrin. These findings might relate to the reduced inflammation in the mouth of the subject cats after treatment with lactoferrin. From these results, they concluded that topical application of lactoferrin to oral mucous membranes was useful as a treatment for intractable stomatitis.

3.2. Parasitic Infection in Fish

In fish, no protein similar to lactoferrin has yet been found. Kakuta *et al.* tested the protective effect of orally administered bovine lactoferrin against a parasitic infection in goldfish. The survival of goldfish experimentally infected with deadly "white spot" disease caused by a protozoan, *Ichthyophthyris multifiliis*, was improved by addition of bovine lactoferrin to the diet at a concentration of 0.4%[39]. In addition, the white blood cells of fish given lactoferrin displayed enhanced phagocytic activity. Bovine lactoferrin has been applied also to the prevention of "white spot" disease in culture farming of red sea bream. In the case of red sea bream, this disease is caused by the protozoan *Cryptocaryon irritans*. The survival rate of red sea bream given a diet supplemented with lactoferrin at a concentration of 1% was significantly higher than that of the control group[40]. They measured some blood parameters of normal sea bream to investigate the mechanism of this defensive effect of lactoferrin. Each group of fish given lactoferrin had significantly higher counts of granulocytes and lymphocytes in the blood compared with the control group[41]. In view of these results, it seems evident that lactoferrin activates the immune system of fish and augments the host defense against parasitic infection. Higher amounts of skin mucus which is important for protection of the body epidermis was observed in both goldfish and red sea bream given lactoferrin[39,41]. These results provide further evidence that lactoferrin may contribute to the host defense of fish through its activation of non-specific defense mechanisms.

4. STIMULATORY EFFECT OF LACTOFERRIN ON PHAGOCYTOSIS BY HUMAN NEUTROPHILS

In humans, lactoferrin is a prominent component of all mucosal secretions and it can be detected in almost all body fluids. Lactoferrin is a major component of the secondary granules of neutrophils and this protein is released from activated neutrophils in the inflammatory response. When infection occurs, neutrophils soon move to the site of infection. There is a possibility that lactoferrin in the secondary granules of neutrophils functions to kill the microbes which are taken into these cells. It has been estimated that the total amount of lactoferrin produced in neutrophils increases 6-fold to 30 g/day during infection and the levels of lactoferrin released into blood plasma increase 20-fold to 10 g/day[42]. The importance of lactoferrin in the host defense is also underlined by studies indicating that patients with defects in lactoferrin production exhibit an abnormal predisposition to recurrent infections by bacterial and fungal pathogens[43].

In addition to direct its antimicrobial action, various immunomodulatory effects of lactoferrin have been reported. Lactoferrin may function as a component of the complex

Table 2. Biological functions of lactoferrin

Biological functions of lactoferrin
Antimicrobial, antiviral, and antiparasitic activities
Growth stimulation of various cell lines
Regulation of intestinal iron absorption
Modulation of the inflammatory-immune response
Up-modulation
Phagocytosis and cytotoxicity (PMN and macrophage)
IL-8 release (PMN)
NK activity (NK cell)
Cell growth and differentiation (leukocytes)
Down-modulation
GM-CSF production (monocyte/macrophage)
RNA transcription (leukocytes)
Release of IL-1, IL-2 and TNF (leukocytes)
Complement activation
Antioxidant activity

Abbreviations: PMN = polymorphonuclear leukocytes, IL = interleukin, NK = natural killer, GM-CSF = granulocyte-macrophage colony-stimulating factor, TNF = tumor necrosis factor.

regulatory network that exists to control the behavior of mature leukocytes, and the proliferation and differentiation of hematopoietic progenitor cells. Lactoferrin up-regulates phagocytosis and cytotoxicity of neutrophils[44] and macrophages[45], IL-8 release from neutrophils[46], natural killer activity[47], and the growth of lymphocytes[48]. On the other hand, lactoferrin had been shown to down-regulate GM-CSF production by macrophages[49], release of IL-1, IL-2 and TNF from leukocytes[50], and complement activation[51]. Table 2 summarizes various biological functions of lactoferrin including immunomodulatory effects observed mainly in in vitro experiments. Due to the limited evidence available from in vivo studies, the physiological effects of lactoferrin remain controversial.

4.1. Stimulatory Effect on Neutrophils in Vitro

Our recent in vitro study revealed that the phagocytic activity of human blood neutrophils were stimulated by addition of bovine lactoferrin[52]. In this study, neutrophils prepared from blood of healthy volunteers were used. Bovine lactoferrin stimulated the phagocytic activity of the neutrophils in a dose dependent manner. Interestingly, the pepsin hydrolysate of bovine lactoferrin or lactoferricin^{B} showed a similar stimulatory effect. These results suggest that the lactoferricin^{B} region near the amino terminus of the lactoferrin molecule plays an important role in stimulating the phagocytic activity of neutrophils.

In another study, Okutomi *et al.* showed that lactoferrin and blood neutrophils act cooperatively to inhibit the growth of *Candida albicans*, an opportunistic fungal pathogen[53]. The growth inhibition caused by human neutrophils was augmented by the addition of bovine lactoferrin at concentrations which did not show any inhibitory effect in the absence of neutrophils. Human lactoferrin showed the same activity as bovine lactoferrin. These findings indicate that lactoferrin acts cooperatively with neutrophils to inhibit pathogenic microorganisms.

4.2. Influence of Orally Administered Bovine Lactoferrin on the Immune System of Healthy Volunteers

We have examined the effects of orally administered bovine lactoferrin on the immune system of healthy volunteers[54]. Ten healthy male volunteers (age range of 31 to 55 years old) were given 2 g of lactoferrin daily for 4 weeks as tablets containing 30% bovine lactoferrin. Using flow cytometry, the phagocytic activity of blood polymorphonuclear leucocytes (PMN) was evaluated from the ratio of cells capable of phagocytizing polymer particles. During the period of lactoferrin administration, the phagocytic activity of PMN was increased in three of ten subjects. In one subject with higher phagocytic activity of PMN before lactoferrin administration, the activity decreased after lactoferrin administration. Obvious changes were not observed in the other 6 subjects. These findings suggest the possibility that orally administered bovine lactoferrin exerts an immunomodulatory effect in humans and acts to up-regulate the phagocytic activity of blood neutrophils, although the effect of lactoferrin differs in individual cases.

5. CONCLUSION

Our findings support the view that orally administered bovine lactoferrin exerts host defensive effects in fish, animals, and humans.

1. Ingested bovine lactoferrin has been shown to exert bacteriostatic effects against *Enterobacteriaceae* and some other kinds of bacteria in the gut of milk-fed mice.
2. Supplementation of infant formula with bovine lactoferrin has been shown to promote the establishment of a bifidobacteria predominant flora in low birth weight infants.
3. Orally administered lactoferrin has been found to exert a protective effect against parasitic infection in fish and to improve the symptoms of intractable stomatitis in cats.
4. In studies designed to elucidate the mechanisms of these protective effects of lactoferrin, it was shown that ingested lactoferrin stimulates the phagocytic activity of white blood cells in cats. In addition, ingested lactoferrin increased the number of lymphocytes and granulocytes in fish and increased the amount of mucus on the epidermis of fish.
5. Oral administration of bovine lactoferrin to healthy volunteers revealed that some immunomodulatory effects of this protein were displayed in humans.

These findings prompt us to consider the utilization of lactoferrin as an active ingredient in dietary products. Further studies will be required to precisely establish the effective dose of orally administered lactoferrin for protection of animals or humans.

Nowadays, various bioactive materials derived from bovine milk are being manufactured on an industrial scale for use as functional ingredients in many fields. We hope that our research efforts concerning lactoferrin might afford some new perspectives on the commercial use of such minor components in milk.

REFERENCES

1. Masson PL, Heremans JF, Dive CH. (1966) An iron-binding protein common to many external secretions. Clin. Chim. Acta 14:735–739.
2. Lash JA, Coates TD, Lafuze J, Baehner RL, Boxer LA. (1983) Plasma lactoferrin reflects granulocyte activation in vivo. Blood 61:885–888.
3. Lönnerdal B, Iyer S. (1995) Lactoferrin: molecular structure and biological function. Annu. Rev. Nutr. 15:93–110.
4. Metz-Boutigue M-H, Jollès J, Mazurier J, Schoentgen F, Legrand D, Spik G, Montreuil J, Jollès P. (1984) Human lactotransferrin: amino acid sequence and structural comparisons with other transferrins. Eur. J. Biochem. 145:659–676.

5. Pierce A, Colavizza D, Benaissa M, Maes P, Tartar A, Montreuil J, Spik G. (1991) Molecular cloning and sequence analysis of bovine lactotransferrin. Eur. J. Biochem. 196:177–184.
6. Anderson BF, Baker HM, Norris GE, Rice DW, Baker EN. (1989) Structure of human lactoferrin: crystallographic structure analysis and refinement at 2.8 Å resolution. J. Mol. Biol. 209:711–734.
7. Norris GE, Anderson BF, Baker EN, Baker HM, Gärtner AL, Ward J, and Rumball SV. (1986) Preliminary crystallographic studies on bovine lactoferrin. J. Mol. Biol. 191:143–145.
8. Shimazaki K, Kawaguchi A, Sato T, Ueda Y, Tomimura T, Shimamura S. (1993) Analysis of human and bovine milk lactoferrins by Rotofor and chromatofocusing. Int. J. Biochem. 25:1653–1658.
9. Arnold RR, Brewer M, Gauthier JJ. (1980) Bactericidal activity of human lactoferrin: sensitivity of a variety of microorganisms. Infect. Immun. 28:893–898.
10. Weinberg ED. (1978) Iron and infection. Microbiol. Rev. 42:45–66.
11. Arnold RR, Russell JE, Champion WJ, Brewer M, Gauthier JJ. (1982) Bactericidal activity of human lactoferrin: differentiation from the stasis of iron deprivation. Infect. Immun. 35:792–799.
12. Ellison III RT, Giehl TJ, LaForce FM. (1988) Damage of the outer membrane of enteric Gram-negative bacteria by lactoferrin and transferrin. Infect. Immun. 56:2774–2781.
13. Erdei J, Forsgren A, Naidu AS. (1994) Lactoferrin binds to porins OmpF and OmpC in *Escherichia coli*. Infect. Immun. 62:1236–1240.
14. Tomita M, Bellamy W, Takase M, Yamauchi K, Wakabayashi H, Kawase K. (1991) Potent antibacterial peptides generated by pepsin digestion of bovine lactoferrin. J. Dairy Sci. 74:4137–4142.
15. Bellamy W, Takase M, Yamauchi K, Wakabayashi H, Kawase K, Tomita M. (1992) Identification of the bactericidal domain of lactoferrin. Biochim. Biophys. Acta 1121:130–136.
16. Teraguchi S, Shin K, Fukuwatari Y, Shimamura S. (1996) Glycans of bovine lactoferrin function as receptors for the type 1 fimbrial lectin of *Escherichia coli*. Infect. Immun. 64:1075–1077.
17. Izhar M, Nuchamowitz Y, Mirelman D. (1982) Adherence of *Shigella flexneri* to guinea pig intestinal cells is mediated by a mucosal adhesin. Infect. Immun. 35:1110–1118.
18. Spik G, Brunet B, Mazurier-Dehaine C, Fontaine G, Montreuil J. (1982) Characterization and properties of the human and bovine lactotransferrins extracted from the faeces of newborn infants. Acta Pediatr. Scand. 71:979–985.
19. Niida Y, Takayanagi N, Terashima H, Horino K. (1988) Lactoferrin in human breast milk and feces of newborn measured by ELISA. J. Jpn. Pediatr. Soc. 92:1496–1501. (in Japanese)
20. Hentges DJ, Marsh WW, Petschow BW, Thal WR, Carter MK. (1992) Influence of infant diets on the ecology of the intestinal tract of human flora-associated mice. J. Pediatr. Gastroent. Nutr. 14:146–152.
21. Roberts AK, Chierici R, Sawatzki G, Hill MJ, Volpato S, Vigi V. (1992) Supplementation of an adapted formula with bovine lactoferrin: 1. Effect on the infant faecal flora. Acta Pediatr. 81:119–124.
22. Balmer SE, Scott PH, Wharton BA. (1989) Diet and faecal flora in the newborn: Lactoferrin. Arch. Dis. Child. 64:1685–1690.
23. Spik G, Legrand D, Leveugle B, Mazurier J, Mikogami,T, Montreuil J, Pierce A, Rochard E. (1993) Binding properties of different lactotransferrins to human lactotransferrin receptor, p. 77–83. In B. Renner and G. Sawatzki (ed.), New Perspectives in Infant Nutrition; symposium Antwerp 1992. Thieme Medical Publishers, New York.
24. Hutchens TW, Henry JF, Yip T-T. (1991) Structurally intact (78-kDa) forms of maternal lactoferrin purified from urine of preterm infants fed human milk: Identification of a trypsin-like proteolytic cleavage event in vivo that does not result in fragment dissociation. Proc. Natl. Acad. Sci. USA 88:2994–2998.
25. Johanson B. (1960) Isolation of an iron-containing red protein from human milk. Acta Chem. Scand. 14:510–512.
26. Groves ML. (1960) The isolation of a red protein from milk. J. Am. Chem. Soc. 82:3345–3350.
27. Teraguchi S, Ozawa K, Yasuda S, Shin K, Fukuwatari Y, Shimamura S. (1994) The bacteriostatic effects of orally administered bovine lactoferrin on intestinal *Enterobacteriaceae* of SPF mice fed bovine milk. Biosci. Biotech. Biochem. 58:482–487.
28. Teraguchi S, Shin K, Ozawa K, Nakamura S, Fukuwatari Y, Tsuyuki S, Namihira H, Shimamura S. (1995) Bacteriostatic effect of orally administered bovine lactoferrin on proliferation of *Clostridium* species in the gut of mice fed bovine milk. Appl. Environ. Microbiol. 61:501–506.
29. Teraguchi S, Ogata T, Shin K, Kingaku M, Fukuwatari Y, Kawase K, Hayasawa H, Tomita M. (1997) The mechanism of in vivo bacteriostasis of bovine lactoferrin. (this refers to the article in the same lactoferrin monograph).
30. Berg RD. (1992) Translocation and the indigenous gut flora. p. 55–85. In R. Fuller (ed.), Probiotics: The Scientific Basis, Chapman & Hall, Cambridge.
31. Teraguchi S, Shin K, Ogata T, Kingaku M, Kaino A, Miyauchi H, Y. Fukuwatari Y, Shimamura S. (1995) Orally administered bovine lactoferrin inhibits bacterial translocation in mice fed bovine milk. Appl. Environ. Microbiol. 61, 4131–4134.

32. Kawaguchi S, Hayashi T, Masano H, Okuyama K, Suzuki T, Kawase K. (1989) Shusankiigaku 19, 125–130. (in Japanese)
33. Zagulski T, Lipinski P, Zagulska A, Broniek S, Jarzabek Z. (1989) Lactoferrin can protect mice against a lethal dose of *Escherichia coli* in experimental infection in vivo. Br. J. Exp. Path. 70:697–704.
34. Machnicki M, Zimecki M, Zagulski T. (1993) Lactoferrin regulates the release of tumor necrosis factor alpha and interleukin 6 *in vivo*. Int. J. Exp. Path. 74:433–439.
35. Lu L, Hangoc G, Oliff A, Chen LT, Shen R-N, Broxmeyer HE. (1987) Protective influence of lactoferrin on mice infected with the polycythemia-inducing strain of Friend virus complex. Cancer Res. 47:4184–4188.
36. Bezault J, Bhimani R, Wiprovnick J, Furmanski P. (1994) Human lactoferrin inhibits growth of solid tumors and development of experimental metastases in mice. Cancer Res. 54:2310–2312.
37. Yoo Y-C, Watanabe S, Watanabe R, Hata K, Shimazaki K, Azuma I. (1997) Bovine lactoferrin and lactoferricin, a peptide derived from bovine lactoferrin, inhibit tumor metastasis in mice. Jpn. J. Cancer Res. 88:184–190.
38. Sato R, Inanami O, Tanaka Y, Takase M, Naito Y. (1996) Oral administration of bovine lactoferrin for treatment of intractable stomatitis in feline immunodeficiency virus (FIV)-positive and FIV-negative cats. Am. J. Vet. Res. 57:1443–1446.
39. Kakuta I. (1996) Protective effect of orally administrated bovine lactoferrin against experimental infection of goldfish *Carassius auratus* with *Ichthyophthirius multifiliis*. Suisanzoshoku 44:427–432. (in Japanese)
40. Kakuta I, Kurokura H. (1995) Defensive effect of orally administered bovine lactoferrin against *Cryptocaryon irritans* infection of red sea bream. Fish Pathol. 30:289–290. (in Japanese)
41. Kakuta I, Kurokura H, Nakamura H, Yamauchi K. (1996) Enhancement of the nonspecific defense activity of the skin mucus of red sea bream by oral administration of bovine lactoferrin. Suisanzoshoku 44:197–202. (in Japanese)
42. Sawatzki G, Rich IN. (1989) Lactoferrin stimulates colony stimulating factor production in vitro and in vivo. Blood Cells 15:371–385.
43. Boxer LA, Coates TD, Haak RA, Wolach JB, Hoffstein S, Baehner RL. (1982) Lactoferrin deficiency associated with altered granulocyte function. New Engl. J. Med. 307:404–409.
44. Gahr M, Speer CP, Damerau B, Sawatzki G. (1991) Influence of lactoferrin on the function of human polymorphonuclear leukocytes and monocytes. J. Leuk. Biol. 49:427–433.
45. Lima MF, Kierszenbaum F. (1985) Lactoferrin effects on phagocytic cell function. I. Increased uptake and killing of an intracellular parasite by murine macrophages and monocytes. J. Immunol. 134:4176–4183.
46. Shinoda I, Takase M, Fukuwatari Y, Shimamura S, Köller M, König W. (1996) Effects of lactoferrin and lactoferricin® on the release of interleukin 8 from human polymorphonuclear leukocytes. Biosci. Biotech. Biochem. 60:521–523.
47. Nishiya K, Horwitz DA. (1982) Contrasting effects of lactoferrin on human lymphocyte and monocyte natural killer activity and antibody-dependent cell-mediated cytotoxicity. J. Immunol. 129:2519–2523.
48. Zimecki M, Mazurier J, Machnicki M, Wieczorek Z, Montreuil J, Spik G. (1991) Immunostimulatory activity of lactotransferrin and maturation of $CD4^- CD8^-$ murine thymocytes. Immunol. Lett. 30:119–124.
49. Broxmeyer HE, Platzer E. (1984) Lactoferrin acts on I-A and I-E/C antigen$^+$ subpopulations of mouse peritoneal macrophages in the absence of T lymphocytes and other cell types to inhibit production of granulocyte-macrophage colony stimulatory factors in vitro. J. Immunol. 133:306–314.
50. Crouch SPM, Slater KJ, Fletcher J. (1992) Regulation of cytokine release from mononuclear cells by the iron-binding protein lactoferrin. Blood 80:235–240.
51. Kijlstra A, Jeurissen SHM. (1982) Modulation of classical C3 convertase of complement by tear lactoferrin. Immunology 47:263–270.
52. Unpublished data
53. Okutomi T, Abe S, Tansho S, Wakabayashi H, Kawase K, Yamaguchi H. (1997) Augmented inhibition of growth of *Candida albicans* by neutrophils in the presence of lactoferrin. FEMS Immunol. Med. Microbiol. (in press)
54. Yamauchi K, Wakabayashi H, Hashimoto S, Teraguchi S, Hayasawa H, Tomita M. (1997) Effects of orally administered bovine lactoferrin on the immune system of healthy volunteers. (this refers to the article in the same lactoferrin monograph)

ANTIVIRAL ACTIVITY OF LACTOFERRIN

Piera Valenti,[1] Magda Marchetti,[2] Fabiana Superti,[3]
Maria Grazia Amendolia,[3] Patrizia Puddu,[4] Sandra Gessani,[5] Paola Borghi,[5]
Filippo Belardelli,[5] Giovanni Antonini,[6] and Lucilla Seganti[2]

[1]Institute of Microbiology
II University of Naples, Italy
[2]Institute of Microbiology
University of Rome "La Sapienza", Italy
[3]Department of Ultrastructures
National Institute of Health, Rome, Italy
[4]Laboratory of Immunology
National Institute of Health, Rome, Italy
[5]Laboratory of Virology
National Institute of Health, Rome, Italy
[6]Department of Basic and Applied Biology
University of l'Aquila, Italy

In 1976, Matthews et al.[19] reported the antiviral activity of milk proteins, underlining their possible clinical importance. Only recently, this effect has been ascribed mainly to lactoferrin (Lf). Lf was initially shown to exert a protective influence in mice infected with Friend leukemia virus[15]. Subsequently, a potent antiviral activity has been attributed to Lf, both *in vitro* towards cytomegalovirus (HCMV)[9,7], herpes simplex virus (HSV)[17], human immunodeficiency virus (HIV)[7,29] and *in vivo* towards HSV-1[6] and HCMV[25]. Our groups provided evidence on the antiviral activity of Lf towards HSV-2[18], rotavirus[28], and HIV[22] infections (Table 1).

In vivo levels of Lf have been shown to be altered in the course of some microbial infections. In bacterial infections an up to 50-fold increase of Lf levels has been described[21] while in viral infections and mostly during HIV infections, markedly decreased Lf levels in tears and plasma down to 0.68 μg/ml have been found[3,5]. The identification of glycosaminoglycans as binding sites for the N-terminus of human and bovine lactoferrin[12,16,30] has led to a preliminary hypothesis on the Lf antiviral mechanism (i.e. competitive interaction with putative virus receptors). In fact, these ubiquitous costituents of most mammmalian cell plasma membranes and extracellular matrices may act as receptors for some enveloped viruses, including HSV[31] and HIV[23]. However, although the antiviral mechanism of Lf has been partially understood, there is still a lot to be elucidated.

Advances in Lactoferrin Research, edited by Spik *et al.*
Plenum Press, New York, 1998.

Table 1. *In vitro* antiviral activity of bovine lactoferrin towards different viruses

Virus	CC_{50}[a] (mg/ml)	IC_{50}[b] (μg/ml)	SI[c]	References
HCMV	>0.250	36.7	>6.0	Harmsen et al. (7)
HSV-1	>10	12.0	833	Marchetti et al. (17)
HSV-2	>10	5.2	1923	Marchetti et al. (18)
HIV-1	>0.250	40.0	>6.3	Swart et al. (29)
HIV-2	>0.250	>250	≈ 1	Swart et al. (29)
HIV-1	>12.5	2.0	>6250	Puddu et al. (22)
Rotavirus SA-11	> 30	50.0	>600	Superti et al. (28)

[a]CC_{50}, cytotoxic concentration, 50%
[b]IC_{50}, inhibiting concentration, 50%
[c]SI, selectivity index ($SI=CC_{50}/IC_{50}$)

To understand the Lf mechanism of action, both the type of utilized virus and the susceptible cells must be taken into account. It is relevant to distinguish the Lf specific binding to identified receptors on cell surfaces from the nonspecific binding to many target cells that Lf, being a basic protein, can perform. These two events could have a different relevance on the Lf antiviral activity. On the other hand, it is worth noting that Lf is active against viruses exhibiting different structural characteristics (both enveloped and naked viruses; DNA and RNA viruses) and replication strategies.

Data reported so far support the following points: bovine Lf is more active than human Lf; apo-Lf is less effective than the iron saturated form, the Lf antiviral activity does not appear to be associated with the iron withholding.

The comparison between bovine and human Lf structures shows that, in spite of their high homology, the aminoacid sequences in the putative regions responsible for antiviral activity, the glycan chains of the molecules and the number of disulphide bridges vary; these variations are likely to contribute to the differences in the antiviral effectiveness. The higher inhibiting activity of iron saturated Lf could be due to the large conformational change occurring after metal ion saturation[26], which might result in a higher affinity for Lf receptors on eukaryotic cells than the apo form[4]. This is supported by the experiments reported by Marchetti et al.[17], in which the preincubation of Vero cells with human Lf in apo or saturated form led to a different antiviral effect: the apo-Lf exerted a lower inhibition of HSV-1 yield (20%) compared with the iron saturated one (60%). This result is also in agreement with a lower affinity of the apo-form for Vero cells, as tested by different assays. A further explanation of the greater antiviral activity of iron saturated Lf could be supplied by its high resistance to denaturation and to proteolytic enzymes[8]. Experiments carried out with 100% iron-saturated Lf incubated throughout the infection demonstrated that the antiviral effect was maintained up to 24–48 hours infection, in culture conditions in which 80% of the immunoreactive form of Lf was still detectable. Data reported by Harmsen et al.[7] showed that when cells were preincubated with native Lf (partially iron saturated), the HCMV inhibition remained stable only for 60 min and then was completely lost. The activation of a protease capable of degrading the cell-bound Lf has been supposed[7]. All together these data strongly support the importance of the metal in the Lf binding sites and allow to rule out that the antiviral effectiveness is linked to the capability of the molecule to withhold iron from the environment.

The putative administration of Lf in antimicrobial therapy, especially in the control of viral infections, should take into account the role played by metal ions eventually released into the cells by the protein, following its interaction with cell surface[4,10]. Although a variety of viral enzymes require metal ions for their activity, literature data concerning the effect of iron chelating drugs on viral replication are greatly controversial[2,11,1]. In particular, it is well known that ribonucleotide reductases (RNRs) from HSV1 and HSV2, like other iron-dependent RNRs, require iron in the R2 subunit[13]; thus, the availability of ferric ions could enhance HSV replication, while iron chelators could inhibit its replication. To add further information about the influence of iron released by Lf on viral replication, experiments were carried out on HSV1 and HSV2 by preincubating the cell monolayers with ferric citrate at the same concentration (10 μM) of the metal present in 400 μg/ml lactoferrin in the fully iron saturated form. The results showed only a weak effect of ferric citrate (10–15% inhibition of viral antigen synthesis), thus indicating that the simple intracellular release of this metal ion neither induce any increase of viral infectivity, nor is sufficient to account for the antiviral activity of iron saturated Lf (94–98% inhibition of viral antigen synthesis) (18). These data further support the hypothesis of a higher antiviral activity of saturated Lf, linked to the conformation and stability of the molecule induced by the binding of ferric ions[8].

A proper approach to distinguish the step of viral infection affected by an antiviral agent requires the synchronization of the viral multiplication cycle by a temperature shift. This experimental procedure consists in : i) the preincubation of the drug with cells for 1 h at 37°C; ii) the incubation of the inhibitor together with virus inoculum during the adsorption step at 4°C; iii) the addition of the drug after the adsorption step for various lenghts of time at 37°C. The study of the Lf antiviral activity on various steps of the different viral infections, even if differently or partially investigated, led to comparable results.

Hasegawa et al.[9] reported that the Lf preincubation with cells resulted in a greater inhibition of HCMV replication in comparison with the preincubation with virus. Similar results were also obtained by us for HSV1[17], HSV2[18] and HIV-1[22] whereas the preincubation of Lf with the enterocyte-like cell line HT-29 did not inhibit the SA-11 rotavirus entry and replication[28]. These different results may be explained by the fact that HCMV, HSV and HIV use glycosaminoglycans as initial cellular binding sites, while rotavirus is known to interact with different carbohydrate moieties of the HT-29 cells[27].

It appears of interest that the complete inhibition of viral infectivity by Lf took place when the protein was added only during the viral attachment step, thus indicating the block of the viral infection in the very early phases of the multiplication cycle. In this experimental condition, Lf can bind to both the cell and virus surfaces. Marchetti et al.[17] demonstrated the Lf capability of interacting with Vero cells and HSV1 particles. Accordingly, Harmsen et al.[7] reported that Lf exerts its effect towards HCMV through the binding to the target cell membranes, even though a direct virus-protein interaction cannot be excluded. The Lf binding to the V3 loop of the gp120 envelope protein of HIV[29] and the Lf interaction with rotavirus particles, determined by the inhibition of the viral hemagglutination and by the virus binding to plastic absorbed Lf[28] have also been demonstrated.

Different results were obtained when Lf was added to infected cells after the viral adsorption step. Harmsen et al.[7] reported that, in postinfection assays the anti-HCMV activity of Lf gradually decreases, whereas Marchetti et al.[18] observed a 60% residual antiviral activity by Lf towards HSV1 and HSV2, probably due to the inhibition of extracellular virus released by permissive cells. In this model, Lf bound to cell surfaces and present in the culture medium, is likely to prevent the infection of neighbouring cells. Moreover, the postincubation of enterocyte-like cells with Lf after rotavirus adsorption step, also resulted in a similar significant antiviral effect. However, this experimental condition allowed a single replication cycle and an interaction of Lf on the rotavirus particles released can be excluded. As specific receptors for Lf and its transepithelial transport in HT-29 cells have

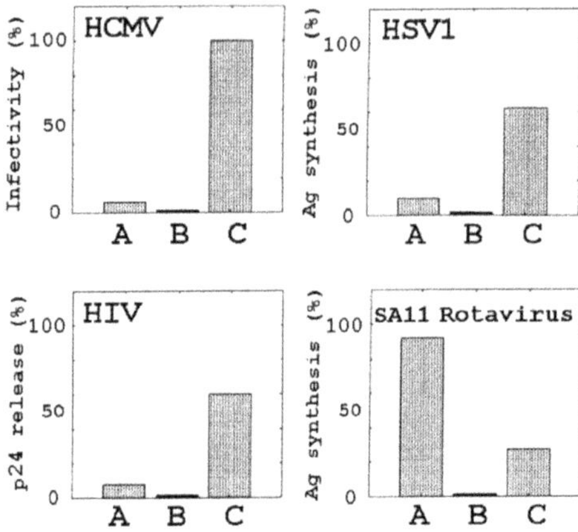

Figure 1. Effect of Lf on different steps of viral replication. A) Cell preincubation with Lf; B) Lf addition during the viral adsorption step; C) Lf incubation during the viral postadsorption period. HCMV[9,7]; HSV1[17]; HIV[22]; SA-11 rotavirus[28].

been well described[24,20] an effect of Lf on the intracellular steps of rotaviral infection can be hypothesized.

All these data, deriving from different studies with several virus-cell models, are summarized in Fig. 1.

In conclusion, Lf appears to play an extended role in the host defence against microbial infections as well as in iron metabolism. The new function ascribed to Lf, consisting in a protection of host cells towards the early interactions with different viruses, can be correlated to the protective effect towards the entry of cells by some facultative intracellular bacteria[14] Indeed, the antiviral activity of Lf is strictly correlated to its competitive interaction with cell receptors for viruses, even though an intracellular effect cannot been excluded. The influence of the immunoreactive or digested forms of Lf in the virus binding to cells, in the DNA and RNA virus replication and the role of iron eventually released from the protein should be considered fundamental topics for further investigations.

ACKNOWLEDGMENTS

Work has been carried out by MURST and CNR grants.

REFERENCES

1. Balogh-Nair V, Brathwaite CE, Chen CX, Vargas J Jr (1995) Synthesis, activity and toxicity of novel macrocyclic ligands against HIV-1 in Jurkat and CEM-SS cell lines. Cell Mol Biol 41 Suppl 1: 9–14
2. Cinatl J Jr, Cinatl J, Rabenau H, Gumbel HO, Kornhuber B, Doerr HW (1994) *In vitro* inhibition of human cytomegalovirus replication by desferrioxamine. Antiviral Res 25: 73–77
3. Comerie-Smith SE, Nunez J, Hosmer M, Farris RL (1994) Tear lactoferrin and ocular bacterial flora in HIV positive patients. Adv Exp Med Biol 350: 339–344
4. Davidson LA, Lonnerdal B (1989) Fe saturation and proteolysis of human lactoferrin: effect on brushborder receptor mediated uptake of Fe^{2+} and Mn^{2+}. Am J Physiol 257: G930-G934
5. Defer MC, Dugas B, Picard O, Damais C (1995) Impairment of circulating lactoferrin in HIV-infection. Cell Mol Biol 41: 417–421
6. Fujihara T and Hayashi K (1995) Lactoferrin inhibits herpes simplex virus type-1 (HSV-1) infection to mouse cornea. Arch Virol 140: 1469–72
7. Harmsen MC, Swart PJ, de Béthune MP, Pawels R, De Clercq E, The TH, Meijer DKF (1995) Antiviral effects of plasma and milk proteins: lactoferrin shows a potent activity against both human immunodeficiency virus and human cytomegalovirus replication in vitro. J Infect Dis 172: 380–388

8. Harrington JP (1992) Spectroscopic analysis of the unfolding of transition metal-ion complexes of human lactoferrin and transferrin. Int J Biochem 24: 275–280
9. Hasegawa K, Motsuchi W, Tanaka S, Dosako S (1994) Inhibition with lactoferrin of in vitro infection with human herpes virus. Jpn. J Med Sci Biol 47: 73–85
10. Hu W-L, Mazurier J, Montreuil J, Spik G (1990) Isolation and partial characterization of a lactotransferrin receptor from mouse intestinal brush border. Biochemistry 29: 535–541
11. Inouye Y, Kanamori T, Yoshida T, Bu X, Shionoya M, Koike T, Kimura E (1994) Inhibition of human immunodeficiency virus proliferation by macrocyclic polyamines and their metal complexes. Biol Pharm Bull 17: 243–250
12. Ji Z.S. and Mahley R.W. (1994) Lactoferrin binding to heparan sulphate proteoglycans and LDL receptor-related protein. Further evidence supporting the importance of direct binding of remnant lipoproteins to HSPG. Arterioscler Thromb 14, 2025–2031
13. Liuzzi M, DÆeziel R, Moss N, Beaulieu P, Bonneau AM, Bousquet C, Chafouleas JG, Garneau M, Jaramillo J, Krogsrud RL, et al. (1994) A potent peptidomimetic inhibitor of HSV ribonucleotide reductase with antiviral activity in vivo. Nature 372: 6507–6508
14. Longhi C, Conte MP, Seganti, Polidoro M, Alfsen A, Valenti P (1993) Influence of lactoferrin on the entry process of *Escherichia coli* HB101 (pRI203) in HeLa cells. Med Microbiol Immunol 182: 25–35
15. Lu L, Hangoc G, Oliff A, Chen L, Shen RN, Broxmeyer HE (1987) Protective influence of lactoferrin on mice infected with the polythemia-inducing strain of Friend virus complex. Cancer Res 47: 4184–4188
16. Mann DM, Romm E, Migliorini M (1994) Delineation of the glycosaminoglycan-binding site in the human inflammatory response protein lactoferrin. J Biol Chem 269: 23661–23667
17. Marchetti M, Longhi C, Conte MP, Pisani S, Valenti P, Seganti L (1996) Lactoferrin inhibits herpes simplex virus type 1 adsorption to Vero cells. Antiviral Res 29: 221–231
18. Marchetti M, Pisani S, Antonini G, Valenti P, Seganti L, Orsi N In vitro inhibition of herpes simplex virus type 1 and 2 by bovine lactoferrin saturated with iron, manganese and zinc ions. Submitted for publication
19. Matthews THJ, Nair CDG, Lawrence MK, Tyrrel DAJ (1976) Antiviral activity in milk of possible clinical importance. Lancet 7: 1387–1389
20. Mikogami T, Heyman M, Spik G, Desjeux JF (1994) Apical-to-basolateral transepithelial transport of human lactoferrin in the intestinal cell line HT-29cl.19A. Am J Physiol G308-G315
21. Nuijens JH, Abbink JJ, Wachtfogel YT, Colman RW, Eerenberg AJ, Dors D, Kamp AJ, Strack van Schijndel RJ, Thijs LG, Hack CE (1992) Plasma elastase alpha 1- antitrypsin and lactoferrin in sepsis: evidence for neutrophils as mediators in fatal sepsis. J Lab Clin Med 119: 159–68
22. Puddu P, Borghi P, Gessani S, Valenti P, Belardelli F, Seganti L Antiviral effect of bovine lactoferrin with divalent and trivalent metal ions on early steps of human immunodeficiency virus type I infection. Submitted for pubblication
23. Roderiquez G, Oravecz T, Yanagishita M, Bou-Habib DC, Mostowski H, Norcross MA (1995) Mediation of human immunodeficiency virus type 1 binding by interaction of cell surface heparan sulfate proteoglycans with the V3 region of envelope gp120-gp41. J Virol 69:2233–2239.
24. Roiron, D, Amouric M, Marvaldi J, Figarella C (1989) Lactoferrin-binding sites at the surface of HT29-D4 cells. Eur J.Biochem 186: 367–373.
25. Shimizu K, Matsuzawa H, Okada K, Tazume S, Dosako S, Kawasaki Y, Hashimoto K, Koga Y (1996) Lactoferrin-mediated protection of the host from murine cytomegalovirus infection by a T-cell-dependent augmentation of natural killer cell activity. Arch Virol 141: 1875–1889
26. Smith CA., Anderson BF, Baker HM, Baker EN (1992) Metal substitution in transferrins: the crystal structure of human copper-lactoferrin at 2.1-A resolution. Biochemistry 31: 4527–4533
27. Superti F, Donelli G (1995) Characterization of SA-11 rotavirus receptorial structures on human colon carcinoma cell line HT-29. J Med Virol 47: 421–428.
28. Superti F, Ammendolia MG, Valenti P, Seganti L Antirotaviral activity of milk proteins: lactoferrin prevents rotavirus replication in the enterocyte-like cell line HT-29. Submitted for publication
29. Swart PJ, Kuipers ME, Smith C, Pawels R, de Béthune MP, De Clercq E, Meijer DKF Huisman JG (1996) Antiviral effects of milk proteins: acylation results in polyanionic compounds with potent activity against human immunodeficiency virus types 1 and 2 in vitro. AIDS Res Human Retrov 12 :769–775
30. Wu HF, Monroe DM, Church FC (1995) Characterization of the glycosaminoglycan-binding region of lactoferrin. Arch Biochem Biophys 317: 85–92
31. WuDunn D, Spear PG (1989) Initial interaction of herpes simplex virus with cells is binding to heparan sulphate. J Virol 63:52–58

24

LACTOFERRIN

Antiviral Activity of Lactoferrin

Pieter J. Swart,[1] E. Mirjam Kuipers,[1] Catharina Smit,[1]
Barry W. A. Van Der Strate,[1] Martin C. Harmsen,[2] and Dirk K. F. Meijer[1]

[1]Department of Pharmacology
Section Pharmacokinetics and Drug Targeting
University of Groningen
A. Deusinglaan 1, 9713 AV Groningen, The Netherlands
[2]Department of Immunology
Hanzeplein 1, 9713 EZ Groningen, The Netherlands

1. SUMMARY

A series of native and chemically derivatized lactoferrins (Lfs) purified from milk and colostrum were assayed *in vitro* for their anti-HIV and anti-HCMV-cytopathic effects in MT4 cells and fibroblasts respectively. All Lfs from bovine and human milk or colostrum were able to completely block HCMV replication as well as inhibited HIV-1 induced cytopathic effects.

Through acylation of the amino function of the lysine residues in Lf, using anhydrides of succinic acid or *cis*-aconitic acid, negatively charged Lf derivatives were obtained that all showed a strong antiviral activity against the HIV-1 *in vitro*. Acylated-Lf exhibited a 4-fold stronger antiviral effect on HIV-1 than the parent compound but the activity on HCMV was abolished.

Peptide scanning studies indicated that the native Lf as well as acylated Lf strongly bind to the V3 domain of the HIV envelope protein gp120, with K_d values in the same concentration range as the *in vitro* IC_{50}. Therefore, shielding of this domain, resulting in inhibition of the virus-cell fusion and entry of the virus in MT4 cells is the likely mechanism underlying the anti-HIV activity.

In contrast, addition of positive charges to Lf through amination of the proteins resulted in an increased anti-HCMV activity and a loss of anti-HIV activity, with anti-HCMV IC_{50} values in the low micromolar concentration range. The N-terminal portion of Lf appeared essential to this anti-HCMV effect. The specific distribution of positively and negatively charged domains in the molecule appears to be important in both the anti-HIV and anti-HCMV effects.

Advances in Lactoferrin Research, edited by Spik *et al.*
Plenum Press, New York, 1998.

2. INTRODUCTION

Human cytomegalovirus belongs to the *Herpesviridae* and remains persistently or latently present in the human host after initial infection[1,2]. In immunosuppressed subjects like organ graft recipients and AIDS patients, virus replication often resumes, which may lead to high morbidity and mortality rates. In AIDS patients HCMV is suggested to act synergistically with HIV on the progression of the disease[3,4]. Between 90 and 100% of all AIDS patients are HCMV seropositive[5]. At present, active HCMV infections in immunocompromised patients are treated intravenously with ganciclovir or with foscarnet[6]. Both drugs inhibit the replication of HCMV in infected cells. However, foscarnet is nephrotoxic[6], and although ganciclovir is the generally preferred drug, it shows bone marrow toxicity[6]. In AIDS patients under ganciclovir maintenance therapy HCMV may develop resistance to the drug. It has been attempted to inhibit HCMV replication by blocking the transcription of immediate early genes with modified complementary nucleic acids like phosphorothioate oligonucleotides[7] or antisense RNAs[8]. A host of other potential antiviral compounds has been reported, however, many of these are unlikely to be applicable in a clinical setting[9].

The above mentioned drawbacks may be overcome by site specific delivery of the particular drug. In this concept, therapeutic agents are covalently attached to or included in macromolecular carriers that are selectively recognized and taken up by the target tissue. We have previously reported that negatively charged albumins (NCAs) inhibit HIV-1 infections and can be used as intrinsically active carriers for AZT like compounds in the treatment of HIV[10,11]. This paper describes the antiviral activity of native lactoferrin against HIV and HCMV and the potentials of Lf as a drug carrier in the treatment of these viral infections.

3. MATERIALS AND METHODS

3.1. Chemicals

Human serum albumin (HSA) was obtained from the Central Laboratory of the Netherlands Red Cross Blood Transfusion Service (Amsterdam, The Netherlands). Bovine milk lactoferrin (bLf) was obtained from DMV (Leeuwarden, The Netherlands). Cis-aconitic anhydride and succinic anhydride were obtained from Janssen Chimica (Beerse, Belgium). Human lactoferrin (hLf) and all other chemicals (analytical reagent grade) were purchased from Sigma (St. Louis, MO, USA).

3.2. Preparation of Succinic- and Cis-Aconitic Treated Proteins

Derivatization of the proteins was done as follows: 10 mg of the protein was dissolved in 10 ml 0.2 M K_2HPO_4 pH 8.0. Solid succinic anhydride or cis-aconitic anhydride (10 mg), was added and the solution was stirred until all anhydride was dissolved. The pH was kept at 8–8.5 using 3 M NaOH. The modified proteins were purified and characterized as described[12].

3.3. Preparation of Cationic Proteins

Proteins were derivatized by the modified method of Purtell *et al.*[13] The modified proteins were purified and characterized as described[12].

3.4. Radiolabeling of Proteins

The (modified) proteins were radiolabeled with ^{125}I by using the chloramine-T method.

3.5. Anti-HIV Screening

3.5.1. Cells and Virus. The HLTV-I infected T4 lymphocyte line, MT-4[14] was used for the anti-HIV-1 assay. The cells were grown in RPMI-1640 medium supplemented with 10% (v/v) heat inactivated fetal calf serum and 20 μg.ml^{-1} gentamicin and maintained at 37°C in a humidified atmosphere of 5% CO_2 in air. Every 3–4 days, cells were seeded at 3.10^5 cells.ml^{-1}. HIV-1 (strain IIIb) was obtained from the culture supernatant of persistently virus infected HUT-78 cells. The virus titer of the supernatant was determined in MT-4 cells. The virus stock was stored at -70°C until used.

3.5.2. Antiviral Assay. Anti-HIV-1 activity of the test compounds was assessed by measuring the inhibitory effect on virus-induced cytopathicity in MT-4 cells and was monitored by the 3-(4,5-dimethyl thiazol-2-yl)-2,5 diphenyl-tetrazolium bromide (MTT) method. The cytotoxicity was also monitored by the MTT assay[15].

3.5.3. HIV Peptides. V3 loop sequences and linear peptide sequences from the V2 and V3 loop of the HIV-1 envelope glycoprotein gp120 used in this study are presented in Table 1. V3-Q17 and V3-160Bal represent the whole circular V3 domain derived from a T-cell tropic SI and a monocytotropic NSI HIV-1 isolate respectively. Both V3 domains were obtained from Cambridge Research Biochemicals (Northchurch Cheshire, UK). Linear V3-Q17 was obtained after reduction of the disulfide bound with dithiothreitol. The reactivity of the SH function was inactivated with iodoacetamide according to Pless and Lennartz[16].

Linear peptides were synthesized using a Milligen 9050 peptide synthesizer according to the procedures provided by the manufacturer using Fmoc-protected activated amino acids (Millipore Corp. Milford, USA). Peptides were purified by liquid chromatography on Sephadex G-15 in 5% acetic acid (Pharmacia). The amino acid composition analysis of

Table 1. Sequences and charges at pH 8 of different gp120 peptides

HIV-peptide	Sequence	Charge pH 8.0
V2 (n°29) (NSI)	KLDVVPIDNDNTSYRLISC	−2.38
V2 (n°30) (sNSI)	KLDVVPIDNTIDNTSYRLISC	−2.38
V2 (n°31) (SI)	KLDVVPIDKDNDNTSYRLISC	−2.39
V3 (Q17) circular (SI)	CTRPNNNTRKRIHIGPGRAFYTTGQIIGN.IRQAHC	+4.38
V3 (160Bal) circular (NSI)	CTRPNNNTRKSIHMGPGRAFYATGQIIGD.IRQAHC	+2.38
V3 (Q17) linear (SI)	CTRPNNNTRKRIHIGPGRAFYTTGQIIGN.IRQAHC	+4.38
V3 (Q17) thrombin digestion	CTRPNNNTRKRIHIGP GRAFYTTGQIIGN.IRQAHC	+4.38
V3 (Q17 n°4)	CTTRPNNNTRK	+1.68
V3 (Q17 n°8)	RKRIHIGPGK	+2.91
V3 (Q17 n°12)	GRAFYTTGQK	−0.03
V3 (Q17 n°14)	YTTGQIIGNK	−0.08
V3 (Q17 n°17)	KIGNIRQAHC	+0.63

the peptides used were in accordance with the sequence claimed (see Table 1 for sequences). The charge of the peptides at pH 8 was computed using the CHARGEPRO routine from PCGENE (Intelligenetics).

3.5.4. Coupling of gp120-Peptides to CNBr-Activated Sepharose. Lyophilized peptides (1 mg) were dissolved in coupling buffer containing 0.1 M $NaHCO_3$ and 0.5 M NaCl, pH 8.3. 50 mg of CNBr-activated Sepharose (Pharmacia) was added and the mixture was incubated overnight at 4°C. After centrifugation the supernatant was decanted and the remaining binding sites were blocked by overnight incubation at 4°C with 0.2 M glycine pH 8.0. Non-covalently bound peptide was removed by a repeated washing procedure using a phosphate buffer with a low and high pH. The peptide-Sepharose was suspended in PBS 0.01% NaN_3 at a final peptide concentration of 40 μg peptide/2mg Sepharose/ml.

3.5.5. HIV Peptide Binding Studies. Several dilutions of the peptide-Sepharose were prepared in assay buffer containing 50 mM Tris.HCl, 100 mM NaCl, 0.2% Tween-20 and 1% BSA pH 8.0 and were incubated with the different proteins (^{125}I-labeled) for 4 hours at room temperature. Thereafter, this the peptide-Sepharose was washed 5 times with PBS containing 0.2% Tween and 1% BSA respectively. The bound protein was measured by counting gamma radiation. Radioactivity was recalculated as the percentage of added label bound to the solid-phase peptide.

3.5.6. Inhibition Studies in HIV Peptide Binding of Lf. The specificity of the binding of native and negatively charged milk proteins to HIV-peptides coupled to Sepharose were studied by competitive inhibition experiments. Iodinated native or charge-modified milk proteins were diluted with various concentrations of the non-radiolabeled inhibitor Aco-HSA. Low peptide-Sepharose suspensions were used in order to guarantee that binding was limited by peptide/protein interaction and that equilibrium conditions were achieved. The remaining procedure was performed as described above.

3.6. Anti-HCMV Screening

3.6.1. Cells and Virus. Human foetal lung fibroblasts (FLFs), between passage 7 and 17, were used as target cells for infection with HCMV strain AD169. Fibroblasts were maintained in Dulbecco's modified Eagle's medium (Gibco-BRL, Gaithersburg, MD, USA) supplemented with 10% fetal calf serum (Gibco) and 60 $\mu g.ml^{-1}$ gentamicin sulphate (Gibco) at 37°C, 100% humidity and 5% CO_2. Infective virus was prepared by infecting subconfluently growing FLFs in 150 cm^2 tissue culture flasks (Costar, Cambridge, UK) at MOI=0.1. Infection was allowed to proceed until the cells showed a maximal cytopathic effect and started lysing; this was usually approximately 10 days post infection. These cells were detached from the flasks and centrifuged together with the medium. The resulting supernatant was frozen in liquid nitrogen and stored at –80°C and used for the infections in the experiments described below.

3.6.2. Anti-HCMV Testing of Antivirals. The anti-HCMV screening was done as described[17]. In brief: antiviral proteins were dissolved in PBS at 2 $mg.ml^{-1}$ immediately prior to the experiments. One day prior to infection with HCMV, subconfluent growing FLFs were detached using trypsin, suspended in culture medium at 100,000 FLFs per ml and seeded into a flat bottom 96 well tissue culture plate (Costar) at approximately 10,000 cells per well. The next day the medium was refreshed and antiviral protein to be tested was added in 50 μl medium. Initially, antivirals were tested at 8 concentrations in twofold concentration dilution series starting at 250 $\mu g.ml^{-1}$ (final concentration after addition of virus). After 30 min incubation 50 μl HCMV diluted in medium was added at MOI=1 and infection was allowed to proceed for 72 h.

4. RESULTS

4.1. Antiviral Activity of Lf

Lf was the only native milk protein capable of inhibiting HIV-1 infection of MT4 cells. The antiviral activity showed minor differences for Lfs obtained from other animal or human sources. IC_{50} values ranged from 40 to 118 $mg.ml^{-1}$ (500–1500 nM). A fourfold increase in inhibitory effect on HIV-1 replication was found after derivatization of Lf with succinic anhydride, producing a negatively charged protein, while an even more pronounced anti-HIV-1 effect was found when bLf was derivatized by cis-aconitic anhydride. Removing of the terminal sialic acids of bLf and hLf, resulting in asialo derivatives had no significant effect on the antiviral activity. In contrast to this, complete pertubation of the secondary structure, by treatment of Lf with dithiothreitol, completely destroyed its antiviral effect. Cationization of bLf resulted in a 5 fold increase of the activity against HCMV, whereas this procedure abolished the activity against HIV-1 (Table 2).

4.2. Binding to gp120 Derived Peptides

The binding of some Lfs and charge modified albumins to the V3 loop of a T cell tropic syncytium inducing HIV-1 isolate (V3-Q17) was studied using a constant amount of protein and a variable amount of V3 loop-immobulized to Sepharose. In recent studies we validated the peptide/Sepharose binding[12]. Native (HSA) showed no interaction with this test system (Fig. 1). Therefore, HSA was used as a negative control. Negatively charged al-

Table 2. *In vitro* 50% inhibitory concentrations of native and modified proteins

	HIV-1		HCMV	
Neo(glyco)protein	IC_{50} ($\mu g.ml^{-1}$)	CC_{50} ($\mu g.ml^{-1}$)	IC_{50} ($\mu g.ml^{-1}$)	CC_{50} ($\mu g.ml^{-1}$)
dextran sulfate 5K	0.04 ± 0.01	> 250	5.5 ± 1.9	> 250
poly-L-lysine	> 250	> 250	6.0 ± 2.0	233
HSA	> 250	> 250	> 250	> 250
Suc-HSA	0.2 ± 0.2	> 250	> 250	> 250
Aco-HSA	0.04 ± 0.01	> 250	> 250	> 250
Cat-HSA	> 250	> 250	> 250	> 250
bLf milk	39.6 ± 21.8	> 250	36.7 ± 7.5	> 250
bLf colostrum	nd	nd	40.8 ± 13.8	> 250
hLf milk	75.2 ± 52.7	> 250	90.2 ± 38.0	> 250
hLf serum	> 62.5	> 62.5	108.6 ± 21.3	> 250
Suc-bLf	11.8 ± 7.6	> 250	> 250	> 250
Aco-bLf	0.17 ± 0.02	> 250	> 250	> 250
Asialo-bLf	46.0 ± 23.0	> 250	36.7 ± 9.2	> 250
Asialo-hLf	nd	nd	110.9 ± 40.8	> 250
Sulfitolyzed-bLf	> 125	> 125	> 250	> 250
Cat-bLf	> 250	> 250	7.1 ± 1.2	> 250

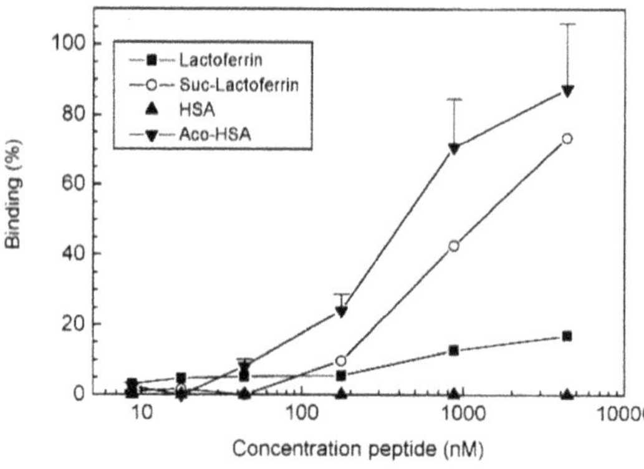

Figure 1. Binding of native and charge modified Lf to the peptide V3-Q17 bound to Sepharose. Serial dilutions of V3 peptide were incubated with ^{125}I-labelled protein.

bumins (NCAs), like Aco-HSA showed a concentration dependent binding of the peptides and therefore we used this modified protein as a positive control and as a potential inhibitor.

Suc-Lf showed a concentration-dependent binding to some HIV-peptides. The binding affinity of this protein was in the same range and as that for the binding of Suc-HSA or Aco-HSA[12]. The concentration of radiolabeled negatively charged protein at which 50% of the peptide was bound (P_{50}-value) was used as parameter indicating the relative affinity.

Interestingly, native Lf showed a weak interaction with the V3 loops V3-Q17 and V3-Bal160. These peptides differed in amino acid sequence and total net charge. A P_{50}-value could not be determined under the experimental conditions used. V2 derived peptides of different HIV-1 isolates showed hardly any binding with Lf. Linearization of V3-Q17 by an irreversible cleavage of the disulfide bond showed a two fold increase in the extent of binding of Lf, whereas cleavage of the peptide in the GPGRAF domain by thrombin resulted in a total loss of Lf/peptide interaction. Based on previous observed interactions between Suc-HSA and some net positively charged linear decapeptides from V3-Q17, we determined the binding affinity between the positively charged Lf and some peptides. Hardly any binding (less that 10%) was found with the net negatively charged peptides (n° 12 and 14), whereas a relatively strong and reversible binding (> 10%) was observed for the net positively charged peptides 4, 8 and 17.

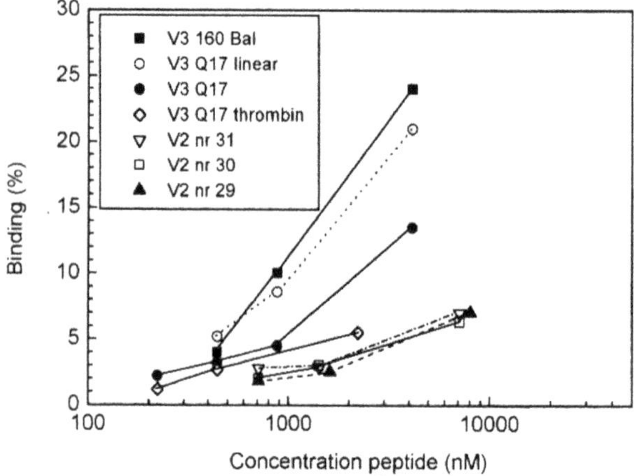

Figure 2. Binding of ^{125}I labeled Lf to different V2 loop peptides, V3 loop peptides as well as peptide fragments of the V3 loop, covalently attached to Sepharose.

4.3. Inhibition Studies

The specificity of the binding of iodinated Lf and Suc-Lf was studied using the protein itself and Aco-HSA as potential inhibitors. For Lf a 1:1 dilution of V3-Q17 bound Sepharose was used, whilst the affinity studies of the NCAs were performed at the P_{50} value, which corresponded to a 1:20 dilution of the peptides/Sepharose solution. All three compounds reversibly bound to the peptide-Sepharose.

The binding characteristics of Aco-HSA to V3-Q17 adapted Sepharose were described by a B_{max} of 4.6 pmol per mg Sepharose and a K_d of 24 nM. Using these data we calculated the K_d's values for the (modified) Lfs. The negatively charged Lf was found to be strong displacer with affinity constants was (K_d = 50 nM) a value that is that was comparable to the affinity constant of Suc-HSA. Lf itself was found to be a relatively weak displacer with a K_d value of 5200 nM.

5. DISCUSSION

In search for new antivirals against the human immunodeficiency virus, we investigated potential macromolecular vectors for cell specific delivery of antiviral drugs. In this framework we discovered protein carriers with an intrinsic antiviral activity. Charge modified plasma proteins with extra negative charge from various animal sources as well as from human origin turned out to be potent anti-HIV-1/2 inhibitors with in vitro IC_{50} values in the low nanomolar concentration range[18]. The anti-HIV effects correlated positively with the number of negatively charges introduced in the various polypeptides.

We recently detected high affinity binding of the negatively charged albumins (NCAs) to the V3 domain of gp120 preventing proteolytic cleavage of the envelope protein that is necessary for exposition of gp41 as part of the virus entry process[12]. Importantly, NCAs exhibited good efficacy against a variety of clinical HIV isolates which differ in syncytium inducing capacity and cellular tropism and showed low toxicity and immunogenicity[18,19].

During the late phase of the asymptomatic stage AIDS patients commonly start to suffer from opportunistic infections such as CMV infection. Bearing this in mind we tested the antiviral capacity of a series of native and charge- modified proteins against both HIV and HCMV. All the negatively charged proteins were active against HIV-1 but failed to inhibit HCMV infection in vitro. Interestingly, among all the native proteins tested unmodified Lf was able to inhibit both HIV-1 and HCMV infection in vitro. The IC_{50} values were about 0.5 μM for both HCMV and HIV-1[17]. In this scope Hasegawa *et al.*[20] reported an inhibitory effect of Lf obtained from bovine and human sources, against the human herpes virus type 1 at concentrations ranging from 0.5–1 mg (12 μM) of protein per ml of medium.

Lf obtained from bovine milk and colostrum, that only differ in the extent of glycosylation, exhibited equipotent effects, while Lf obtained from bovine milk was about two times more effective on HIV and HCMV replication than the human protein. The amino acid sequences of these proteins are 70% identical[21]. This difference in antiviral activity against HCMV could be explained by its minor differences in net charge of the protein.

The anti-HIV-1 activity of Lf was somehow unexpected taken into account its net positive charge as well as the positive charge of the V3 domain of HIV gp120. Nevertheless, we detected significant binding to the V3 domain of gp120. Such binding can be well explained on the basis of clustered negatively charged amino acids in the residues 210 to

240 as determined by the distribution of aspartic acid and glutamic acid[22]. This peptide sequence is part of the antiparallel backbone strands contributing to β sheets with an interruption in the form of a negatively charged loop. This loop is easily accessible and therefore may play an important role in the interaction with HIV-1.

Matthews *et al.*[23] in 1976 already reported distinct anti-viral effects of skimmed human and cow milk against arboviruses, rhinoviruses and influenza A/B viruses. However the activity was attributed to the milk IgA. Other researchers reported that undenaturated whey protein concentrates administrated to HIV infected individuals elevated their blood mononuclear cells and therefore, had an positive effect on the sense of well being of the seropositive HIV patient[24]. In relation to the latter observation, the Italian National Registry of AIDS described that breastfed HIV-1 infected children had a longer median incubation time (19 months) than bottlefed infants (9.7 months). Breastfed children also had a slower progression to AIDS[25]. Newburg *et al.* described that milk contains a factor that inhibits binding of HIV-1 to the CD4 receptor[26].

During bacterial infections, often a significant rise in plasma levels of Lf is observed[27] and plasma concentrations reach values close to the IC_{50} as was observed by us in vitro. Recently, we and others[28] also observed that plasma Lf concentrations in HIV infected individuals and AIDS patients decline with the progression of the disease. As mentioned before, opportunistic infections develop during the course of AIDS. Among others this might be facilitated by the decreased levels of Lf by which an important part of the non-specific host defense against microorganisms is lost. The protective function of Lf may gradually be reduced during progression of the disease.

Thus, our finding that Lf has a combined anti-HIV-1 and anti-HCMV activity may point to a physiological anti-viral activity of Lf in the host response against viral infections. Therefore we anticipate that Lf has a broad therapeutic value if administrated to patients with combined infections with HIV and CMV.

ACKNOWLEDGMENTS

This work is supported by grants from the Programma Coordinatie AIDS Onderzoek (PccAo, project number 95026) the Netherlands and from Nutricia Research, Zoetermeer, the Netherlands.

REFERENCES

1. Plummer G: Cytomegaloviruses of man and animals. Prog Med Virol 1973;9:302–340.
2. Weller TH: The cytomegaloviruses: ubiquitous agents with protean clinical manifestations. I. N Engl J Med 1971;285:203–214.
3. Skolnik PR, Kosloff BR, and Hirsch MS: Bidirectional interactions between human immunodeficiency virus type 1 and cytomegalovirus. J Infect Dis 1988;157:508–514.
4. Davis MG, Kenney SC, Kamine J, Pagano JS, and Huang ES: Immediate-early gene region of human cytomegalovirus trans-activates the promotor of human immunodeficiency virus. Proc Natl Acad Sci USA 1987;84:8642–8646.
5. Drew WL, Mintz L, Miner RC, Sands M, and Kettener B: Prevalence of cytomegalovirus infection in homosexual men. J Infect Dis 1981;143:188–192.
6. Bowden RA: Antivirals for cytomegalovirus. In: Multidisciplinary approach to understanding cytomegalovirus disease. MIchelson S, and Plotkin SA (Eds.). Exerpta Medica, Amsterdam, 1993,pp. 241–249.
7. Azad RF, Driver VB, Tanaka K, Crooke RM, and Anderson KP: Antiviral activity of a phosphorothioate oligonucleotide complementary to RNA of the human cytomegalovirus major immediate-early region. Antimicrob Agents Chemother 1993;37:1945–1954.

8. Bryant LA, and Sinclair JH: Inhibition of human cytomegalovirus major immediate early gene expression by antisense RNA expression vectors. J Gen Virol 1993;74:1965–1967.
9. Snoeck R, Neyts J and De Clercq E: Strategies for the treatment of cytomegalovirus infections. In: Multidisciplinary approach to understanding cytomegalovirus disease. MIchelson S, and Plotkin SA (Eds.). Excerpta Medica, Elsevier Science Publishers B.V. Amsterdam, 1993,pp. 269–278.
10. Haas M, Meijer DKF, Moolenaar F, De Jong PE and De Zeeuw D: Renal drug targeting: optimalisation of renal pharmacotherapeutics. In: International Yearbook of Nephrology 1996. Andreucci VE, and Fine LG (Eds.). Oxford University Press, 1996,pp. 3–11.
11. Mayer U, Wagenaar E, Beijnen JH, Smit JW, Meijer DKF, van Asperen J, Borst P, and Schinkel AH: Substantial excretion of digoxin via the intestinal mucosa and prevention of long-term digoxin accumulation in the brain by the mdr1a P-glycoprotein. Br J Pharmacol 1996;119:1038–1044.
12. Kuipers ME, Huisman JG, Swart PJ, de Béthune M-P, Pauwels R, De Clercq E, Schuitemaker H, and Meijer DKF: Mechanism of anti-HIV activity of negatively charged albumins: biomolecular interaction with the HIV-1 envelope protein gp120. J Acq Immun Defic Synd Hum R 1996;11:419–429.
13. Purtell JN, Pesce AJ, Clyne DH, Miller WC, and Pollak VE: Isoelectric point of albumin: Effect on renal handling of albumin. Kidney Int 1979:16:366–376.
14. Miyoshi I, Taguchi H, Kubonishi I, Yoshimoto S, Ohtsuki Y, Shiraishi Y, and Akagi T: Type C virus-producing cell lines derived from adult T cell leukemia. Gann Monogr Cancer Res 1982;28:219–228.
15. Pauwels R, Balzarini J, Baba M, Snoeck R, Schols D, Herdewijn P, Desmyter J, and De Clercq E: Rapid and automated tetrazolium-based colorimetric assay for the detection of anti-HIV compounds. J Virol Methods 1988;20:309–321.
16. Pless DP, and Lennarz WJ: Enzymatic conversion of proteins to glycoproteins. Proc Natl Acad Sci USA 1977;74:134–138.
17. Harmsen MC, Swart PJ, de Béthune M-P, Pauwels R, De Clercq E, The TH, and Meijer DKF: Antiviral effects of plasma and milk proteins: lactoferrin shows potent antiviral activity on both human immunodeficiency virus and human cytomegalovirus. J Infect Dis 1995;172:380–388.
18. Jansen RW, Schols D, Pauwels R, De Clercq E, and Meijer DKF: Novel, negatively charged, human serum albumins display potent and selective in vitro anti-human immunodeficiency virus type 1 activity. Mol Pharmacol 1993;44:1003–1007.
19. Swart PJ, and Meijer DKF: Negatively-charged albumins: A novel class of polyanionic proteins with a potent anti-HIV activity. Antiviral News 1994;2:69–70.
20. Hasegawa K, Motsuchi W, Tanaka S, and Dosako S: Inhibition with Lactoferrin of In Vitro Infection with Human Herpes Virus. Jpn J Med Sci Biol 1994;47:73–85.
21. Metz-Boutigue MH, Jollès J, Mazurier J, Schoentgen F, Legrand D, Spik G, Montreuil J, Jollès P: Human lactotransferrin: amino acid sequence and structural comparisons with other transferrins. Eur J Biochem 1984;145:659–676.
22. Day CL, Anderson BF, Tweedie JW, and Baker EN: Structure of the Recombinant N-Terminal Lobe of Human Lactoferrin at 2.0 Å Resolution. J Mol Biol 1993;232:1084–1100.
23. Matthews THJ, Lawrence MK, Nair CDG, and Tyrrell DAJ: Antiviral acitivity in milk of possible clinical importance. Lancet 1976;2:1387–1389.
24. Gold P: Method of treatment of HIV-seropositive individuals with dietary whey proteins. 1994:1–18; PCT/CA93/00107 (Application number);WO 93/20831 (Patent number).
25. Mok J: HIV-1 Infection. Breast Milk and HIV-1 Transmission. Lancet 1993;341:930–931.
26. Newburg DS, Viscidi RP, Ruff A, and Yolken RH: A human milk factor inhibits binding of human immunodeficiency virus to the CD4 receptor. Pediatr Res 1992;31:22–28.
27. Lash JA, Coates TD, Lafuze J, Baehner RL, and Boxer LA: Plasma lactoferrin reflects granulocyte activation in vivo. Blood 1983;61:885–888.
28. Boyle MJ, Connors M, Flanigan ME, Geiger SP, Ford H, Baseler M, Adelsberger J, Davey RT, and Lane HC: The human HIV/peripheral blood lymphocyte (PBL)-SCID mouse—A modified human PBL-SCID model for the study of HIV pathogenesis and therapy. J Immunol 1995;154:6612–6623.

25

A HELICAL REGION ON HUMAN LACTOFERRIN

Its Role in Antibacterial Pathogenesis

D. S. Chapple,[1-3] C. L. Joannou,[1] D. J. Mason,[3] J. K. Shergill,[1] E. W. Odell,[2] V. Gant,[3] and R. W. Evans[1]

[1]Division of Biochemistry and Molecular Biology
United Medical and Dental Schools of Guy's and St Thomas's Hospitals
Guys Hospital Medical School
London Bridge, London SE1 9RT,United Kingdom
[2]Department of Oral Medicine and Pathology
United Medical and Dental Schools of Guy's and St Thomas's Hospitals
Guys Tower
London Bridge, London SE1 9RT, United Kingdom
[3]Division of Infectious Diseases
United Medical and Dental Schools of Guy's and St Thomas's Hospitals
St Thomas's Hospital
London SE1 9EH, United Kingdom

1. SUMMARY

Human lactoferrin contains a 47 amino acid peptide, named lactoferricin H, which is thought to be responsible for its antimicrobial activity. Lactoferricin includes a loop region, which resides on the outer surface of the N-lobe of lactoferrin, adopting an alpha helix with a hydrophobic tail. Peptides have been synthesised corresponding to the highly charged alpha helix (HLP 2) and hydrophobic tail region (HLP 5). HLP 2 has potent antibacterial activity whereas HLP 5 had no activity. To investigate the relationship between structure and function of HLP 2, HLP 6 was synthesised with a proline replacing methionine. This substitution was predicted to disrupt the helical region of the peptide and the orientation of the positively charged residues. Antibacterial activity was significantly reduced when tested against *Escherichia coli* serotype 0111, NCTC 8007. The mode of action of HLP 2 against the bacterial membrane was investigated by flow cytometric analysis, using *Escherichia coli,* NCTC 8007. Membrane potential and integrity were monitored using the fluorescent probes, bis 1,3-(dibutylbarbituric acid) trimethine oxonol

Advances in Lactoferrin Research, edited by Spik *et al.*
Plenum Press, New York, 1998.

and propidium iodide respectively. HLP 2 caused complete loss of membrane potential and integrity, with irreversible damage to the cell as shown by rapid loss of viability. We conclude that HLP 2 causes membrane disruption and that helicity is an important factor for antibacterial activity.

2. INTRODUCTION

The mechanism by which lactoferrin exerts its antimicrobial activity has been the subject of much controversy[1–2,4–5,13–14]. The discovery of a potent antimicrobial peptide, released upon degradation with gastric pepsin,[2,16] indicated that the bacteriostatic/cidal activity of lactoferrin was not simply due to iron deprivation. We and others have previously shown that the liberated 47 amino acid peptide, named lactoferricin, contains a loop region which is exposed on the outer surface of the N-lobe, an area distinct from iron binding and available for bacterial binding[13]. The loop region of lactoferricin consists of a highly charged alpha helical region and hydrophobic tail, held together by a disulphide bridge. Based on studies of other naturally occurring bactericidal alpha helical peptides, such as cecropins[15] and magainins[11], we proposed that the alpha helical portion of the loop region of lactoferricin is an important factor in the killing of bacteria by lactoferrin and lactoferricin. To test this proposal, we have synthesised peptides corresponding to the loop region of lactoferricin, minus the disulphide bridge (HLP 1), the alpha helix (HLP 2), and the hydrophobic tail (HLP 5).

Incubation of Gram-negative bacteria with lactoferrin leads to the release of lipopolysaccharide, indicating a direct interaction between lactoferrin and the bacteria[8]. Other naturally occurring peptides have been shown to cause pore formation in artificial membranes[6–7], and it is suggested that these peptides cause bacterial death through membrane interactions. To investigate whether this mechanism applies to HLP 2 we have used flow cytometry to measure membrane potential and integrity of bacterial cells exposed to the peptide in the presence of selected fluorescent probes. The advantage of flow cytometry is that it is a fast and sensitive technique allowing the use of whole bacterial cells rather than artificial membranes, mutant bacteria or membrane constituents.

3. MATERIALS AND METHODS

3.1. Lactoferrin Purification, Peptide Synthesis, and Bacterial Strains

The bacterial strain used was *Escherichia coli*, serotype 0111, NCTC 8007 (Collindale, United Kingdom). Bacteria were grown overnight on Columbia blood Agar, and incubated to log phase in 1% proteose peptone. Lactoferrin was purified from human milk whey using the method of Blackberg and Hernell[3] . Peptide HLP2, corresponding to the helical charged portion of the loop region within lactoferricin (residues 20–30; NH_2-FQWQRNMRKVR-C00H) was synthesised commercially (Neosystem Laboratoire, Strasbourg). Peptides HLP 1, corresponding to the loop region within human lactoferricin (residues 20–35; NH_2-fqwqrnmrkvrgppvs-cooh), HLP 5, corresponding to the hydrophobic tail region (residues 31–35; NH_2-GPPVS-COOH) and HLP 6, corresponding to the proline substituted analogue (NH_2-FQWQRNPRKVR-COOH) were synthesised using Fmoc chemistry on a Synergy Peptide Synthesiser (Perkin Elmer, UK) and purified using high pressure liquid chromatography.

3.2. Antibacterial Activity

Escherichia coli NCTC 8007 was grown to log phase in 1% proteose peptone and incubated with lactoferrin or peptides at 37°C. After two hours, the cells were washed twice and viability assessed by serial dilution and drop counting on nutrient agar plates.

3.3. Peptide Modelling

Peptides were modelled using Silicon Graphics and Insight II software (Molecular Simulations).

3.4. Flow Cytometry

Bacteria were incubated in the presence of HLP 2 for two hours at 37°C, washed twice in fresh filtered 1% proteose peptone and propidium iodide (PI) or bis-(1.3-dibutyl-barbituric acid) [DiBAC$_4$(3)] used as fluorescent probes. DiBAC$_4$(3) was used to monitor membrane potential, and PI used to measure membrane integrity. Both dyes were added to the bacterial suspension to give a final concentration of 10 μg/ml and fluorescence monitored using a HS Bryte flow cytometer (Bio-Rad, UK)[12]].

4. RESULTS AND DISCUSSION

4.1. Antimicrobial Activity

Peptides HLP 1 (whole loop) and HLP 2 (positively charged alpha helical portion of the loop) exhibited potent anti-bacterial activity against *E. coli* 8007. In contrast, native human apolactoferrin and HLP 5 (hydrophobic tail portion of the loop region) had no activity under the same conditions (Fig. 1). HLP 6 (methionine substituted analogue of HLP 2) also had no activity against *E. coli* 8007.

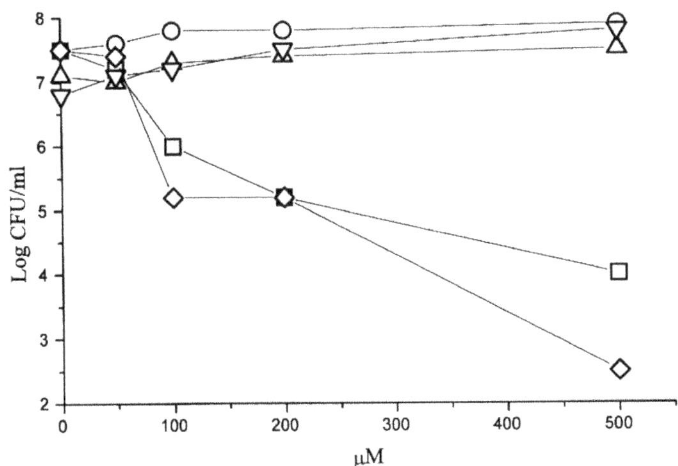

Figure 1. Antibacterial activity of lactoferrin and synthetic peptides against *E. coli* 8007. Mean viability derived from a minimum of 4 experiments; lactoferrin (△), HLP 1(◇), HLP 2 (□), HLP 5(○), and control(▽).

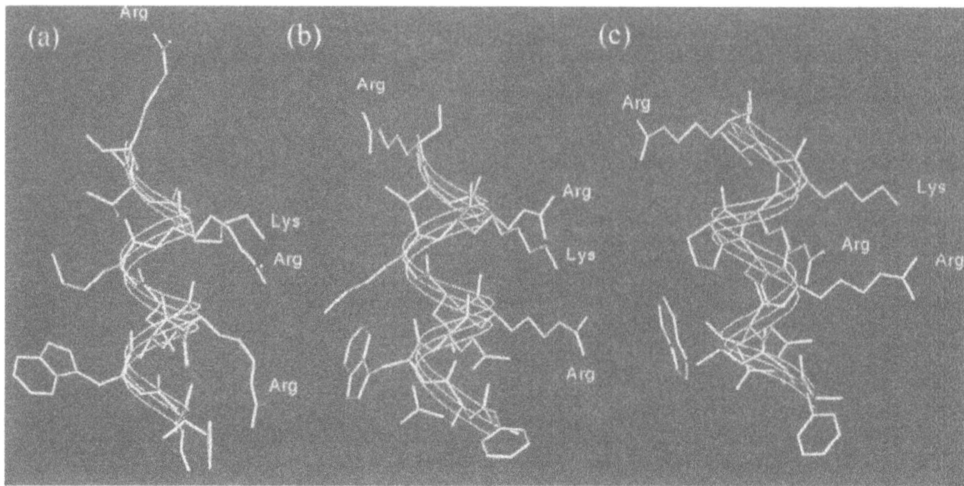

Figure 2. The helical structure of the loop region showing the charged amino acids and their orientation on lactoferricin (a). Peptides HLP 2 (b) and HLP 6 (c) are predicted structures derived using the molecular modelling program Insight II.

4.2. Peptide Modelling

A ribbon model of lactoferricin showing the helical conformation adopted in native human lactoferrin is shown in Fig. 2a. The charged arginine and lysine sidechains within the helix lie in one plane on the outer surface of the protein. The structure predicted for HLP 2 (Fig. 2b) again has these three residues lying in one plane, whereas that predicted for HLP 6 (Fig. 2c) shows that substitution of the methionine by a proline disrupts this arrangement. These observations, taken together with the results of the antimicrobial studies, indicate that orientation of the charged sidechains within the helical region of peptide is important for antibacterial activity and that even subtle changes in the orientation of the sidechains can have a dramatic effect on this activity.

4.3. Mechanism of Action

Flow cytometry enables both light scatter and fluorescence of individual bacteria to be measured simultaneously as they pass through an excited beam of light at a rate in excess of one thousand cells per second. $DiBAC_4(3)$ will only fluoresce if the membrane potential of the bacteria has collapsed and an increase in the percentage of the bacterial population which fluoresce will result in a shift in peaks[12]. PI will only fluoresce when bound to nucleic acids which are found in the cytoplasm of the bacterial cell. Although PI is a small molecule dye, it is unable to pass through the bacterial membrane, so fluorescence indicates pore formation, resulting in a collapse of membrane integrity and a shift in fluorescence[9]. Exposure of the bacteria to HLP 2 for 2 hours resulted in a shift in both $DiBAC_4(3)$ and PI fluorescence(Fig. 3). This indicates that HLP 2 interacts with the bacterial membrane resulting in a collapse of membrane potential and integrity, due to pore formation, leading to cell death.

The mechanism of action of other cationic peptides has been well characterised; they interact with negatively charged divalent cation binding sites on surface lipopolysaccharide, disrupting these sites and leading to uptake of peptide across the outer membrane.

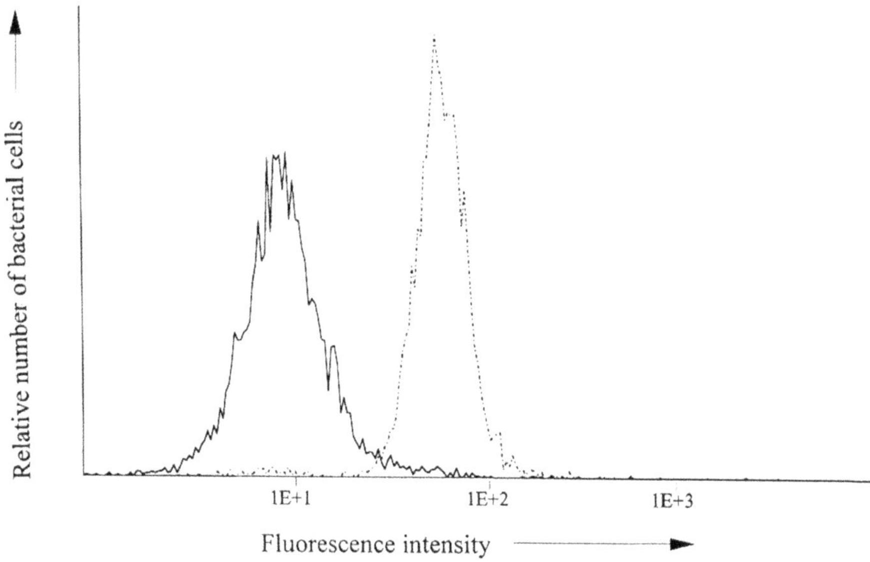

Figure 3. Comparison of HLP 2 treated bacteria (——) with a control culture of *E. coli* 8007 (-----) stained with DiBAC4(3). Data are displayed as flow cytometric histograms of 5,000 bacterial events in which the axis represent cell number on a linear scale (y axis) and associated cell fluorescence on a logarithmic scale (x axis). A similar histogram is seen when bacteria are stained with PI.

The affected membrane is thought to form channels leading to leakage of cytoplasmic molecules and cell death[10]. The results of our own studies, using flow cytometry, indicate that HLP2 exerts its antimicrobial effect by a similar mechanism. Although the relevance of these results to the antibacterial activity of intact lactoferrin remains to be determined, these studies go some way to elucidating its antimicrobial mode of action.

ACKNOWLEDGMENTS

We are grateful to the Special Trustees for St Thomas' Hospital for financial support. We wish to thank Gillian Weaver (Queen Charlotte's Hospital, London) and Ann Billington (King's College Hospital, London) for provision of human milk.

REFERENCES

1. Arnold. R.R., R.M.Cole. J.R.McGhee. 1977. A bactericidal effect for human lactoferrin. Science. 197: 263–265
2. Bellamy. W., M.Takase. K.Yamauchi, H.Wakabayashi, K.Kawase, and M.Tomita. 1992. Identification of the bactericidal domain of lactoferrin. Biochim. Biophys. Acta. 1121: 130–136
3. Blackberg. L. and Hernell. O. 1980. Isolation of Lactoferrin from human whey by a single chromatographic step. FEBS Lett. 109. 180–183.
4. Brock. J.H. 1980. Lactoferrin in human milk: its role in iron absorption and protection against enteric infection in the newborn infant. Arch. Dis.Child. 55: 417–421
5. Bullen. J.J.. H.J.Rogers, and L.Leigh. 1972. Iron-binding proteins in milk and resistance to *Escherichia coli* infection in infants. B M J. 1: 69–75

6. Christensen, B., J.Fink., R.B.Merrifield, and D.Mauzerall. 1988. Channel-forming properties of cecropins and related model compounds incorporated into planar lipid membranes. Proc. Natl. Acad. Sci. USA. 85: 5072–5076
7. Duclohier, H., G.Molle and G.Spach. 1989. Antimicrobial peptide magainin I from *Xenopus* skin forms anion-permeable channels in planar lipid bilayers. Biophys. J. 56:1017–1021
8. Ellison III, R.T., T.J.Giehl, and F.M.LaForce. 1988. Damage of the outer membrane of enteric Gram-negative bacteria by lactoferrin and transferrin. Infect.Immun. 56: 2774–2780
9. Gant, V.A., G.Warnes, I.Phillips, and G.F.Savage. 1993. The application of flow cytometry to the study of bacterial responses to antibiotics. J. Med. Microbiol. 23: 83–88
10. Hancock, R.E.W. 1997. Peptide Antibiotics (Review). Lancet. 349: 418–422
11. Marion, D., M.Zasloff, and A.Bax. 1988. A two-dimensional NMR study of the antimicrobial peptide magainin 2. FEBS Lett. 227: 21–26
12. Mason, D.J., R.Allman, J.M.Stark, and D.Lloyd. 1994. Rapid estimation of bacterial antibiotic susceptibility. J. Microsc. 176: 8–16
13. Odell, E.W., R. Sarra, M. Foxworthy, D.S.Chapple, R.W.Evans. 1996. Antibacterial activity of peptides homologous to a loop region in human lactoferrin. FEBS Lett. 382: 175–178
14. Oram, J.D., and B.Reiter. 1968. Inhibition of bacteria by lactoferrin and other iron-chelating agents. Biochim. Biophys. Acta. 170: 351–365
15. Varra M & T.Varra. Ability of cecropin B to penetrate the enterobacterial outer wall. Antimicrob. Agent Chemother. 38(10): 2498–501
16. Yamauchi, K., M.Tomita, T.J.Giehl, and R.T.Ellison III. 1993. Antibacterial activity of lactoferrin and a pepsin-derived lactoferrin peptide fragment. Infect. Immun. 61(2): 719–728.

INTERACTION OF LACTOFERRIN WITH *Micrococcus spp.* AND ITS ROLE IN ANTIMICROBIAL ACTIVITY

A. de Lillo,[1,3] R. Cernuda,[2] and J. H. Brock[3]

[1]Department of Functional Biology (Microbiology)
[2]Department of Cellular Biology and Morphology
Medical School
University of Oviedo, Spain
[3]Department of Immunology
Western Infirmary
Glasgow G11 6NT, Scotland, United Kingdom

1. INTRODUCTION

The host non-specific defence mechanisms are of major importance in those areas first encountered by invading microorganisms and influence the initial success of bacterial colonization of the host, especially in mucosal areas. Secretions bathing the mucosal surfaces contain several non-specific antibacterial components such as lysozyme, lactoperoxidase and lactoferrin (McNab and Tomasi, 1981) and specific ones such as secretory immunoglobulin A (sIgA) that control the growth of pathogens or commensal microbiota on mucosal surfaces.

Lactoferrin contributes to non-specific immunity by sequestering the available iron and limiting microbial growth (Reiter, 1983), and by a direct bactericidal activity which is independent of iron binding and involves the N-terminal cationic peptide of the molecule (Bellamy *et al.*, 1992). Furthermore, Tomita *et al.* (1991) found an increased antibacterial activity of both bovine and human lactoferrin after enzymatic digestion with pepsin. How lactoferrin exerts its bactericidal effect is still unclear but may involve blockage of microbial carbohydrate metabolism (Arnold *et al.*, 1982) or destabilization of the bacterial cell envelope (Ellison *et al.*, 1990).

The bactericidal action of lactoferrin implies binding to the cell surface (Dalmastri *et al.*, 1988). Lactoferrin can bind to various bacteria including *Escherichia coli*, *Aeromonas hydrophila* and *Shigella flexneri* (Naidu and Arnold, 1997).

Advances in Lactoferrin Research, edited by Spik *et al.*
Plenum Press, New York, 1998.

In Gram-positive bacteria, lactoferrin, like defensins and other cationic components from neutrophils, can increase membrane permeability making the cell more susceptible to complement and lysozyme attack.

To provide further insight into the mode of action lactoferrin against Gram-positive bacteria, we have studied the interaction of lactoferrin with four *Micrococcus spp.* and related this to the susceptibility to killing by lactoferrin.

2. MATERIALS AND METHODS

2.1. Lactoferrin

Human-apolactoferrin (apoLf) (Sigma) was checked for purity by FPLC and SDS-PAGE, and labelled with ^{125}I by the chloramine-T method (Greenwood *et al.*, 1963). For some experiments lactoferrin fragments with antibacterial activity, generated by the heat-treatment method of Saito *et al.* (1991), were used.

2.2. Bactericidal Assays

Mid log phase *Micrococcus luteus* ATCC 4698, *Micrococcus radiophilus* NCTC 27603, *Micrococcus roseus* NCTC 186, *Micrococcus varians* NCTC 15306 were washed three times in sterile saline (pH 7.9), containing 100 μM of the iron-chelator ethylen-diamine di-*0*-hydroxyphenylacetic acid (EDDA), once in saline (pH 7.4), suspended at 10^8 c.f.u./ml in sterile saline and incubated for 4h with or without lactoferrin. Killing and lysis were monitored at intervals by plate counting and absorption at 660 nm.

2.3. Lactoferrin Binding Assay

M. luteus and *M. varians* (100 μl of A_{660} = 0.9–1.0) were incubated with 100 μl of ^{125}I-Lf (0.2–8 μg) in PBS pH 7.4, for 1h at 37°C, washed three times with ice-cold PBS containing 0.1% of Tween 20 and radioactivity in the pellet measured. Non-specific binding was estimated by incubating the cells with a 50 fold excess of unlabelled lactoferrin.

2.4. Interaction of Lipomannan from *M. luteus* with Lactoferrin

Lipomannan (LM) from *M. luteus* was prepared by the hot phenol-water method (Westphal and Jann, 1965), and purified by hydrophobic interaction chromatography using essentially the method described by Fisher (1991). Interaction with lactoferrin was detected by the chromatographic method of Ohno and Morrison (1989).

2.5. Transmission Electron Microscopy (TEM) and Immunocytochemistry

Bacteria were embedded in Araldite 502, sectioned at 70 nm and stained with aqueous solutions of uranyl acetate and lead citrate. TEM was performed on a Jeol 2000 EX II electron microscope.

For the immunolocalization of lactoferrin associated with microorganisms, cells were embedded in LR White resin. Ultrafine sections were incubated with 1% BSA in PBS, pH 7.2 at room temperature for 3h followed by incubation with goat anti-human lactoferrin (1:500–1:1000; Immunoresearch Labs, Avondale, USA) in PBS-BSA at 4°C for 24 h. After washing, sections were incubated with 1:40 rabbit anti-IgG conjugated with colloidal gold and finally stained with 2% uranyl acetate.

2.6. Trypsin Treatment of Lf-LM Complexes

^{125}I-Lf (0.4 μg) was incubated with varying lipomannan concentrations in 36 μl of 50 mM Tris-HCl pH 7.0 at 37°C for 1h. Trypsin-EDTA 10x solution (Sigma) was added to give a final concentration of 0.5 mg/ml trypsin/0.2 mg/ml EDTA, and incubation continued for another 2½h. Samples were analyzed by SDS-PAGE and autoradiographed.

3. RESULTS

3.1. Bactericidal Effect of Human Lactoferrin on *Micrococcus spp.*

Apo-lactoferrin had a dose dependent bactericidal effect on *M. luteus* cells as shown by a decrease in the viability of the bacterial suspension, while iron-saturated lactoferrin had no effect. This effect was irreversible and needed the continuous presence of free lactoferrin in the medium since changing the cells to new medium, without lactoferrin after 45 min stabilized the number of viable bacteria (Figure 1a).

Lactoferrin decreased the absorbance of the cultures (Figure 1b) indicating that cell killing was accompanied by lysis. This was confirmed by TEM studies, which showed that lactoferrin caused a progressive change in cell morphology, affecting the cell wall as well as the division septum and the membrane, culminating in cell lysis (Figure 2: micrographs 1,2,3).

The other three species were resistant to the lytic action of lactoferrin, but susceptible to a peptide mixture generated by acid hydrolysis of the whole molecule (Figure 3). The effect of lactoferrin hydrolyzate on *M. luteus,* was greater than that of the intact molecule after 2½ h.

3.2. Lactoferrin Binding to *M. luteus* and *M. varians*

There were clear differences in the affinity of lactoferrin binding to *M. luteus* and *M. varians.* Specific binding of ^{125}I-Lf to *M. luteus* reached saturation at 0.34 μM (Figure 4a).

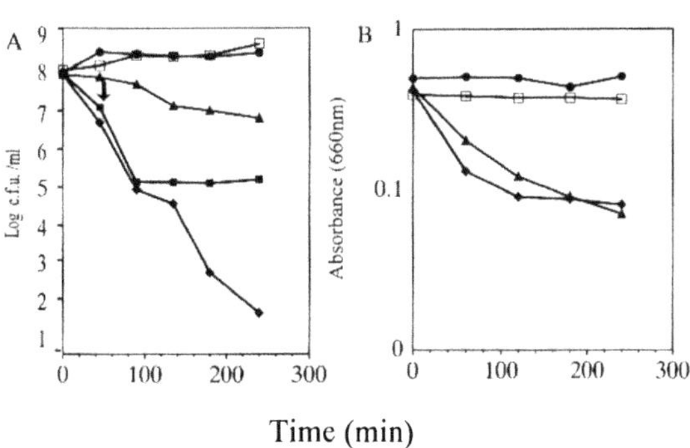

Figure 1. Antimicrobial effect of human lactoferrin on *M. luteus* viability. (A) Dose-dependent bactericidal effect. 10^8 c.f.u./ml were incubated in sterile saline pH 7.4 at 37°C for 4 h in the presence of (□) control without Lf, (■) 20 μM Fe_2Lf, (▲) 15 μM apoLf, (◆) 20 μM apoLf, (●) 20 μM apoLf, removed at 45 min (arrow). (B) A_{660} of the suspensions during the experiment.

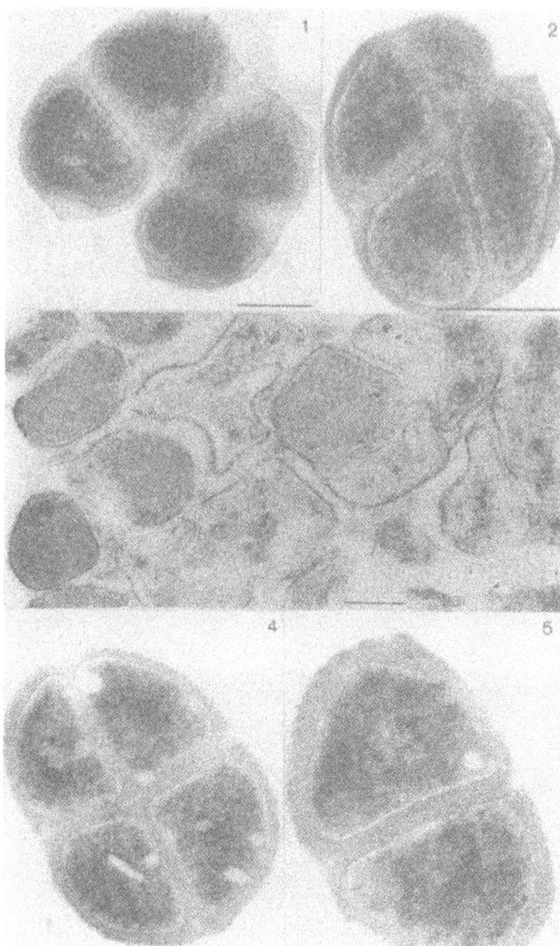

Figure 2. Effect of lactoferrin on *M. luteus* and *M. varians* morphology. Micrograph 1: *M. luteus* cells in saline; 2: *M. luteus* cells incubated with 20 μM of Fe_2Lf; 3: *M. luteus* cells incubated with 20 μM apoLf; 4: *M. varians* cells incubated in saline; 5: *M. varians* cells incubated with 20 μM apoLf. All observations were made after 4 h of incubation with lactoferrin.

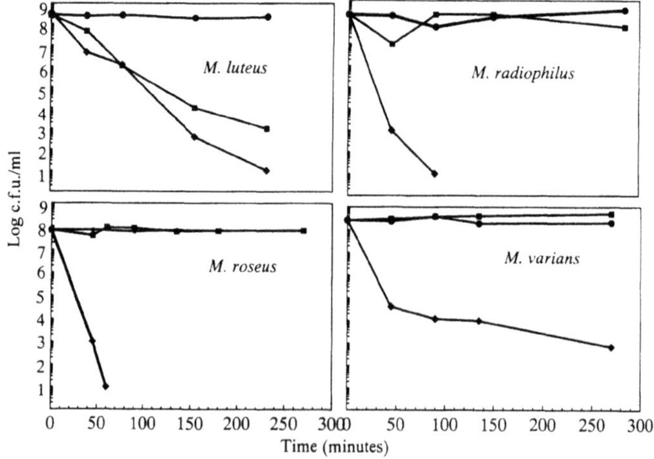

Figure 3. Effect of heat-treated lactoferrin on *Micrococcus spp.* Bacteria were incubated with heat-treated lactoferrin or the intact molecule. (●) Control without lactoferrin; (■) 20 μM untreated lactoferrin; (◆) 250 μg/ml lactoferrin hydrolyzate.

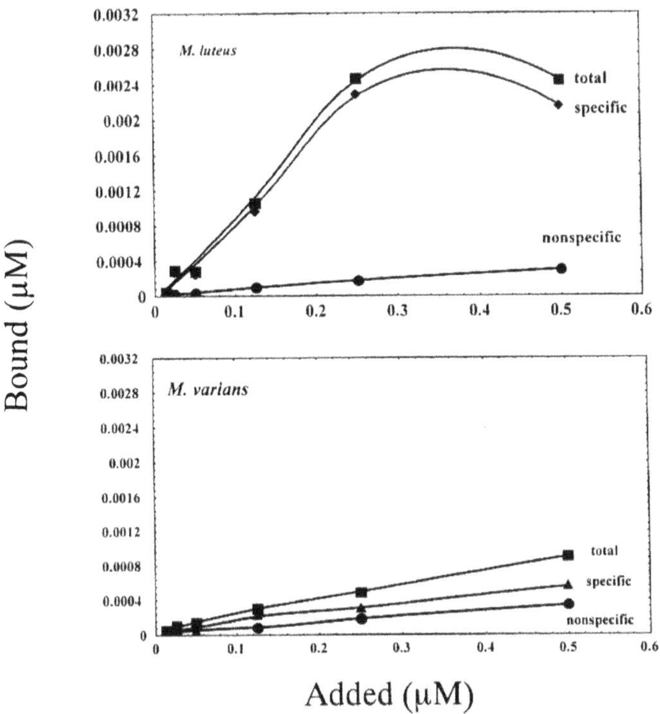

Figure 4. Binding of ^{125}I-Lf to *M. luteus* and *M. varians* cells. Binding saturation curves are shown. Non-specific binding was estimated by incubation with 50 fold excess of unlabelled lactoferrin.

Binding to *M. varians* cells seemed to be non-specific since it did not reach saturation even at high concentrations of labelled lactoferrin (Figure 4b).

3.3. Interaction of Lactoferrin with Lm from *M. luteus*

In order to study whether lipomannan could be the first binding site for lactoferrin on *M. luteus*, advantage was taken of the significant differences in molecular weight of these two molecules. As shown in Figure 5, when lipomannan was chromatographed on

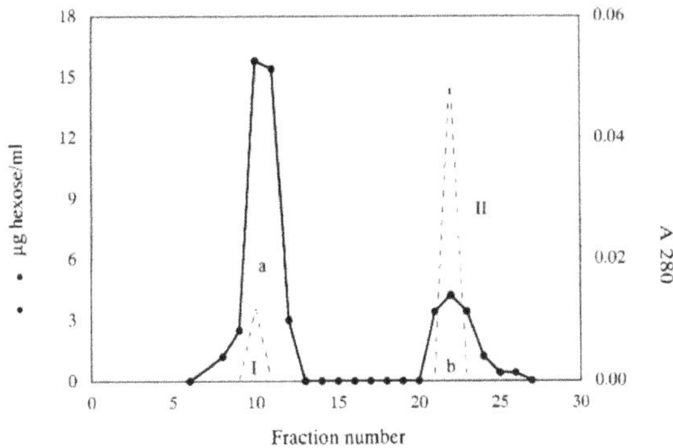

Figure 5. Gel permeation chromatography of lipomannan and lactoferrin on Superose 12. Lactoferrin was detected by A_{280} and normally eluted in fraction 21. Lipomannan was detected by hexose analysis and eluted in fraction 10. The elution peak profile of a preincubated mixture of apoLf and LM was as follows: peak I: A_{280} of apoLf-LM complex; peak II: A_{280} of free lactoferrin: peak 1: hexose content of LM and apoLf in the complex: peak 2: hexose due to free lactoferrin.

Superose 12, it eluted in the void volume (Vo). Apo-lactoferrin eluted in the included volume (Vi) in fraction 21 corresponding to a molecular mass of 80 kDa. When a preincubated mixture of both molecules was chromatographed, there was a significant shift in the 280 nm Lf peak from the included volume to the void volume (fraction 10) indicating lactoferrin-lipomannan complex formation.

3.4. Immunolocalization of Lactoferrin

Immunogold staining showed the existence of an antigenic determinant reacting with anti-lactoferrin antibody inside *M. luteus* cells following incubation with apoLf but not Fe_2Lf (Figure 6). *M. varians* gave negative result with apoLf.

3.5. Effect of Trypsin on Lm-Lf Complexes

When lactoferrin-lipomannan complexes were digested with trypsin, a different fragment pattern was produced compared with digestion of lactoferrin alone (Figure 7). A band of 5.4 kDa progressively replaced a smaller one with increasing lipomannan concentrations.

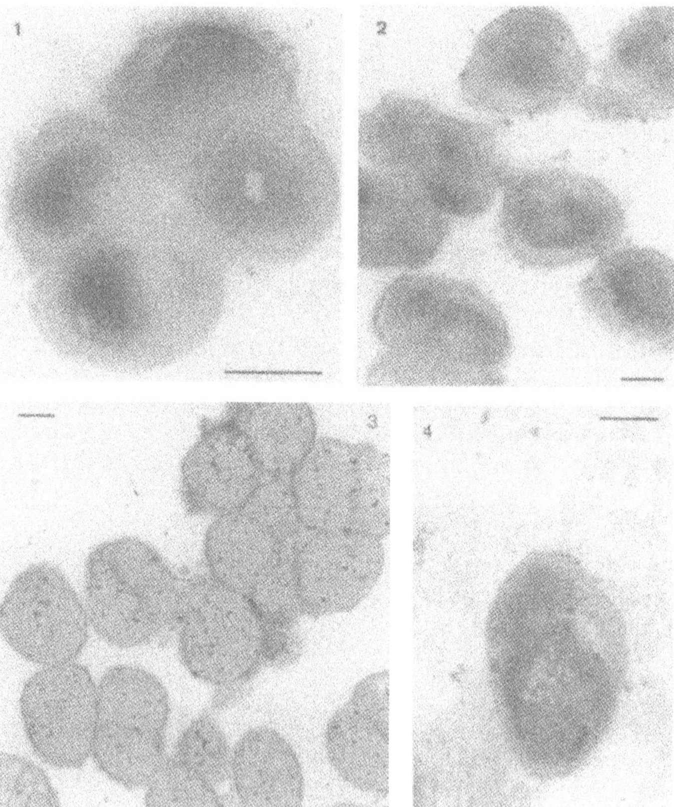

Figure 6. TEM immunolocalization of lactoferrin associated with *M. luteus* and *M. varians* cells. Micrograph 1: control, *M. luteus* in sterile saline; 2: *M. luteus* with 20 μM Fe_2Lf; 3: *M. luteus* with 20 μM apoLf; 4: *M. varians* with apoLf. Bar indicates 0.5 μm. All samples were incubated for 4 h.

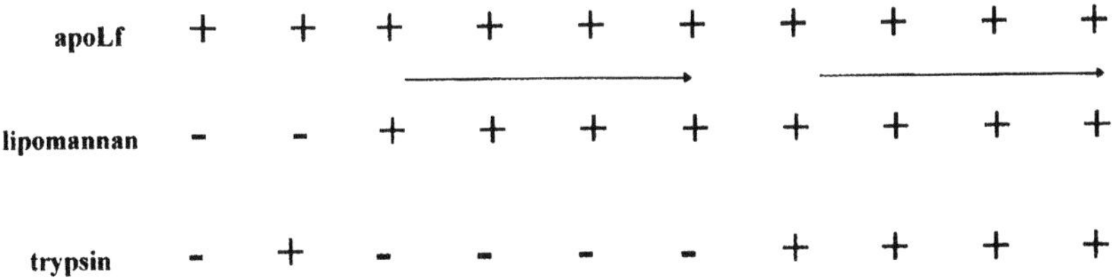

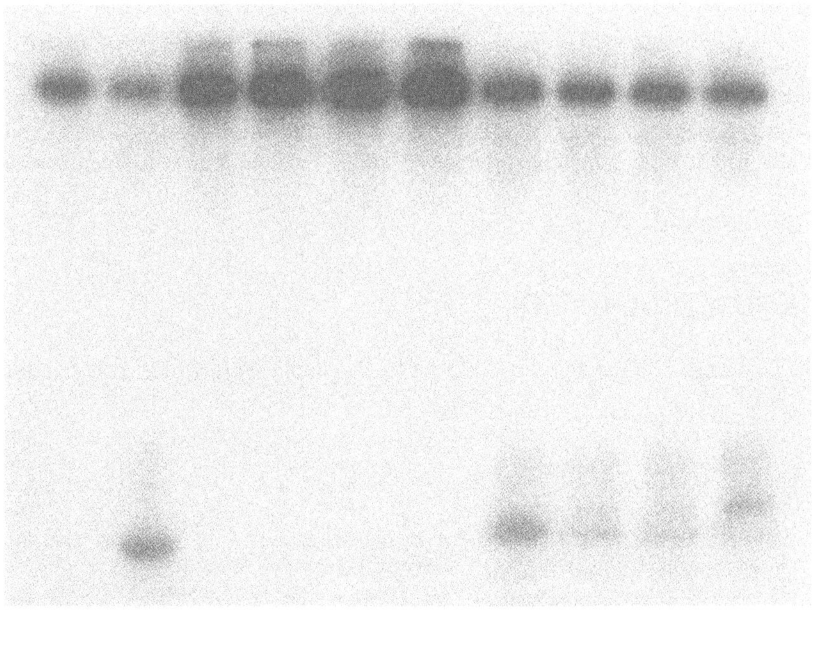

1 2 3 4 5 6 7 8 9 10

Figure 7. Autoradiograph of SDS-PAGE of trypsin-treated Lf-Lm complexes. Lane 1: lactoferrin incubated in buffer; lane 2: trypsin-digested lactoferrin; lanes 3 to 6: Lf incubated with increasing concentrations of LM; lanes 7 to 10: incubation of Lf with increasing concentrations of LM for 1h followed by trypsin digestion.

4. DISCUSSION

Using *Micrococcus spp* as a model for the action of lactoferrin on Gram-positive bacteria, we have demonstrated that lactoferrin, in its iron-free state, is bacteriolytic against *M. luteus*, while the other three species of the same genus *M. radiophilus*, *M. roseus* and *M. varians* were resistant to the effect of lactoferrin. However, these three resistant species were sensitive to a lactoferrin hydrolyzate, which also had a more potent effect on *M. luteus* compared to the intact molecule. This indicates that the target site for lactoferrin lytic action was present in all species tested and that differential accessibility to the cell surface and the site of action could explain the differences in susceptibility between these microorganisms of the same genus.

Since binding is the first step for the bactericidal action, we looked for differences in binding affinities of lactoferrin for *M. luteus* and *M. varians*. The results suggested that there was a correlation between binding and susceptibility to lactoferrin action, since binding to *M. luteus* was greater, saturable and specific, while binding to *M. varians* seemed to be lower and non-specific. Different cell wall structural components may be responsible for the variations in lactoferrin binding and susceptibility, as the sensitive *M. luteus* contains lipomannan (Powell *et al.*, 1975) which was found to bind lactoferrin, while the other three species contained lipoteichoic acids.

Immunogold techniques demonstrated that a lactoferrin antigenic determinant can enter *M. luteus* cells. Since lactoferrin has a molecular mass of 80 kDa it is unlikely to penetrate through the cell wall and membrane unless in a fragmented form. This might involve a cationic fragment of the lactoferricin type, produced by microbial proteases which we have detected in *M. luteus* membranes (unpublished data). Since interaction with lipomannan was found to alter the proteolytic fragment pattern of lactoferrin, this interaction could be an important factor in this mechanism. More work including identification of potentially bactericidal fragments produced "in vivo" will be required to confirm these proposals.

ACKNOWLEDGMENTS

This work was supported by DENTAID, S.A and a short term EMBO fellowship.

REFERENCES

1. Arnold, R.R., Russell, J.E., Champion, W.J., Brewer, M. and Gauthier, J.J. 1982 Bactericidal activity of human lactoferrin: differentiation from the stasis of iron deprivation. Infect.Immun. 35:792–799.
2. Bellamy, W., Takase, M., Yamauchi, K. Wakabayashi, H., Kawase, K. and Tomita M. 1992. Identification of the bactericidal domain of lactoferrin. Biochim Bipophys Acta 1121:130–136.
3. Dalmastri, C., Valenti, P., Visca, P., Vittorioso, P. and Orsi N. 1988. Enhanced antimicrobial activity of lactoferrin by binding to the bacterial surface. Microbiologica 11: 225–230.
4. Ellison, R.T. III, LaForce, F.M., Giehl, T.J., Boose, D.S. and Dunn B.E. 1990. Lactoferrin and transferrin damage of the Gram-negative outer membrane is modulated by $Ca^{2+}$and Mg^{2+}.J. Gen. Microbiol. 136:1437–1446.
5. Fischer, W. 1991. One step purification of bacterial lipid macroamphiphiles by hydrophobic interaction chromatography. Anal. Biochem. 194:353–358.
6. - Greenwood, F.C., Hunter W.M. and Glover, J.S. 1963. The preparation of ^{131}I-labelled human growth hormone of high specific radioactivity. Biochem. J. 89: 114–123.
7. McNab, P.C. and Tomasi, T.B. 1981. Host defence mechanisms at mucosal surfaces. Annu. Rev. Microbiol. 35:477.
8. Naidu, S. and Arnold, R.R. 1997. Influence of lactoferrin on host-microbe interactions. In: Lactoferrin. Interactions and Biological Functions. Ed T.W. Hutchens and B. Lönnerdal. Humana Press. New Jersey pp. 259–276.
9. Ohno, N. and Morrison D.C. 1989. Lipopolysaccharide interaction with lysozyme. J. Biol. Chem. 264:4434–4441.
10. Powell, D.A., Duckworth, M. and Baddiley J. 1975. A membrane-associated lipomannan in *Micrococci*. Biochem J. 151:387–397
11. Reiter, B. 1983. The biological significance of lactoferrin. Int. J. Tiss. Reac. 5:87–96
12. Saito, H., Miyakawa, H., Tamura, Y., Shimamura, S. and Tomita, M. 1991. Potent bactericidal activity of bovine lactoferrin hydrolysate produced by heat treatment at acidic pH. J. Dairy Sci 74: 3724–3730.
13. Tomita, M., W. Bellamy, M. Takase, K. Yamamuchi, H. Wakabayashi and Kawase, K. 1991. Potent antibacterial peptides generated by pepsin digestion of lactoferrin. J. Dairy Sci. 74:4137–4142.
14. Westphal, O. and Jann K. 1965. Bacterial lipopolysaccharides. Methods Carbohydr. Chem. 5:83–91.

ENHANCED ANTI-*CANDIDA* ACTIVITY OF NEUTROPHILS AND AZOLE ANTIFUNGAL AGENTS IN THE PRESENCE OF LACTOFERRIN-RELATED COMPOUNDS

Hiroyuki Wakabayashi,[1] Takafumi Okutomi,[2] Shigeru Abe,[2]
Hirotoshi Hayasawa,[1] Mamoru Tomita,[1] and Hideyo Yamaguchi[2]

[1]Nutritional Science Laboratory
Morinaga Milk Industry Co., Ltd.
Zama, Kanagawa 228, Japan
[2]Department of Microbiology and Immunology
Teikyo University School of Medicine
Itabashi-ku, Tokyo 173, Japan

1. SUMMARY

We investigated the effects of lactoferrin (Lf)-related compounds on growth inhibition of *Candida albicans* by neutrophils or antifungal agents *in vitro*. Human neutrophils partially inhibited the growth of *C. albicans*. The growth inhibition caused by human neutrophils was augmented by the addition of human Lf at concentrations which did not show any inhibitory effect in the absence of neutrophils. Similar observations were obtained also with the following combinations: human neutrophils + bovine Lf, murine neutrophils + bovine Lf, and murine neutrophils + iron saturated bovine Lf, but not in the case of murine neutrophils + human transferrin. The minimum inhibitory concentration (MIC) of azole antifungal agents, clotrimazole, ketoconazole, fluconazole, and itraconazole was reduced by 1/4 to 1/16 in the presence of a sub-MIC level of each of bovine Lf, bovine Lf pepsin hydrolysate, and the antimicrobial peptide "lactoferricin® B" (Lfcin B). Other types of antifungal agents, amphotericin B, nystatin, and flucytosine did not show such combined effects with these Lf-related compounds. The anti-*Candida* activity of bovine Lf or Lfcin B in combination with clotrimazole was shown to be synergistic by checkerboard analysis. Clinically isolated azole-resistant *C. albicans* strains were more susceptible to bovine Lf or Lfcin B than azole-susceptible strains. Trailing growth of an azole-resistant strain in the presence of fluconazole was reduced by the addition of sub-MIC levels of bo-

Advances in Lactoferrin Research, edited by Spik *et al.*
Plenum Press, New York, 1998.

vine Lf or Lfcin B. These results suggest that Lf-related compounds even at relatively low concentrations may function as an antifungal effector in combination with neutrophils thereby modulating azole antifungal efficacies *in vivo*.

2. INTRODUCTION

Lactoferrin (Lf) is considered to play an important role in the host defense against microbial infections. However, its antimicrobial activity *in vitro* is weak or frequently undetectable at physiological concentrations. It is likely that Lf acts cooperatively with other host defense factors or antimicrobial agents administered in chemotherapy *in vivo*. Cooperative action of Lf in combinations with IgA[18] and lysozyme[6,16] has been demonstrated. Lf and some β-lactam antibiotics were reported to have synergistic antimicrobial effects against Gram-negative bacteria including *Klebsiella pneumoniae*, *Escherichia coli*, *Pseudomonas aeruginosa* and *Salmonellae* spp., and Gram-positive bacteria such as *Staphylococcus aureus in vitro* and *in vivo*[4,10,11].

Candida albicans is an opportunistic fungal pathogen that has become an increasingly common cause of disease in immunocompromised hosts[14]. Recently, repeated treatments with fluconazole, the azole antifungal agent mainly used in severely immunocompromised patients, such as those with late-stage AIDS or chronic mucocutaneous candidiasis, have led to the appearance of *Candida* isolates resistant to these agents[7,17]. Lf shows anti-*Candida* activity in salt solution of low concentration or minimal medium[12,21,22]. We attempted to elucidate the effects of Lf-related compounds on the anti-*Candida* activity of neutrophils and antifungal agents *in vitro*. *C. albicans* is a dimorphic organism that grows in yeast form or in hyphal form depending on the growth conditions. The hyphal form is considered to be important in pathogenicity[13]. For the studies on neutrophils, we used a crystal violet (CV) staining method by which hyphal form cells could be detected in the presence of leukocytes[1]. For the studies on antifungal agents, we checked combination effects with Lf-related compounds against yeast form cells and then tested the combinations against azole-resistant strains by the CV staining method.

3. METHODS

3.1. Preparation of Lf-Related Compounds

Bovine Lf and bovine Lf pepsin hydrolysate (Lfhyd) were produced by Morinaga Milk Industry Co., Ltd. (Tokyo, Japan). The apo form and iron-saturated holo form of bovine Lf were obtained by the method previously reported[9]. Lactoferricin B (LFcin B) was produced as described previously[2]. Human Lf was isolated from healthy volunteers according to the previous report[20]. Human transferrin was purchased from Sigma Chemical Co. (St. Louis, Mo).

3.2. Culture of *C. albicans*

Azole-susceptible strains were ATCC90028 (American Type Culture Collection, Rockville, Maryland) and TIMM1768 (Teikyo University Institute of Medical Mycology, Tokyo, Japan). Clinically isolated azole-resistant strains were TIMM3164, TIMM3315, and TIMM3317 (Teikyo University Institute of Medical Mycology). The cells were cultivated on fresh Sabouraud glucose agar, which contained 1% Bactopeptone (Difco Laboratories Co., Detroit, Mich.), 2% glucose, and 1.5% agar, at 28°C before use.

3.3. Preparation of Murine and Human Neutrophils

Murine neutrophils were prepared from peritoneal exudate as described previously[15]. C3H/He N mice, or C3H/He J mice when indicated otherwise, were injected intraperitoneally with 3 ml of 8% casein sodium (Tokyo Kasei, Tokyo, Japan) saline. Six hours later peritoneal cells were collected and contaminating erythrocytes were lysed by the addition of hypotonic PBS diluted to 1/3 with distilled water. After being washed with PBS, they were resuspended in 2 ml of RPMI 1640 (Nissui Pharmaceutical Co., Tokyo, Japan) medium containing 2.5% heat-inactivated (60°C, 30 min) fetal caLf serum with 20 mM HEPES, 2 mM L-glutamine, and 16 mM sodium hydrogen carbonate (RP-medium) and then layered on 10 ml of 90% Ficoll-Hypaque solution (Pharmacia Fine Chemicals, N. J.). After centrifugation at 300×g for 30 min. at room temperature, the cells from the bottom phase were washed with PBS, and were shown to be more than 95% neutrophils by Giemsa staining.

Human peripheral blood neutrophils were obtained as follows[15]. Heparinized venous blood obtained from healthy male volunteers was mixed with dextran 70 and allowed to stand at room temperature for 30 min. The leukocyte-rich supernatant was collected and then centrifuged on a Ficoll-Hypaque density gradient. The neutrophil-rich layer at the bottom was washed with PBS, and contaminating erythrocytes were lysed by the addition of 0.83% NH_4Cl in Tris-HCl buffer (pH 7.7). The mixture was suspended for 3 min. and an equal volume of cold PBS was added. The residual leukocytes were washed and resuspended in RP-medium. More than 95% of the suspension was neutrophils.

3.4. Assay for Growth Inhibition of *C. albicans* in Hyphal Form

For hyphal growth of *C. albicans*, yeast cells picked up from the agar slants (1×10^4 cells/ml) were put into 200 μl of RP-medium with or without Lf-related compounds, neutrophils, and antifungal agents in a 96 well flatbottom microplate and incubated at 37°C for 15 hr. in a 5% CO_2 atmosphere. To determine hyphal growth content of *C. albicans*, the CV staining assay was performed as described[1,15]. After incubation, the medium in the wells was discarded by inverting the microplate. The adhesive *Candida* mycelia in the wells were sterilized by immersion in 70% ethanol for 1 min. and, when neutrophils were present, the adherent neutrophils were washed out with 100 μl of 0.25% sodium dodecyl suLfate (SDS). The plates were washed twice by immersing them in distilled water. The mycelia were stained with 0.02% CV in 100 μl of PBS for 15 min. and washed 3 times with water. After drying the microplates, 150 μl of isopropanol containing 0.04 N HCl and 50 μl of 0.25% SDS were added to the wells and the samples were mixed by a plate mixer for 30 sec. in order to extract CV from the mycelia. The absorbance at 550–630 nm of triplicate samples was measured spectrophotometrically.

3.5. Assay for Growth Inhibition of *C. albicans* in Yeast Form

For yeast growth of *C. albicans*, yeast-form cells were collected from agar slants, suspended (1×10^4 cells/ml) in 200 μl of Sabouraud glucose broth (1% Bactopeptone, 2% glucose) with or without Lf-related compounds and antifungal agents, and incubated at 37°C for 17 hr. After incubation, the absorbance at 630 nm of triplicate samples was measured spectrophotometrically.

3.6. Checkerboard Analysis

To ascertain the combination effects between Lf-related compounds and antifungal agents, fractional inhibitory concentration (FIC) index by the checkerboard method[5] was calculated as follows: $[(A)/MIC_A]+[(B)/MIC_B]$ = FIC index, where MIC_A and MIC_B are the MICs of drug A and B, determined separately, and (A) and (B) are the MICs of drug A and B when determined in combination. Drugs were interpreted to act in synergy or with indifference when the FIC index was ≤0.5 or 0.5–4, respectively.

4. RESULTS AND DISCUSSION

4.1. Effects of Lf on the Anti-*Candida* Activity of Neutrophils

Cooperative activities of LF and neutrophils on the inhibition of *C. albicans* TIMM1768 were investigated using the CV staining method as reported previously[15]. Human Lf had a strong inhibitory effect on *Candida* growth when used alone. Complete inhibition of hyphal growth was observed at the concentration of about 30 μg/ml. Human neutrophils at 50 times higher numbers than *Candida* cells, i.e. effector to target ratio of 50 (E/T=50), inhibited *Candida* growth by about 50%. When human LF was present at 10 μg/ml, human neutrophils at E/T=50 completely inhibited the growth of *Candida*.

Interspecies combinations of Lf and neutrophils were investigated. Effects of bovine Lf on the anti-*Candida* activity of murine or human neutrophils were studied. Bovine Lf alone showed only slight activity even at 100 μg/ml (Fig. 1). Relatively low concentra-

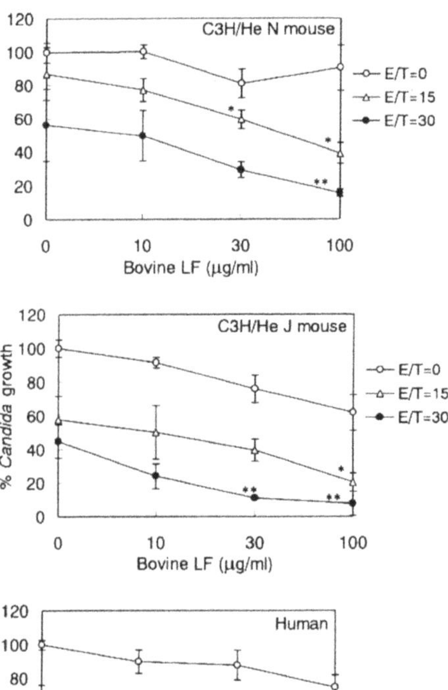

Figure 1. Augmentation of the anti-*Candida* activity of murine or human neutrophils by bovine Lf. *C. albicans* TIMM1768 was cultured with neutrophils of C3H/He N mouse, C3H/He J mouse, or human, in the presence of various concentrations of bovine Lf for 15 hr. The results represent the mean of three separate experiments ± SD. *,**; Statistically significant difference ($p < 0.01$) by Student's t test from each control in the absence of bovine Lf.

tions of bovine Lf in the presence of C3H/He N murine neutrophils at E/T=15 or 30 clearly augmented the growth inhibition of *Candida*. Similar augmentation effects were also observed with the combination of bovine Lf and neutrophils obtained from C3H/He J mice which are known to display a low response to LPS. Enhanced anti-*Candida* activity was also seen in the co-culture of human neutrophils in the presence of bovine Lf. These results indicate that bovine Lf is effective to stimulate neutrophils derived from other mammalian species.

In order to clarify the role of iron ions, the effects of the apo form and iron-saturated holo form of bovine Lf on the anti-*Candida* activity of murine neutrophils were compared. The combination of murine neutrophils and apo-Lf displayed potent anti-*Candida* activity. Holo-Lf at concentrations up to 1000 μg/ml did not show any effect on the growth of *Candida* in the absence of neutrophils but it augmented the anti-*Candida* activity of murine neutrophils. Thus Lf in combination with neutrophils appears to augment the growth inhibition of *C. albicans* independently of its iron status.

We also examined the effect of human transferrin on the anti-*Candida* activity of murine neutrophils. Human transferrin inhibited *Candida* growth in a dose-dependent manner, but the effect of transferrin in combination with neutrophils seemed to be only additive.

Under physiological conditions, Lf is present in secretions such as saliva, tears and nasal secretions, and the accumulation of neutrophils is observed in mucosal lesions caused by *Candida* infection[8]. These observations indicate that neutrophils and Lf may coexist at the site of an inflammatory lesion in mucosal candidiasis. Therefore, it can be assumed that their combined effects play an important role in mucosal defense against *Candida* invasion.

4.2. Effects of Lf-Related Compounds on Anti-*Candida* Activities of Antifungal Agents

Combination effects of Lf-related compounds and clinically used antifungal agents were investigated[24]. *Candida* cells were incubated in Sabouraud glucose broth for 17 hr. *Candida* was grown in yeast form and the extent of growth was determined from the increase in turbidity.

We compared several types of antifungal agents in combination with the Lf-related compounds (Fig. 2). The MICs of the azole antifungal agents, clotrimazole, ketoconazole, fluconazole, and itraconazole, were reduced by the addition of a sub-MIC level of bovine Lf, Lfhyd, or Lfcin B. On the other hand, other types of antifungal agents, amphotericin B, nystatin (polyene type), and flucytosine (fluoropyrimidine type) did not show such significant effects in combination with the Lf-related compounds. These observed activities were specific to azole type antifungal agents.

We have analyzed the combination effects of Lf-related compounds and azole antifungal agents by the checkerboard method. The FIC index of the combination of bovine Lf and clotrimazole was 0.187 indicating synergy. In the case of Lfcin B and clotrimazole, the FIC index was 0.190 interpreted to be synergy.

4.3. Effects of Lf-Related Compounds on Inhibition of Azole-Resistant *C. albicans* by Fluconazole

We have compared the susceptibilities of azole-resistant and azole-susceptible strains of *C. albicans* to bovine Lf or Lfcin B and examined the combination effects of Lf-related compounds and fluconazole in azole-resistant *C. albicans*. In this study, we monitored the hyphal growth of *Candida* measured by the CV staining method.

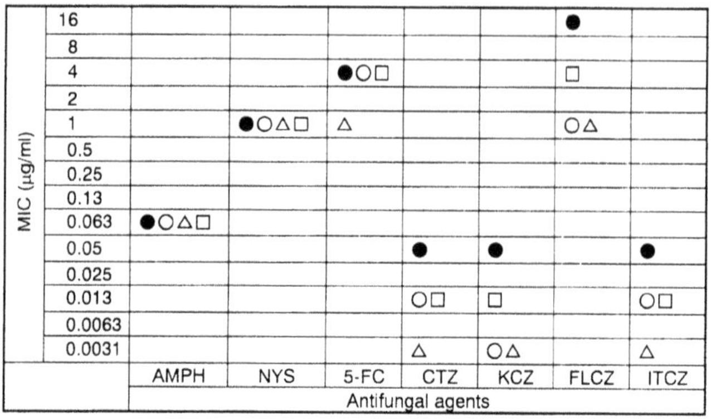

MIC (μg/ml)	AMPH	NYS	5-FC	CTZ	KCZ	FLCZ	ITCZ
16						●	
8							
4			●○□			□	
2							
1		●○△□	△			○△	
0.5							
0.25							
0.13							
0.063	●○△□						
0.05				●	●		●
0.025							
0.013				○□	□		○□
0.0063							
0.0031				△	○△		△
	Antifungal agents						

Figure 2. MIC of antifungal agents against *C. albicans* TIMM1768 in the presence of Lf-related compounds as determined by the increase in turbidity in Sabouraud glucose broth. Antifungal agents tested were amphotericin B (AMPH), nystatin (NYS), flucytosine (5-FC), clotrimazole (CTZ), ketoconazole (KCZ), fluconazole (FLCZ), and itraconazole (ITCZ). Symbols used are ●: control; ○: with bovine Lf (100 μg/ml); ▲: with Lfhyd (200 μg/ml); □: with Lfcin B (3.1 μg/ml).

The MICs of antifungal agents and Lf-related compounds against two azole-susceptible strains and three azole-resistant strains are shown in Table 1. Since strain TIMM3164 was defective in hyphae formation, we determined the extent of growth of this strain by measuring the increase in turbidity of the culture. TIMM3164 was moderately resistant to fluconazole and itraconazole. Strains TIMM3315 and TIMM3317 were highly resistant to these agents. On the other hand, these three azole-resistant strains were more susceptible to bovine Lf than the azole-susceptible strains. Two highly azole-resistant strains also showed lower MIC values for LFcin B than those of the azole-susceptible strains.

The azole-resistant strain TIMM3317 showed trailing growth in the presence of fluconazole (Fig. 3). This growth was reduced by the addition of a relatively low concentration of bovine Lf. In the case of Lfcin B, although the peptide did not show any inhibitory activity at concentrations up to 100 μg/ml, it significantly decreased the trailing growth in the presence of fluconazole. Especially, the combination of Lfcin B at 100 μg/ml and fluconazole completely inhibited the hyphal growth of this strain.

Table 1. MIC of antifungal agents and Lf-related compounds against azole-susceptible or azole-resistant *C. albicans* as determined by the CV staining method

	MIC (μg/ml)[a]				
C. albicans	AMPH	FLCZ	ITCZ	Bovine Lf	Lfcin B
Azole-susceptible					
ATCC90028	0.13	0.25	0.0063	>6400	400
TIMM1768	0.13	0.25	0.0063	6400	400
Azole-resistant					
TIMM3164[b]	0.13	1	0.05	1600	400
TIMM3315	0.25	128	>51	200	50
TIMM3317	0.13	>256	>51	1600	200

[a]Antifungal agents tested were amphotericin B (AMPH), fluconazole (FLCZ), and itraconazole (ITCZ).
[b]MIC was determined from the increase in OD630 of the culture.

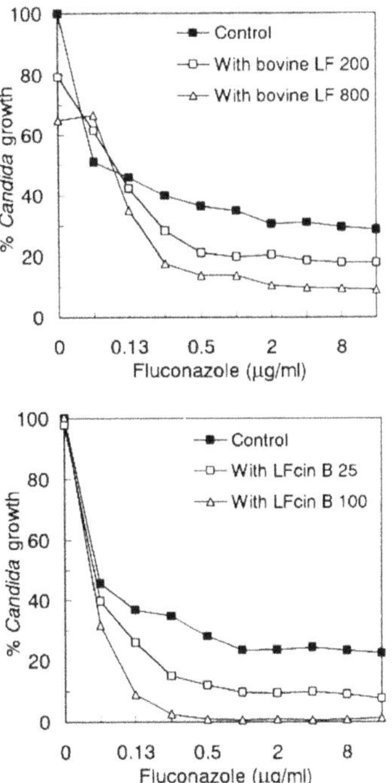

Figure 3. Inhibition of hyphal growth of *C. albicans* TIMM3317 by fluconazole in the presence of bovine Lf (200, 800 μg/ml) or Lfcin B (25, 100 μg/ml) as determined by the CV staining method.

Recently it was demonstrated that the majority of azole-resistance in the case of *C. albicans* could be correlated with multidrug efflux transporters, which are energized by proton motive force or ATP[19,23]. Lf induces cell surface alterations including leakage of proteins and formation of surface blebs in *Candida* spp.[12] Lfcin B induces disruption of the proton gradient across the cell membrane of *C. albicans*[25] and inhibits uptake of glucose by *Trichophyton rubrum*[3], and consequently, it may reduce ATP production by fungi. Lf or Lfcin B may deprive *C. albicans* of the potential to develop azole-resistance by inhibiting the activity of the multidrug efflux transporters or by facilitating azole uptake through the membrane.

5. CONCLUSIONS

1. Relatively low concentrations of Lf cooperatively acted with neutrophils to inhibit *C. albicans*, and this action was evident in the case of interspecies combinations of neutrophils and Lf.
2. Combinations of bovine Lf or Lfcin B with azole antifungal agents synergistically inhibited the growth of *C. albicans*.
3. Azole-resistant *C. albicans* strains were more susceptible to bovine Lf or Lfcin B than azole-susceptible strains.
4. Trailing growth of an azole-resistant strain in the presence of fluconazole was reduced by the addition of sub-MIC levels of bovine Lf or Lfcin B.
5. Overall, Lf-related compounds even at relatively low concentrations may function as an antifungal effector in combination with neutrophils thereby modulating azole antifungal efficacies *in vivo*.

REFERENCES

1. Abe, S., Satoh, T., Tokuda, Y., Tansho, S., and Yamaguchi, H. (1994) A rapid colorimetric assay for determination of leukocyte-mediated inhibition of mycelial growth of Candida albicans. Microbiol. Immunol. 38, 385–388.
2. Bellamy, W., Takase, M., Yamauchi, K., Wakabayashi, H., Kawase, K., and Tomita, M. (1992) Identification of bactericidal domain of lactoferrin. Biochim. Biophys. Acta 121, 130–136.
3. Bellamy, W., Yamauchi, K., Wakabayashi, H., Takase, M., Takakura, N., Shimamura, S., and Tomita, M. (1994) Antifungal properties of lactoferricin B, a peptide derived from the N-terminal region of bovine lactoferrin. Lett. Appl. Microbiol. 18:230–233.
4. Chimura, T., Hirayama, T., and Nakahara, M. (1993) In vitro antimicrobial activities of lactoferrin, its concomitant use with cefpodoxime proxetil and clinical effect of cefpodoxime proxetil. Jap. J. Antibiotics 46, 482–485.
5. Eliopoulos, G. M. and Moellering, R. C. (1991) Antimicrobial combinations, in Antibiotics in laboratory medicine, (Lorian, V., eds.), Williams & Wilkins, Baltimore, pp. 432–492.
6. Ellison III, R. T. and Giehl, T. J. (1991) Killing of gram-negative bacteria by lactoferrin and lysozyme. J. Clin. Invest. 88, 1080–1091.
7. Johnson, E. M., Warnock, D. W., Luker, J., Porter, S. R., and Scully, C. (1995) Emergence of azole drug resistance in Candida species from HIV-infected patients receiving prolonged fluconazole therapy for oral candidiasis. J. Antimicrob. Chemother. 35, 103–114.
8. Lacasse, M., Fortier, C., Trudel, L., Collet, A. J., and Deslauriers, N. (1994) Experimental oral candidosis in mouse: microbiologic and histologic aspects. J. Oral Pathol. Med. 19, 136–141.
9. Mazurier, J. and Spik, G. (1980) Comparative study of the iron-binding properties of human transferrins. I. Complete and sequential iron saturation and desaturation of lactotransferrin. Biochim. Biophys. Acta 629, 399–408.
10. Miyazaki, S., Harada, Y., Tsuji, A., and Goto, S. (1991) In vivo combined effects of lactoferrin (Lf) and drugs on bacterial infections in mice. Chemotherapy(Tokyo) 39, 829–835.
11. Naidu, A. S. and Arnold, R. R. (1994) Lactoferrin interaction with Salmonellae potentiates antibiotic susceptibility in vitro. Diagn. Microbiol. Infect. Dis. 20, 69–75.
12. Nikawa, H., Samaranayake, L. P., Tenovuo, J., Pang, K. M., and Hamada T. (1993) The fungicidal effect of human lactoferrin on Candida albicans and Candida krusei. Arch. Oral Biol. 38, 1057–1063.
13. Odds, F. C. (1988) Candida and candidosis: a review and bibliography, 2nd ed., Balliere Tindall, London.
14. Odds, F. C., Schmid, J., and Soll, D. R. (1990) Epidemiology of Candida infections in AIDS, in Mycoses in AIDS patients, (Vanden Bossche, H., et al., eds.), Plenum, New York, pp. 67–74.
15. Okutomi, T., Abe, S., Tansho, S., Wakabayashi, H., Kawase, K., and Yamaguchi, H. (1997) Augmented inhibition of growth of Candida albicans by neutrophils in the presence of lactoferrin. FEMS Immunol. Med. Microbiol. (in press)
16. Perraudin, J. P. and Prieels, J. P. (1982) Lactoferrin binding to lysozyme-treated Micrococcus luteus. Biochim. Biophys. Acta 718, 42–48.
17. Rex, J. H., Rinaldi, M. G., and Pfaller, M. A. (1995) Resistance of Candida species to fluconazole. Antimicrob. Agents Chemother. 39, 1–8.
18. Rogers, H. J. and Synge, C. (1978) Bacteriostatic effect of human milk on Escherichia coli: the role of IgA. Immunology 34, 19–28.
19. Sanglard, D., Kuchler, K., Pagani, J.-L., Monod, M., and Bille, J. (1995) Mechanisms of resistance of azole antifungal agents in Candida albicans isolates from AIDS patients involve specific multidrug transporters. Antimicrob. Agents Chemother. 39, 2378–2386.
20. Sawatzki, G. and Kubanek B. (1983) Isolation and ELISA of mouse and human lactoferrin, in Structure and function of iron storage and transport protein, (Urushizaki, I., et al., eds.), Elsevier, Amsterdam, pp. 441–443.
21. Soukka, T., Tenovuo, J., and Lenander-Lumikari, M. (1992) Fungicidal effect of human lactoferrin against Candida albicans. FEMS Microbiol. Lett. 90, 223–228.
22. Valenti, P., Visca, P., Antonini, G., and Orsi, N. (1986) Interaction between lactoferrin and ovotransferrin and Candida cells. FEMS Microbiol. Lett. 33, 271–275.
23. Venkateswarlu, K., Denning, D. W., Manning, N. J., and Kelly, S. L. (1995) Resistance to fluconazole in Candida albicans from AIDS patients correlated with reduced intracellular accumulation of drug. FEMS Microbiol. Lett. 131:337–341.

24. Wakabayashi, H., Abe, S., Okutomi, T., Tansho, S., Kawase, K., and Yamaguchi, H. (1996) Cooperative anti-Candida effects of lactoferrin or its peptides in combination with azole antifungal agents. Microbiol. Immunol. 40, 821–825.
25. Wakabayashi, H., Hiratani, T., Uchida, K., and Yamaguchi, H. (1996) Antifungal spectrum and fungicidal mechanism of an N-terminal peptide of bovine lactoferrin. J. Infect. Chemother. 1:185–189.

THE MECHANISM OF IN VIVO BACTERIOSTASIS OF BOVINE LACTOFERRIN

Tomohiro Ogata, Susumu Teraguchi, Kouichirou Shin, Michiko Kingaku, Yasuo Fukuwatari, Kouzou Kawase, Hirotoshi Hayasawa, and Mamoru Tomita

Nutritional Science Laboratory
Morinaga Milk Industry Co.Ltd.
Zama-City, Kanagawa 228, Japan

Recently we have reported that orally administered bovine Lf(bLf) exerts bacteriostatic effects against bacterial overgrowth in the intestine of specific-pathogen-free (SPF) mice fed milk. In this animal model, the in vivo bacteriostatic effect of bLf against the proliferation of intestinal *Enterobacteriaceae*, the bacteria most sensitive to bLf, was independent of the iron-chelating ability of bLf. In addition various proteolytic hydrolysates of bLf (with differing antibacterial activities in vitro) showed the same bacteriostatic effect as undigested bLf. These results suggest that the mechanism of in vivo bacteriostasis of Lf differs from the in vitro mechanism reported.

In SPF mice fed milk differing in concentrations of lactose, glucose and galactose, the proliferation of intestinal *Enterobacteriaceae* was dependent on the carbohydrate concentration in the diet. The addition of 2% bLf to the diets significantly suppressed this carbohydrate-dependent proliferation of bacteria except in the case of diets containing excess carbohydrate. In germ-free mice fed sterile milk, the addition of 2% bLf to milk resulted in a significant decrease in concentrations of lactose, glucose and galactose in the cecal contents. In an in vitro assay system using everted sacs of the small intestine of SPF mice, both bLf and its pepsin hydrolysate apparently stimulated glucose absorption. Based on these findings, we propose that the in vivo mechanism of action of ingested bLf involves the stimulation of carbohydrate absorption resulting in a bacteriostatic effect against *Enterobacteriaceae* in the intestine of mice fed milk.

1. INTRODUCTION

Lactoferrin (Lf) is prominently found in mammalian milk and is particularly abundant in human milk[7]. Although it is well recognized that breast feeding offers protection to newborn infants[3–5], the contribution of ingested Lf remains still unproved. We have re-

Advances in Lactoferrin Research, edited by Spik *et al.*
Plenum Press, New York, 1998.

cently reported that orally administered bovine lactoferrin (bLf) is effective to suppress proliferation of various bacteria[8,9] and bacterial translocation[10] in the intestine of specific-pathogen-free (SPF) mice fed bovine milk. In this animal model, the in vivo bacteriostatic effects of ingested bLf responsible for suppression of bacterial overgrowth were found to be independent of the iron-chelating activity of bLf which is required for bacteriostasis in vitro[8,9]. In addition, the in vivo effect of bLf differed among species of intestinal bacteria[8,9]. Various proteolytic hydrolysates of bLf showed the same in vivo bacteriostatic effect as occurred with undigested bLf[8,9], indicating that ingested bLf may exert this activity even after it has been digested to some extent. Considering these results, the in vivo mechanism seems to differ in certain respects from the in vitro mechanism reported.

In this study the in vivo mechanism of bacteriostasis of bLf was investigated with the aim of further elucidating the biological role of lactoferrin in mammalian milk. The bacteriostasis induced by ingested bLf in milk-fed mice is evident as a suppressive effect on the bacterial overgrowth that occurs in response to milk feeding[8–10]. It is speculated that the bacterial proliferation in the gut of milk-fed mice may be dependent on lactose, and the ingestion of bLf may result in a suppression of bacterial use of lactose or a decrease in the concentration of lactose available in the intestine of mice fed milk. To investigate these hypotheses, we examined the influence of administered carbohydrate and bLf on the proliferation of intestinal bacteria in the gut of SPF mice fed a milk diet. Then we examined whether bLf stimulates the absorption of carbohydrates in vivo and in vitro. In this paper, we propose an explanation for the in vivo mechanism by which orally administered bLf exerts bacteriostasis against intestinal bacteria.

2. MATERIALS AND METHODS

2.1. Animals

Four-week-old female BALB/c SPF mice were kept in a specific-pathogen-free room. They were initially fed with free access to a commercial pelleted diet (F-2, Funabashi Farms Co., Chiba, Japan) and tap water for 7 days. Germ-free (GF) BALB/c mice were obtained from the Animal Facilities of the Institute of Public Health (Tokyo, Japan) and then they were bred. Four-week-old GF mice were housed in a vinyl isolator. Male and female GF mice were kept in separate cages. The GF mice of both sexes were initially fed with free access to a commercial sterile pelleted diet (F-1, Funabashi Farms Co., Chiba, Japan) and sterile water for 7 days. Their maintenance and checking of the GF state was performed by the method of Ueda et al.[11]. Mice at 5 weeks of age were used in the experiments. The composition of tested diets and the feeding schedules employed are described below. Food intake and body weight were measured at intervals during the experiments in the case of SPF mice but only at the end of the experiment in the case of GF mice. Other conditions employed in the animal experiments using SPF mice were the same as those described previously[8].

2.2. Influence of Carbohydrates and bLf on Intestinal Bacteria

The influence of carbohydrates and bLf on proliferation of *Enterobacteriaceae* and bifidobacteria in the intestine was examined as follows. SPF mice were randomly divided into eight groups of five mice each. Four of the eight groups were fed the following diets: bovine milk (4% lactose), milk treated with β-D-galactosidase (EC 3.2.1.23; 250–500 units/mg; Sigma Chemical Co., St. Louis, U.S.A.) (2% glucose and 2% galactose), a mixture of the above two milk preparations (1:1) (2% lactose, 1% glucose and 1% galactose), or milk supplemented with lactose (7% lactose). The other four groups were fed the same diets as indicated above, supplemented with 2% bLf in each instance. Milk treated with β-D-galactosidase was prepared as follows. Filter-sterilized enzyme was added to commercial pasteurized bovine milk at a concentration of 0.008% (wt/wt) and the mixture was incubated at 37°C for 3 h. The reaction was terminated by heating at 90°C for 15 min. After feeding the test diets for 7 days, fresh feces were collected separately from each mouse and the bacterial numbers in the feces were assayed using DHL agar (Eiken Chemical Co., Tokyo, Japan) for *Enterobacteriaceae* and BL agar (Eiken Chemical Co.) for bifidobacteria as described previously[9].

2.3. Carbohydrate Absorption in Germ-Free Mice

The influence of bLf on carbohydrate absorption in the intestine of mice was examined using two groups of 12 GF mice each. One group was fed sterile milk and the other group was fed milk supplemented with 2% bLf. Both diets were prepared by mixing commercial whole sterilized bovine milk (150°C, 2.4 sec) with filter-sterilized distilled water or 20% bLf solution at a volume ratio of 9:1. Before feeding, the diets were incubated at 37°C for 4 days and it was confirmed that there was no bacterial contamination. After feeding each diet for 7 days, the mice were anesthetized by exposure to chloroform. Then the small intestine and the cecum were removed using sterile technique. Concentrations of lactose, glucose and galactose present in the inner contents of the cecum were assayed using F-Kits (Boehringer Mannheim Co., Mannheim, Germany). Mucosal disaccharidase activities were measured using specimens of the small intestine.

2.4. Carbohydrate Absorption in Vitro

Specific-pathogen-free mice fed pellets were anesthetized by exposure to chloroform, and then the small intestine was removed and the inner contents were washed out with ice-cold phosphate buffered saline (PBS, pH 7.4). For assay of the glucose absorption rate, sacs of everted small intestine (about 5–10 cm from the pylorus) were prepared by the method of Wilson and Wiseman (12). 0.5 ml of PBS was injected into each sac. At first, the influence of NaCl on the glucose absorption rate was examined. The sac was incubated in a reaction mixture (4 ml) containing 10 mM potassium phosphate buffer (pH7.4), 10 mM glucose, 0–150 mM NaCl and 150–0 mM choline chloride. The total concentration of NaCl and choline chloride was adjusted to 150 mM to provide the same osmotic pressure. After incubation at 37°C for 30 min, the glucose concentration both in the PBS inside the sac and in the reaction mixture outside the sac was assayed. The glucose absorption rate was estimated from the increase in glucose concentration in the PBS or its decrease in the reaction mixture. The rate was expressed as μmol glucose per gram of tissue per h. The effects of bLf and its pepsin hydrolysate (bLfH) on the glucose absorption rate were examined using reaction mixtures containing 10 mM potassium phosphate buffer (pH7.4), 10 mM glucose, 20 mM NaCl, 40 mM KCl, 90 mM choline chloride and bLf or bLfH at a concentration of 0–1 mg/ml. After incubation at 37°C for 30 min, the glucose absorption rate was assayed.

2.5. Statistical Analysis

Data were expressed as the mean value ± SD_{n-1} of samples. The data were analyzed statistically by Student's *t* test.

3. RESULTS

3.1. Influence of Carbohydrate and bLf on Intestinal Bacteria

Table 1 shows the effect of administered bLf on the numbers of *Enterobacteriaceae* and bifidobacteria in the feces of SPF mice fed milk containing carbohydrates at different concentrations for 7 days. Before feeding the milk diets, the number of *Enterobacteriaceae* in the feces of mice was 6.1 ± 0.4 log_{10}CFU/g of feces (n=5) whereas bifidobacteria were not detected in the feces of any group. The extent of proliferation of *Enterobacteriaceae* in the gut of mice fed milk without bLf appeared to be consistent regardless of the kind and concentrations of carbohydrate tested. The addition of 2% bLf to milk significantly suppressed the in vivo proliferation of *Enterobacteriaceae* in each group, except for the group fed milk containing 7% lactose, and the bacterial numbers decreased to a level about 1/100 of that in the feces of mice fed milk only ($P < 0.01$). On the other hand, the in vivo proliferation of bifidobacteria was dependent on the presence of lactose in the diets. In the three groups fed milk containing lactose (2–7%) but no bLf, the numbers of bifidobacteria in the feces of the mice were about 11 log_{10}CFU/g of feces indicating that bifidobacteria were the most predominant species. The addition of 2% bLf to the diets significantly suppressed the in vivo proliferation of bifidobacteria in the group fed milk containing 2% lactose but did not in the group fed milk containing 4% lactose or 7% lactose. In the group fed milk containing glucose and galactose, each at 2%, but no lactose, bifidobacteria were detected in the feces of one of the five mice and their numbers were much lower than in the case of the other groups tested regardless of the addition of bLF to the diet.

3.2. Carbohydrate Absorption in Germ-Free Mice

Table 2 shows concentrations of lactose, glucose and galactose in the cecal contents of GF mice fed sterile milk or sterile milk containing 2% bLf for 7 days. There was no significant difference in food intake and body weight of GF mice between the two groups ($P < 0.05$). The sterile milk used contained lactose alone at a concentration of 36.5 mg/g. However, lactose, galactose and glucose were detected in the cecal contents of GF mice at different concentrations after feeding for 7 days. The carbohydrate predominantly detected in the cecal contents was lactose. Concentrations of these three carbohydrates present in

Table 1. Effects of administered bovine lactoferrin (bLf) on the numbers of *Enterobacteriaceae* and bifidobacteria in feces of SPF mice fed milk containing carbohydrates at different concentrations

			Mean log_{10} CFU/g of feces ± SD_{n-1}			
Carbohydrate (%)			Enterobacteriaceae		Bifidobacteria	
Lactose	Glucose	Galactose	Milk	Milk + 2% bLf	Milk	Milk + 2% bLf
0	2	2	9.3 ± 0.9	6.7 ± 1.3*	8.4	8.6
2	1	1	9.0 ± 1.3	6.6 ± 0.4*	11.6 ± 0.2	9.2 ± 0.9*
4	0	0	8.6 ± 0.9	6.6 ± 0.8*	10.8 ± 0.2	10.8 ± 0.7
7	0	0	9.5 ± 0.2	9.1 ± 0.5	11.0 ± 0.2	11.7 ± 0.1*

*Significantly different ($P < 0.01$) versus the value for mice fed milk only (n=5).

Table 2. Influence of administered bLf on the concentration of carbohydrate in the cecal contents of germ-free mice fed sterile milk (lactose: 36.5 mg/g)

	Mean mg/g of cecal contents ± SD_{n-1}	
Carbohydrate	Milk	Milk + 2% bLf
Glucose	2.1 ± 1.2	0.5 ± 0.4*
Galactose	3.4 ± 1.0	1.5 ± 0.7*
Lactose	26.6 ± 1.4	16.5 ± 2.0*
Total carbohydrate	32.1 ± 2.2	18.5 ± 2.9*

*Significantly different (P < 0.01) versus the value for mice fed milk only (n=12).

the cecal contents were significantly lower in the group fed sterile milk containing 2% bLf than in the group fed sterile milk only ($P < 0.01$).

3.3. Carbohydrate Absorption in Vitro

To confirm the effect of bLf on carbohydrate absorption, in vitro experiments using sacs of everted small intestine of mice were performed. Figure 1 shows the influence of NaCl on the glucose absorption rate in an everted jejunal sac of SPF mice. The glucose absorption rate was assayed by determining the increase in glucose concentration in the serosal solution inside the everted sac (serosal side) or the decrease in glucose concentration in the mucosal solution outside the sac (mucosal side). The latter rate was about 6–8 times higher than the former rate. Both of the rates appeared to be dependent on the concentration of NaCl present outside the everted sac. Compared with the rates in the absence of NaCl, significantly higher rates were observed in the presence of 150 mM NaCl ($P < 0.05$ or 0.01). Figure 2 shows the influence of bLF or bLFH on the glucose absorption rate in the everted sac. The glucose absorption rate was assayed from the decrease in glucose concentration in the mucosal solution supplemented with 20 mM NaCl and 40 mM KCl which are similar to the concentrations of sodium and potassium in bovine milk. The influence of bLf or bLfH on the glucose absorption rate was compared within the range of 1–1000 μg/ml. The glucose absorption rate in the presence of bLf or bLfH at 10 μg/ml was maximum and significant in comparison with that in the absence of bLf or bLfH ($P < 0.05$ or 0.01).

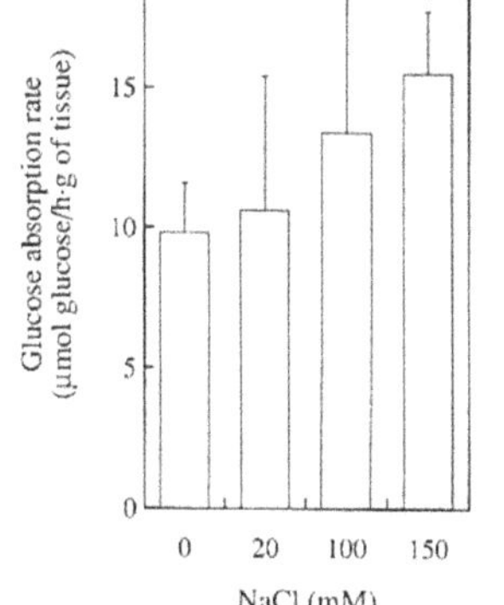

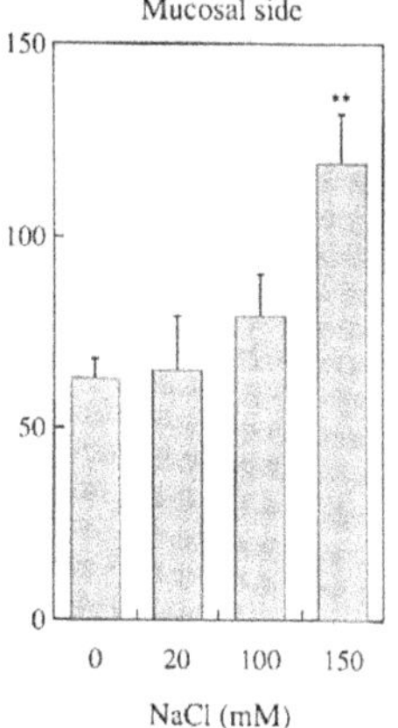

Figure 1. Influence of NaCl concentration on the glucose absorption rate in everted jejunal sacs of SPF mice. Values with asterisks are significantly different (*:$P < 0.05$, **:$P < 0.01$) versus the value in the absence of NaCl (n=3).

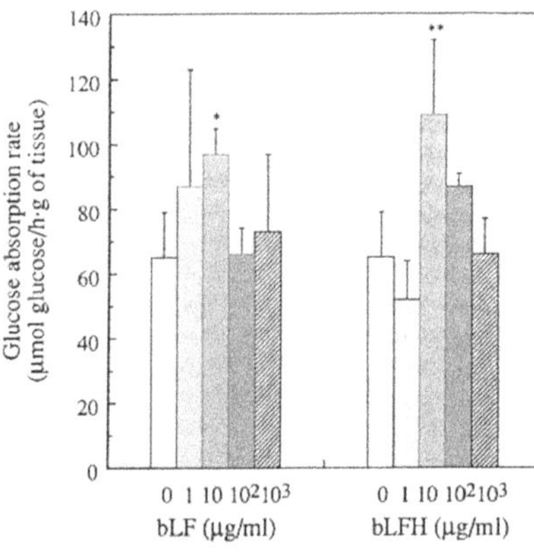

Figure 2. Influence of bLf or pepsin hydrolysate of bLf (bLfH) on the glucose absorption rate in everted jejunal sacs of SPF mice. Values with asterisks are significantly different (*:$P < 0.05$, **:$P < 0.01$) versus the value in the absence of bLF or bLFH (n=3).

4. DISCUSSION

When SPF mice were fed milk containing 4% carbohydrate consisting of different combinations of lactose, glucose and galactose or milk containing 7% lactose, an overgrowth of *Enterobacteriaceae* was consistently observed in the feces. This result indicates that proliferation of *Enterobacteriaceae* in the intestine is not dependent on lactose alone present in milk. This overgrowth of *Enterobacteriaceae* was suppressed by the addition of 2% bLf to milk, except in the case of milk containing 7% lactose. This result suggests that proliferation of intestinal *Enterobacteriaceae* may be suppressed by bLf except in the case of diets containing excess carbohydrate. In contrast, the proliferation of intestinal bifidobacteria was dependent on lactose present in milk. The addition of 2% bLf to milk suppressed the proliferation of bifidobacteria in mice given milk containing 2% lactose but not at all in mice given milk containing 4% lactose or 7% lactose. This result suggests that proliferation of intestinal bifidobacteria may be suppressed by bLf in the case of milk containing low concentrations of lactose. It seems that the bacteriostatic effect of bLf against intestinal bacteria is not expressed in mice fed diets containing low or excessive amounts of carbohydrate. However, it is likely that in mice given the diet containing the low amount of carbohydrate the effect of bLf was not apparent because the number of intestinal bacteria was already maintained at a low level. Thus the proliferation of intestinal bacteria was influenced by the ingestion of carbohydrate and bLf in milk-fed mice. In addition, the response of intestinal bacteria to ingested bLf differed among species of bacteria.

In GF mice fed sterile milk, the addition of 2% bLf to milk resulted in a significant decrease in concentrations of lactose, glucose and galactose in the cecal contents. This result strongly suggests that bLf acts to stimulate the absorption of these carbohydrates in the intestine of mice, that is, lactose digestion or the following absorption of glucose and galactose, or both. Lactose is absorbed in the small intestine of mammals after being hydrolyzed by β-D-galactosidase located on the brush border to galactose and glucose which are transferred into the enterocytes by two types of glucose transporters: Na^+-glucose cotransporters and facilitated glucose transporters[13]. Administered bLf did not influence the amount of mucosal β-D-galactosidase activity expressed, or that of the other disaccharidases examined in GF mice fed milk (data not shown). However, a positive correlation was observed between the concentrations of lactose and monosaccharides present in the cecal contents of GF mice fed milk or milk containing 2% bLf. These results suggest that lactose hydrolysis by mucosal β-D-galactosidase may be inhibited by the accumulation of monosaccharides derived from lactose. Since the concentrations of monosaccharides in the cecal contents significantly decreased in GF mice given bLf, the lower concentrations of monosaccharides induced by ingested bLf may lead to the stimulation of lactose digestion.

In an in vitro assay system using everted jejunal sacs from SPF mice, the glucose absorption rate appeared to be dependent on the concentration of NaCl. Both bLf and bLfH at a concentration of 10 μg/ml stimulated the glucose absorption rate in the presence of 20 mM NaCl, a level similar to that in bovine milk. The concentration range of bLf effective in stimulation of glucose absorption may be narrow since the effect of bLf was not observed at concentrations above 100 μg/ml. From these in vitro and in vivo data, it is considered that bLf functions by some mechanism to stimulate the absorption of glucose and galactose in the intestine of the milk-fed mice. The glucose concentration in the serum of GF mice of the two groups was assayed and found to be around 2 mg/g. Considering the difference between the glucose concentrations present in the cecal contents and in the blood, there is a possibility that bLf stimulates the active transport of glucose by a mechanism involving a Na^+-glucose cotransporter. Further studies are required to elucidate the active domain of bLf and the mechanism by which it promotes carbohydrate absorption.

In the cecal contents of SPF mice fed milk or milk containing 2% bLf for 7 days, lactose, glucose and galactose were not detected. Apparently, all of the unabsorbed carbohydrates were used by intestinal bacteria in SPF mice. On the other hand, substantial amounts of lactose still remained in the cecum of GF mice fed milk containing 2% bLf. Such a substantial amount of lactose may have induced the overgrowth of bifidobacteria but not *Enterobacteriaceae* in SPF mice fed milk containing 2% bLf. In order to examine the effect of carbohydrates on the proliferation of *Enterobacteriaceae*, *Escherichia coli* EC2 , a predominant strain isolated from the feces of SPF mice, was incubated in a broth medium containing lactose and glucose at the same concentrations as found in the cecal contents of GF mice. The growth curves monitored at 660 nm are shown in Fig. 3. *E. coli* EC2 exhibited rapid growth in the medium containing 2.1 g of glucose per liter regardless of the presence of lactose. In the medium containing 16.5 g of lactose and 0.5 g of glucose per liter, glucose was preferentially used for bacterial growth followed by the additional use of lactose after a lag. This phenomenon is well-known as the glucose effect[2] or catabolite repression[6]. Glucose has a broader effect inhibiting the induction of many operons in bacteria concerned with the utilization of various other carbohydrates such as lactose and galactose[1]. From these results, it is considered that glucose present in the intestine exerted a strong positive effect on the intestinal overgrowth of *Enterobacteriaceae* in SPF mice.

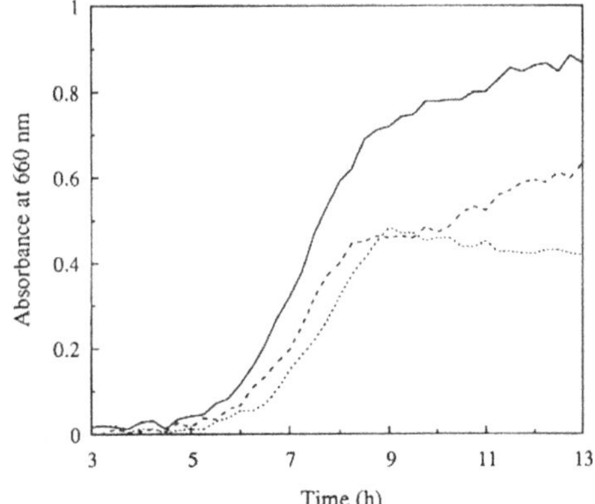

Figure 3. Effect of glucose on the proliferation of *E. coli* EC2 *in vitro*. *E. coli* EC2 was incubated with shaking at 37°C in a medium containing 0.88 g of KH_2PO_4, 1.43 g of Na_2HPO_4, 1.00 g of $(NH_4)_2SO_4$, 0.13 g of $MgSO_47H_2O$, 0.005 g of yeast extract per liter. As carbohydrate source, the medium contained 2.1 g of glucose and 26.6 g of lactose (——), 0.5 g of glucose and 16.5 g of lactose (-----), or 0.5 g of glucose (·······) per liter.

The results of this study suggest that mainly unabsorbed glucose but not lactose and galactose present in the intestine would be responsible for promoting the proliferation of intestinal *Enterobacteriaceae* in SPF mice fed milk. Presumably the addition of bLf to milk would decrease the intestinal concentrations of glucose as well as lactose and galactose in SPF mice as observed in GF mice. Lesser amounts of glucose would limit the proliferation of *Enterobacteriaceae.* Our proposed bacteriostatic mechanism of action of bLf against intestinal *Enterobacteriaceae* in SPF mice fed milk is summarized as follows: (i) orally administered bLf stimulates the absorption of glucose and galactose produced by lactose hydrolysis in the small intestine, and (ii) proliferation of *Enterobacteriaceae* in the intestine is restricted by the resulting reduced amounts of intestinal glucose. Unabsorbed lactose may be used by anaerobic bacteria such as bifidobacteria in the intestine of mice. Thus ingested bLf may also affect the composition of the intestinal flora in the milk-fed mice.

Milk is the natural first food and it contains ingredients essential for infant mammals. Considering the results of this study, lactoferrin present in mammalian milk may contribute both to the rapid growth of newborn animals by stimulating the absorption of lactose and also it may help to protect infant animals from intestinal infections through its bacteriostatic effect.

REFERENCES

1. Davis, B. D., and E. Z. Ron. 1980. Metabolic regulation, p.257–287. *In* B. D. Davis, R. Dulbecco, H. N. Eisen, and H. S. Ginsberg (ed.), Microbiology, third edition. Harper & Row, Publishers, Inc., Philadelphia.
2. Epps, H. M., and E. F. Gale. 1942. The influence of the presence of glucose during growth on the enzyme activities of E. coli: Comparison of the effect with that produced by fermentation acid. Biochem. J., 36:619–623.
3. France, G. L., D. J. Marmer, and R. W. Steel. 1980. Breast feeding and *Salmonella* infection. Am. J. Dis. Child. 134:147–152.
4. Gunn, R. A., A. M. Kimball, R. A. Pollard, J. C. Feeley, and R. A. Feldman. 1979. Bottle feeding as a risk factor for cholera in infants. Lancet. 11:730–732.
5. Howie, P. W., J. S. Forsyth, S. A. Ogston, A. Clark, and C. du V. Florey. 1990. Protective effect of breast feeding against infection. Br. Med. J. 300:11–16.
6. Magsanik, B. 1961. Catabolite repression. Cold Harbor Symp. Quant. Biol. 26: 249–256.
7. Masson, P. L., and J. F. Heremans. 1971. Lactoferrin in milk from different species. Comp. Biochem. Physiol. 39:119–129.
8. Teraguchi, S., K. Ozawa, S. Yasuda, K. Shin, Y. Fukuwatari, and S. Shimamura. 1994. The bacteriostatic effects of orally administered bovine lactoferrin on intestinal *Enterobacteriaceae* of SPF mice fed bovine milk. Biosci. Biotech. Biochem. 58:482–487.
9. Teraguchi, S., K. Shin, K. Ozawa, S. Nakamura, Y. Fukuwatari, S. Tsuyuki, H. Namihira, and S. Shimamura. 1995. Bacteriostatic effect of orally administered bovine lactoferrin on proliferation of *Clostridium* species in the gut of mice fed bovine milk. Appl. Environ. Microbiol. 61:501–506.
10. Teraguchi, S., K. Shin, T. Ogata, M. Kingaku, A. Kaino, H. Miyauchi, Y. Fukuwatari, and S. Shimamura. 1995. Orally administered bovine lactoferrin inhibits bacterial translocation in mice fed bovine milk. Appl. Environ. Microbiol. 61:4131–4134.
11. Ueda, K., S. Yamazaki, and S. Someya. 1972. Studies on tubercle bacillus infection in germ-free mice. J. Reticuloendothel. Soc. 12:545–563.
12. Wilson, T. H., and G. Wiseman. 1954. The use of sacs of everted small intestine for the study of the transference of substances from the mucosal to the serosal surface. J. Physiol. 123:116–125.
13. Wright, E. M. 1993. The intestinal Na^{+}/glucose cotransporter. Annu. Rev. Physiol. 55: 575–589.

THE MAIN SYSTEMIC, HIGHLY EFFECTIVE, AND QUICKLY ACTING ANTIMICROBIAL MECHANISMS GENERATED BY LACTOFERRIN IN MAMMALS *IN VIVO*

Activity in Health and Disease

Tadeusz Zagulski,[1] Zofia Jarzabek,[1] Alina Zagulska,[1] and Michael Zimecki[2]

[1]Polish Academy of Sciences
Institute of Genetics and Animal Breeding
Jastrzebiec, 05-551 Mroków, Poland
[2]Polish Academy of Sciences
Institute of Immunology and Experimental Therapy
53-114 Wroclaw, Poland

1. THE IMPORTANCE OF IRON FOR LIFE AND TWO FACES OF IRON IN LIVING ORGANISMS

The function of iron ions in life processes is unusual. Because its property of easy changes of its oxidation state from plus two to plus three and vice-versa, iron has been used by Nature for the creation of life. All organisms from microorganisms to mammals, utilising oxygen, need iron. Iron is necessary for all living cells, as a crucial ion in many enzymes, particulary in enzymes acting in cell energetic machinery.

But there also exists another, harmful effect of iron; iron, like many "heavy metals" in ionic form is highly toxic in living organisms, first of all because of the generation by ionic iron in body fluids and tissues of extremely toxic oxygen free radicals but also because of the non-specific binding of this ion to many important biological molecules resulting in their inactivation. Because of this toxicity, mammals have developed during evolution a chain of protein carriers of iron ions. The function of these proteins is the transport of iron in the biological fluids and tissues to the places of synthesis of heme groups and to the places of iron storage in the Reticulo-Endothelial System (RES) cells in a manner safe for the mammal. Another important function of the iron-binding proteins is its participation in the generation of important defence systems connected with iron metabolism in mammals. One of these proteins is lactoferrin.

Advances in Lactoferrin Research, edited by Spik *et al.*
Plenum Press, New York, 1998.

2. SYSTEMIC, MICROBIOSTATIC MECHANISM GENERATED BY LACTOFERRIN (AND TRANSFERRIN) *IN VIVO* IN HEALTH AND DISEASE

Lactoferrin (Lf), is an iron metabolism associated glycoprotein of mammalian tissues and their secretions of ectodermal origin[1,2]. Lf has many biological functions. According to our current knowledge, a substantial part of the biological role of Lf already appears to be connected with natural defence of mammals.

One very important, already well-explained function of Lf is the ability of this protein alone or together with another iron-binding protein, serum transferrin to generate the microbiostatic defence mechanism: both of these proteins in the forms free of iron (apoproteins) or not fully saturated with iron, scavenge iron from the physiological environment which can deprive microorganisms of this metal, and inhibit their growth *in vitro* and *in vivo*[3–9]. This microbiostatic system acts in normal health because at the normal physiological conditions serum transferrin is only about in 30% saturated with iron, and lactoferrin *in vivo* is free of iron, and also during infection; at the time of infection iron level in serum is very low, and sometimes for short times practically disappears. Such a reduction of serum iron levels has also been shown after intravenous injection of lactoferrin, lipopolysaccharide (LPS) or interleukin-1 (IL-1). In all these cases the reduction of iron ion levels was probably induced by lactoferrin released from polymorphonuclear leukocytes (PMN) or synthesised *de novo* in the inflammed tissues[10–13].

We suggest that the mechanism of generation of the decrease of iron in serum during infection is very simple: lactoferrin released from destroyed PMN is taken up by the RES and it blocks the release of iron from RES cells to the blood. Iron from serum transferrin is normally used for heme group synthesis. This quickly decreases serum iron levels. (A mechanism like this could also partially regulate the serum iron level under normal physiological conditions, but this is a hypothesis only).

This microbiostatic mechanism eliminates from the pathogenic or potentially pathogenic groups all microoganisms which cannot compete with the iron-binding proteins of mammals for iron in physiological conditions. More than ninety percent of all microorganisms belong to this group; therefore we think it is not a wild exaggeration to say that this defence system enables co-existence between microorganisms and mammals.

3. SYSTEMIC BACTERICIDAL MECHANISM GENERATED BY LACTOFERRIN IN HEALTHY AND DIABETIC ANIMALS *IN VIVO*

Unfortunately, some microorganisms are able to acquire the iron ions needed for their growth in physiological conditions. They are able do this in different ways. During growth in iron-limiting conditions, some microorganisms of this group are able to synthesize and release into the body fluids siderophores, special, low-molecular weight chelators of iron ions, which can compete for this ion with the iron-binding proteins of the host *in vivo*[14–17]. Others release into the body fluids of mammals the hemolytic enzymes, which release hemoglobin[18,19] and heme groups, from the host cells, for use by these microorganisms as a source of iron[20–23]. And at least some of them are able to bind the iron-binding proteins (lactoferrin and transferrin) to specific receptors for these proteins expressed on the outer membrane of these microorganisms, and use iron bound to these proteins for their own growth[24–27].

All these microorganisms which are able to acquire iron for their growth in physiological conditions, are pathogenic or potentially pathogenic in mammals and the above described microbiostatic defence system can in some cases only delay their growth *in vivo*. For elimination of these microorganisms the host organism requires (an) other, killing defence mechanism(s).

We discovered a protective effect of lactoferrin during bacterial infection *in vivo*. We have shown that bovine lactoferrin bLf given i.v. to mice during *E. coli* lethal experimental infections *in vivo* protects a substantial percentage of the animals against death[28]. This effect, protecting animals against septic death, was only in part dependent on iron.

In another series of experiments we have shown that the above-described protecting activity generated by lactoferrin *in vivo* is primarily the killing system and consists of two parts: The first part is the clearing activity—Lf if given intravenously (i.v.) or subcutaneously (s.c.) (or probably orally?) produces a hundred-fold increase in the clearing rate of *E. coli* cells from the blood of experimental animals by the expression of receptors for *E. coli* on the RES cells. The second part of this protecting activity is the stimulation by lactoferrin by hundreds to a thousand times increase in the killing rate of the phagocytosed *E. coli* cells in the organs in which mainly RES macrophages exist.

We have therefore examined also the activity of the killing system generated by bLf in animals with experimentally induced diabetes. The results showed that only in a short initial period after diabetes induction (48 h), the disease substantially downregulated killing rate of bacterial cells in the investigated organs and clearing rate of *E.coli* cells from the blood of experimental animals. bLf given i.v. abolished this unfavourable effect of diabetes. In the later period of disease, until 20 days, the diabetes alone stimulated killing activity but down regulated clearing activity of bacterial cells from the blood of experimental animals.

Lactoferrin substantially increased both killing and clearing activity in these longer-lived diabetic animals. In some cases this stimulating effect of bLf was very high, suggesting the concerted action of bLf and diabetes in animals at 10 and 20 days of diabetes. Despite this beneficial effect of bLf and diabetes (if acting together) in *in vivo* stimulation of killing cells of the investigated organs, in survival experiments *in vivo*, animals from all three of these diabetic groups (48 h, 10 and 20 days) were not protected by bLf. These effects were not dependent on the levels of glucose in diabetic animals. These data suggest that lactoferrin stimulates in mice also another protecting activity than that described both previously and here, of a killing and clearing activity and that diabetes down regulates this unexplained protecting system(s).

The results described in this short review showed that lactoferrin can generate in mammals *in vivo* simultaneously two important, systemic defence mechanisms: described above, a microbiostatic mechanism well established already by many scientists and a microbicidal mechanism(s) discovered by us.

According to our knowledge these are the first results clearly demonstrating that native lactoferrin can generate a powerful, systemic, nonspecific, killing, antibacterial system in mammals *in vivo*.

Both of these protecting systems, microbiostatic and microbicidal can act, we believe, in health and during disease, but exogenous lactoferrin given i.v. can accelerate their activity. Some diseases (for example, diabetes) can abolish the effectiveness of the microbicidal system(s) generated by bLf by the downregulation of another, not yet identified, but extremely important (for survival of animals) protecting system(s). For a full explanation of the mechanisms of action of the bactericidal defence system generated by Lf *in vivo* at the molecular level future research will be needed.

REFERENCES

1. Masson PL, Heremans JF, & Dive C. (1966) An iron binding protein common to many external secretion. Clin.Chim.Acta 14:735–739.
2. Masson PL, & Heremans JF. (1967) Studies on lactoferrin, the iron-binding protein of secretions. Prot.Biol.Fluids 14:115–124.
3. Spik G, & Montreuil J. (1983) Role de la lactotransferrine dans les mecanismes moleculaires de la defense antibacterienne. Bull.Europ.Physiopath.Resp. 19:123–130.
4. Weinberg ED. (1978) Iron and infection. Microbiol.Rev.42: 45–66.
5. Weinberg ED. (1984) Iron withholding: a defence against infection and neoplasia. Physiol.Rev.64: 65–102.
6. Weinberg ED. (1992) Iron depletion: A defence against intracellular infection and neoplasia. Life Sci.50: 1289–1297.
7. Bullen JJ, Ward CG, Rogers HJ. (1991) The critical role of iron in some clinical infections Eur.J.Clin.Microbiol.Infect.Dis.10: 613–617.
8. Letendre ED. (1985) The importance of iron in the pathogenesis of infection and neoplasia. TIBS.10: 166–168.
9. Emery T. (1980) Iron deprivation as a biological defence mechanism. Nature 287:776–777.
10. Kampschmidt RF, & Mesecher M. (1985) Interleukin-1 from P388D1: Effects upon neutrophils, plasma iron, and fibrinogen in rats, mice and rabbits. Proc. Soc. Exp. Med.179: 197--200.
11. Pekarek RS, Wannemacher RW, Beisel WR. (1972) The effect of leukocytic endogenous mediator (LEM) on the tissue distribution of zinc and iron. Proc.Soc.Exp.Biol.Med.140: 685–88.
12. Zagulski T, Zagulska A, Judra M, Jarzebek Z. (1985) Rabbit plasma iron level after in vivo administration of lactoferrin. Prace i Mater.Zoot.36: 95–105.
13. Gutteberg TJ, Rokke O, Andersen O, Jorgensen T. (1989) Early fall of circulating iron and rapid rise of lactoferrin in septicemia and endotoxemia: An early defence mechanism. Scan.J.Infect.Dis.21: 709–715.
14. Brock JH, Pickering MG, McDowall MC, & Deacon AG. (1983) Role antibody and enterobactin in controlling growth of Escherichia coli in human milk and acquisition of lactoferrin- and transferrin-bound iron by Escherichia coli. Infect.Immun.40: 453–459.
15. Ford S, Copper RA, Evans RW, Evans RW, Hider RC, & Williams PH. (1988) Domain preference in iron removal from human transferrin by the bacterial siderophores aerobactin and enterochelin. Eur.J.Biochem.178: 477–481.
16. Ichihara S, & Mizushima S. (1977) Involvement of other membrane proteins in enterochelin-mediated iron uptake in Escherichia coli. J.Biochem.81: 749–756.
17. Yang H, Kooi CD, & Sokol PA. (1993) Ability of Pseudomonas pseudomallei malleobactin to acquire transferrin-bound lactoferrin-bound and cell-derived iron. Infect.Immun.61: 656–662.
18. Stoebner JA, & Payne SM. (1988) Iron-regulated hemolysin production and utilization of heme and hemoglobin by Vibrio cholerae. Infect. Immunity 56: 2891–2895.
19. Pickett CL, Auffenberg T, Pesci EC, Sheen VL, & Jusuf Sri SD. (1992) Iron aquisition and hemolysin production by Campylobacter jejuni. Infect.Immun.60: 3872–3877.
20. Lee BC, & Hill P. (1992) Identification of an outer-membrane haemoglobin-binding protein in Nisseria meningitidis. J.Gen.Microbiol.138: 2647–2656.
21. Nishina Y, Miyoahi SI, Nagase A, & Shinoda S. (1992) Significant role of an exocellular protease in utilization of heme by Vibrio vulnificus. Infect. Immunity 60:2128–2132.
22. Otto BR, Verweij-van Vught AMJJ, & MacLaren MD. (1992) Transferrins and heme-compounds as iron sources for pathogenic bacteria. Critical Rev. in Microbiol.18: 217–233.
23. Smalley JW, Birss AJ, McKee AS, & Marsh PD. (1993) Haemin-binding proteins of Porphyromonas gingivalis W50 grown in a chaemostat under haemin-limitation. J.Gen.Microbiol.139: 2145–2150.
24. Lee BC, & Schryvers AB. (1988) Specifity of the lactoferrin and transferrrin receptors to Neisseria gonorrhoeae. Molec.Microbiol.2: 827–29.
25. Schryvers AB. (1989) Identification of the transferrin- and lactoferrin- binding proteins in Haemophilus influenzae. J.Med.Microbiol.29: 121–130.
26. Tigyi Z, Kishore AR, Maeland JA, Forsgren A, & Naidu AS. (1992) Lactoferrin-binding proteins in Shigella flexneri. Infect. Immunity 60: 2619–2626.
27. Alcantara J, Padda JS, & Schryvers AB. (1992) The N-linked oligosaccharides of human lactoferrin are not required for binding to bactericidal lactoferrin receptors. Can.J.Microbiol.38: 1202–1205.
28. Zagulski T, Lipilski P, Zagulska A, Broniek S, & Jarzebek Z. (1989) Lactoferrin can protect mice against a lethal dose of Escherichia coli in experimental infection in vivo. Br.J.Exp.Path.70: 697–704.

REGULATION BY LACTOFERRIN OF EPIDERMAL LANGERHANS CELL MIGRATION

Ian Kimber,[1] Marie Cumberbatch,[1] Rebecca J. Dearman,[1] Pauline Ward,[2] Denis R. Headon,[3] and Orla M. Conneely[2]

[1]Zeneca Central Toxicology Laboratory
Alderley Park, Macclesfield, United Kingdom
[2]Department of Cell Biology
Baylor College of Medicine
Houston, Texas
[3]Agennix Incorporated
Houston, Texas

1. INTRODUCTION

Lactoferrin is an iron-binding glycoprotein that shares structural homology with transferrin and which is found in milk, other epithelial secretions and the secondary granules of neutrophils[1]. One function ascribed to lactoferrin is that of antibacterial activity which is effected primarily through the sequestration of iron necessary for microbial growth[1,2]. It is recognized, however, that in addition to its bacteriostatic properties, lactoferrin may serve as a modulator of immune and inflammatory responses. Of particular interest is the observation that this glycoprotein may influence the production of some cytokines, including proinflammatory cytokines[1–3]. Among the cytokines shown to be regulated negatively by lactoferrin are tumour necrosis factor α (TNF-α), interleukin 1β (IL-1β) and granulocyte/macrophage colony-stimulating factor (GM-CSF)[2–4]. It has been demonstrated that (bovine) lactoferrin administered intravenously to mice prior to an injection of bacterial lipopolysaccharide (LPS) inhibited significantly the production of TNF-α normally induced by LPS[5]. It has been argued, however, that the ability of lactoferrin to inhibit the stimulation of TNF-α by LPS may be secondary to its capacity to bind LPS and thereby compromise the signal for cytokine production[3].

With the objective of investigating further the regulation of TNF-α by lactoferrin, we have here examined the effect of this molecule on the migration from the skin of epidermal Langerhans cells (LC); a process that is considered to be dependent upon the *de novo* synthesis of TNF-α by keratinocytes.

Langerhans cells reside in the epidermis where they form a semi-contiguous network that serves as a trap for external antigens[6]. Following topical exposure to chemical

Advances in Lactoferrin Research, edited by Spik *et al.*
Plenum Press, New York, 1998.

allergens, or in response to other forms of dermal trauma, LC migrate from the epidermis and travel, via afferent lymphatics, to draining lymph nodes where they accumulate as immunostimulatory dendritic cells (DC)[7,8]. This process is central to the initiation of cutaneous immune responses; a proportion of the LC that leave the skin bears high levels of antigen and when they reach the lymph nodes they have acquired the characteristics necessary for the effective presentation of that antigen to responsive T lymphocytes[6].

The important signals that mediate the migration and functional maturation of LC are provided by cytokines. The skin is an immunologically active tissue and epidermal cells, both LC themselves and keratinocytes, are a rich source of both constitutively expressed and inducible cytokines. It has been reported previously that TNF-α, an inducible product of keratinocytes in the epidermis, may play a particularly important role in the stimulation of LC migration. This cytokine is induced rapidly following skin sensitization[9]. Intradermal injection of mice with homologous recombinant TNF-α has been found to cause a rapid reduction in the frequency of epidermal LC and this is paralleled, somewhat later, by an accumulation of DC within draining lymph nodes.[10,11] It has been demonstrated also that the systemic (intraperitoneal) injection of mice with neutralizing anti-TNF-α antibody serves to inhibit the accumulation of DC in draining nodes which is normally induced by skin sensitization with chemical allergens such as oxazolone.[12]

We have now examined in mice the ability of homologous recombinant lactoferrin to influence the accumulation of DC in draining lymph nodes induced either by topical administration of oxazolone or by the intradermal injection of murine TNF-α.

2. EXPERIMENTAL

In the experiments described here acid-washed (iron-free) recombinant mouse lactoferrin has been used. Mouse lactoferrin was produced in *Aspergillus awamori* using the same expression system as that employed to produce human lactoferrin.[13] Similar results (not presented) were obtained with iron-saturated lactoferrin. In the first series of experiments, the ability of mouse lactoferrin administered intradermally to influence the accumulation of DC in draining lymph nodes induced by oxazolone was investigated. Treatment with lactoferrin was performed two hours prior to skin sensitization and the numbers of DC within lymph nodes measured, by morphological criteria[8], after a further 18 hours. The results of a representative experiment are illustrated in Figure 1. As this figure demonstrates, lactoferrin caused a dose-dependent inhibition of oxazolone-induced DC accumulation compared with controls which had received 0.02% bovine serum albumin (BSA) in place of lactoferrin. At the highest concentration tested (0.02%), lactoferrin resulted in a 61% reduction in the frequency of lymph node DC.

In subsequent experiments the influence of lactoferrin on oxazolone induced migration of DC was compared with its affect on migration stimulated by intradermal injection of TNF-α. A representative experiment is shown in Figure 2. As reported previously,[10] the intradermal administration to mice of homologous TNF-α induced an increased frequency of DC within draining lymph nodes measured 4 hours later. Prior dermal exposure of mice to lactoferrin had no effect on the integrity of DC accumulation compared with control animals that received BSA (Figure 2A). In concurrent experiments, the influence of lactoferrin on oxazolone-induced DC accumulation was determined. Consistent with the data displayed in Figure 1, intradermal exposure of mice to 0.02% lactoferrin caused a substantial reduction in the number of DC found in draining lymph nodes 18 hours following sensitization (Figure 2B). In neither experiment did treatment of mice with BSA cause any

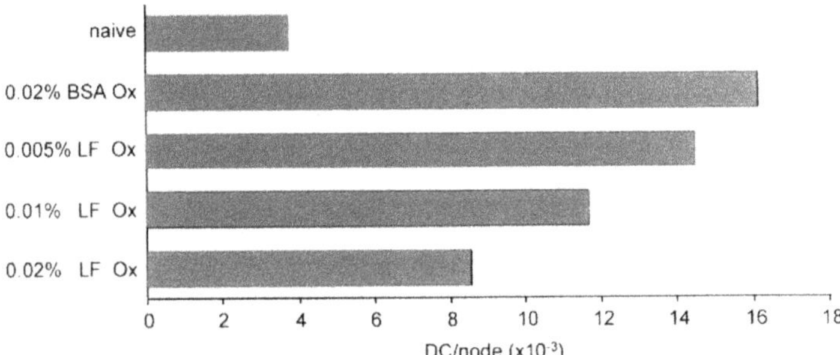

Figure 1. Influence of homologous lactoferrin on the induction in mice of DC accumulation in draining lymph nodes by oxazolone. Groups of BALB/c strain mice (n=10) received intradermal injections (30μl) into both ear pinnae of acid washed homologous recombinant lactoferrin (Lf), or control protein (bovine serum albumin; BSA) each suspended, at the concentrations indicated, in phosphate buffered saline. Two hours later mice were exposed on the dorsum of each ear to 25 μl of 0.5% oxazolone (Ox) dissolved in 4:1 acetone:olive oil. Additional control mice were untreated. Draining auricular lymph nodes were removed 18 hours following sensitization, pooled for each experimental group, and the frequency of DC was measured by direct morphological examination using phase contrast microscopy.

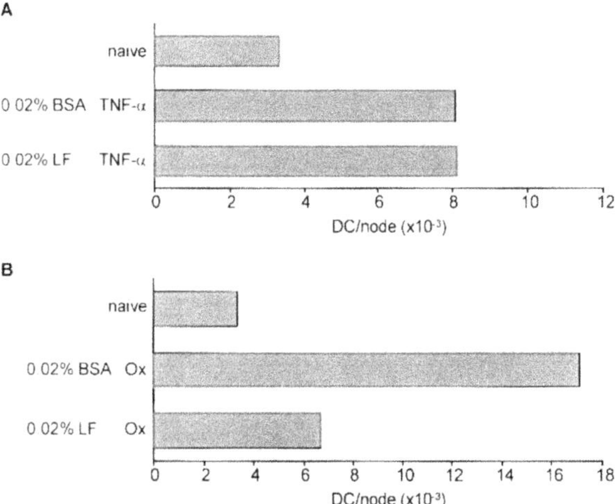

Figure 2. Influence of homologous lactoferrin on the induction in mice of DC accumulation in draining lymph nodes by oxazolone or TNF-α. Groups of BALB/c strain mice (n=10) received intradermal injections (30 μl) into both ear pinnae of acid washed homologous recombinant lactoferrin (Lf), or control protein (bovine serum albumin; BSA) at the same concentration (0.02%) each suspended in phosphate buffered saline. Two hours later mice either received a second intradermal injection (30μl) into the same site of homologous recombinant tumour necrosis factor α (TNF-α; 50 ng) or were exposed topically on the dorsum of each ear to 25 μl of 0.5% oxazolone (Ox) dissolved in 4:1 acetone:olive oil. Draining auricular lymph nodes were removed 4 hours (TNF-α), or 18 hours (oxazolone) later, pooled for each experimental group, and the frequency of DC measured by direct morphological examination using phase contrast microscopy.

inhibition of either oxazolone- or TNF-α induced arrival of DC in draining lymph nodes (data not presented).

3. DISCUSSION

There is compelling evidence that the integrity of induced LC migration from the epidermis and the subsequent accumulation of DC in draining lymph nodes are dependent upon the availability locally of TNF-α and the *de novo* synthesis of this cytokine within the skin.[10–12] The results described here reveal that intradermal injection of homologous lactoferrin caused a marked and dose-dependent inhibition of the accumulation within draining lymph nodes of DC that is normally associated with skin sensitization. In contrast, lactoferrin failed to influence the arrival of DC within draining lymph nodes stimulated by the intradermal administration of homologous recombinant TNF-α. This latter observation indicates that lactoferrin does not pervert the process of LC migration to draining lymph nodes *per se* and suggests that the effect of this molecule is to reduce the availability of TNF-α which provides a necessary signal for LC migration. The mechanism of action of lactoferrin in this context is not known. It is possible, however, that lactoferrin administered locally serves to inhibit or alter the synthesis of this cytokine. In this respect it is of considerable interest that lactoferrin may be internalized by cells and transported to the nucleus where it influences directly gene expression via interaction with distinct sequences.[14] Such a mechanism has been proposed as the basis for the secondary immune and inflammatory regulatory activities of lactoferrin which include influences on cytokine expression.[14] The assumption is that in the epidermis the effects on induced TNF-α expression by keratinocytes are mediated following interaction of lactoferrin with specific receptors on these cells, the existence of which are suggested by preliminary evidence (data not presented).

In conclusion, the results reported here demonstrate that lactoferrin is able *in vivo* to inhibit a biological process for which there is a requirement for TNF-α under conditions where the induced expression of this cytokine is independent of LPS. These data provide further evidence that lactoferrin is able directly to regulate the expression of cytokines and, as a consequence, may exert important influences on immune responses and inflammatory reactions.

REFERENCES

1. Lonnerdal, B, Iyer S. (1995) Lactoferrin: molecular structure and biological function. Ann. Rev. Nutr. 15: 93–110.
2. Sanchez, L, Calvo, M, Brock, JH. (1992) Biological role of lactoferrin. Arch. Dis. Childhood 67: 697–661.
3. Brock, J. (1995) Lactoferrin: a multifunctional immunoregulatory protein? Immunol. Today 16: 417–419.
4. Penco, S, Pastorino, S, Bianchi-Scarra, G, Garre C. (1995) Lactoferrin down-modulates the activity of the granulocyte macrophage colony-stimulating factor promoter in interleukin-1β-stimulated cells. J. Biol. Chem. 270: 12263–12268.
5. Machnicki, M, Zimecki, M, Zagulski, T. (1993) Lactoferrin regulates the release of tumour necrosis factor alpha and interleukin 6 in vivo. Int. J. Exp. Path. 74: 433–439.
6. Kimber, I, Cumberbatch, M. (1992) Dendritic cells and cutaneous immune responses to chemical allergens. Toxicol. Appl. Pharmacol. 117: 137–146.
7. Kinnaird, A, Peters, SW, Foster, JR, Kimber, I. (1989) Dendritic cell accumulation in draining lymph nodes during the induction phase of contact allergy in mice. Int. Arch. Allergy Appl. Immunol. 89: 202–210.

8. Cumberbatch, M, Kimber, I. (1990) Phenotypic characteristics of antigen-bearing cells in the draining lymph nodes of contact sensitized mice. Immunology 71: 404–410.
9. Enk, AH, Katz, SI. (1992) Early molecular events in the induction phase of contact sensitivity. Proc. Natl. Acad. Sci USA 89: 1398–1402.
10. Cumberbatch, M, Kimber, I. (1992) Dermal tumour necrosis factor-α induces dendritic cell migration to draining lymph nodes and possibly provides one stimulus for Langerhans cell migration. Immunology 75: 257–263.
11. Cumberbatch, M, Fielding, I, Kimber, I. (1994) Modulation of epidermal Langerhans cell frequency by tumour necrosis factor-α. Immunology 81: 395–401.
12. Cumberbatch, M, Kimber, I. (1995) Tumour necrosis factor-α is required for accumulation of dendritic cells in draining lymph nodes and for optimal contact sensitization. Immunology 84: 31–35.
13. Ward, PP, Piddington, CS, Cunningham, GA, Zhou, X, Wyatt, RD, Conneely, OM. (1995) A system for production of commercial quantities of human lactoferrin: a broad spectrum antibiotic. Bio/Technology 13: 498–502.
14. Ho, J, Furmanski, P. (1995) Sequence specificity and transcriptional activation in the binding of lactoferrin to DNA. Nature 373: 721–724.

LACTOFERRIN STIMULATES THE MITOGEN-ACTIVATED PROTEIN KINASE IN THE HUMAN LYMPHOBLASTIC T JURKAT CELL LINE

Isabelle Duthille, Maryse Masson, Geneviève Spik, and Joël Mazurier

Centre National de la Recherche Scientifique
Unité Mixte de Recherche n°111
Laboratoire de Chimie Biologique
Université des Sciences et Technologies de Lille
59655 Villeneuve d'Ascq cedex, France

1. INTRODUCTION

Lactoferrin[11] (also called lactotransferrin) has been shown to be involved in numerous inflammatory and immune response functions such as regulation of granulocyte monocyte colony stimulating factor[6,7,12,17] and of interleukin[9] synthesis, activation of NK cell activity[13], inhibition of metastasis[2] and maturation of T- and B-cells[15,16]. Recently, it has also been demonstrated the activation of a nuclear reporter gene harbouring the specific DNA sequence that lactoferrin recognizes[8]. Therefore, lactoferrin could be considered as a transcriptional factor[1]. It requires that lactoferrin is internalized and imported into the nucleus of target cells where it acts as a transcriptional factor[1]. However, there is no evidence that cell-surface lactoferrin receptors can transport lactoferrin either directly or via a new trafficking pathway to the nucleus. In order to investigate how lactoferrin affects the immune system, we have developed an experimental model using the human lymphoblastic T Jurkat cell line which upon lactoferrin treatment slowly undergoes a maturation process[5]. Jurkat cells express the lactoferrin receptor[10] at the cell surface in coated-pit like domains, associated into clusters, and in different subsets of intracellular vesicles[3]. Therefore, we have investigated the first intracellular events following the binding of lactoferrin to lymphocytes: internalization process of lactoferrin and activation of a transduction pathway.

Advances in Lactoferrin Research, edited by Spik *et al.*
Plenum Press, New York, 1998.

2. THE EFFECT OF LACTOFERRIN ON CELL DIFFERENTIATION

We have studied the modulation of CD4 antigen expression in parallel with the expression of serum transferrin receptor (CD71) over a period of 3 weeks culture in presence of 10% fetal calf serum and in the continuous presence of iron-saturated lactoferrin[5]. In these conditions, Jurkat cells after one week of treatment showed an upregulation of CD4 antigen expression. The cell surface density of the CD4 antigen reached a plateau after 2 weeks of treatment. Simultaneously, a progressive down regulation of CD71 with accompanying reduction of cell volume were observed. These modifications in the expression of CD4 and CD71 antigens were not observed in cells growing in the absence of lactoferrin. These effects were dependent on both lactoferrin concentration and presence of iron.

3. INTERNALIZATION PROCESS OF LACTOFERRIN

We have shown[4] using both a biochemical approach and confocal and electron microscopy that after binding to the cell surface, lactoferrin enters T-cells via receptor mediated endocytosis. Both iron-free and iron-saturated lactoferrins were internalized, albeit at different rates, with their endocytic pathway being identical. The demonstration that lactoferrin was readily internalized by a human lymphoblastic cell line favors the notion that there exists 2 binding components: a specific one which allowed internalization of lactoferrin and one less specific component allowing only binding and involving mainly ionic interactions between cationic residues of lactoferrin and membrane anionic components. The specific component could involve the 105 kDa molecular mass membrane glycoprotein that we have described[10] and the lipoprotein receptor-related protein[14]. The volume of the low affinity component, which could be related to the amount of proteoglycans, depends on the cell studied: small in the case of lymphocytes and high in the case of hepatocytes, monocytes and epithelial tumor cells. The internalization of lactoferrin was followed by an accumulation of degraded lactoferrin into cells and into the culture medium. Confocal and electron microscopy have shown a complex pattern of localisation for the lactoferrin molecule. Fluorescein carbohydrate labeled lactoferrin accumulated in vesicles and electron microscopy showed the presence of lactoferrin in endosome compartments, however we have, up to now, no direct evidence that lactoferrin is internalized up to lysosomes. The presence of vesicular intact lactoferrin and the persistent level of 60–70% of undegraded lactoferrin in the culture medium strongly suggest that lactoferrin is recycled, and that a fraction (40–30%) of the ligand is degraded at each round of endocytosis. Confocal microscopy confirmed the colocalisation of these 2 proteins at early stages of ligand internalization. No lactoferrin had been observed to reach the nucleus, suggesting that lactoferrin can act by activating transduction pathways.

4. PHOSPHORYLATION OF INTRACELLULAR PROTEINS

Using anti-phosphotyrosine and anti-phosphothreonine antibodies we demonstrated that lactoferrin induced a cascade of phosphorylations with tyrosine phosphorylations as primary events. Ten clearly separated tyrosine-phosphorylated proteins were induced or enhanced after incubation with lactoferrin. The tyrosine-phosphorylation of the polypeptides was transient, reaching a maximum between 5 and 10 min. The P-Tyr proteins could be grouped into three, based on the time at which phosphorylation reached a maximum.

Some polypeptides (p44, p52 and p57) were rapidly phosphorylated, peaking at 5 min. However, most of the proteins had a peak phosphorylation at 10 min. Whereas all the P-Tyr polypeptides returned to baseline levels by 20 min, one protein was phosphorylated more slowly, reaching a maximum at about 20 min.

The comparison of the pattern of lactoferrin induced-phosphorylated polypeptides to those obtained by activating TCR or IL-2 receptor led to suggest that $p56^{lck}$, $p60^{shc}$ and MAP Kinase was activated by lactoferrin.

5. ACTIVATION OF THE MITOGEN-ACTIVATED PROTEIN KINASE

A polypeptide, with a maximal tyrosine-phosphorylation after 5 min of incubation with lactoferrin, presented the electrophoretic properties of the MAP Kinase protein. To identify this protein as MAP kinase, Jurkat T-cells were first incubated with lactoferrin for specific periods of time and lysates were analysed by Western blotting with anti-MAP Kinase or anti-P-Tyr antibody. We investigated the possible activation of this enzyme by analysing the ability of MAP kinase to phosphorylate the MBP, one of its synthetic substrates. Stimulation of MAP Kinase activity was analysed using the in-gel kinase assay. Cellular extracts were indeed subjected to kinase assay in MBP-containing polyacrylamide gels. After SDS-PAGE, the gel was incubated with [γ-^{32}P]ATP, then dried and analysed by autoradiography. During treatment with lactoferrin, the activity of a polypeptide, migrating at 44 kDa, presented variation, with a maximum following 5–7 min of treatment of lactoferrin.

6. CONCLUSION

The present investigation showed that lactoferrin, which promote differentiation but not proliferation of Jurkat lymphoblastic cells, acts by activating a transduction pathway. In fact, the binding of lactoferrin on Jurkat T-cells induces one or several transduction pathway(s) involving protein tyrosine kinases and MAP kinase. This hypothesis agrees with the internalization study in Jurkat T-cells which suggested that no secreted-lactoferrin is imported into the nucleus of these lymphoblastic cells.

ACKNOWLEDGMENTS

This work was supported by the Université des Sciences et Technologies de Lille, the Centre National de la Recherche Scientifique (UMR n°111 CNRS; Director: Prof. A. Verbert) and the Association pour la Recherche sur le Cancer (Grant 3040).

REFERENCES

1. Baeurle, P.: Enter a polypeptide messenger. (1995) Nature 373, 661–662
2. Bezault, J., R. Bhimani, J. Wiprovnick, P. Furmanski: Human lactoferrin inhibits growth of solid tumors and development of experimental metastases in mice. (1994) Cancer Res. 54, 2310–2312
3. Bi, B.Y., B. Leveugle, J.L. Liu, A. Collard, P. Coppe, A.C. Roche, N. Nillesse, M. Capron, G. Spik, J. Mazurier: Immunolocalization of the lactoferrin receptor on the human T lymphoblastic cell line Jurkat. (1994) Eur. J. Cell Biol. 65, 164–171

4. Bi, B. Y., J. L. Liu, D. Legrand, A.C. Roche, M. Capron, G. Spik, J. Mazurier: Internalization of human lactoferrin by Jurkat a human lymphoblastic T-cell line. (1996) Eur. J. Cell Biol. 69, 288–296
5. Bi, B. Y., A.M. Lefebvre, D. Dus, G. Spik, J. Mazurier: Effect of lactoferrin on proliferation and differentiation of the Jurkat human lymphoblastic T-cell line. (1997) Arch. Immunol. Therap. Exp., 45, 315–320.
6. Broxmeyer, H.E., A. Smithyman, R.R. Eger, P.A. Meyers, M. de Sousa: Identification of lactoferrin as the granulocyte derived inhibitor of colony stimulating activity (CSA-production). (1978) J. Exp. Med. 148, 1052–1067
7. Garré, C., G. Bianchi-Scarra, M. Sirito, M. Musso, R. Ravazzalo: Lactotransferrin binding sites and nuclear localization in K 562 cells (1992) J. Cell. Physiol. 153, 477–482
8. He, J., P. Furmanski: Sequence specificity and transcriptional activation in the binding of lactoferrin to DNA. (1995) Nature 373, 721–724
9. Machnicki, M., M. Zimecki, T. Zagulski: Lactoferrin regulates the release of tumour necrosis factor alpha and interleukin 6 *in vivo*. (1993) Int. J. Exp. Path. 74, 433–439
10. Mazurier, J., D. Legrand, W.L. Hu, J. Montreuil, G. Spik: Expression of human lactoferrin receptors in phytohemagglutinin-stimulated human peripheral blood lymphocytes. Eur. J. Biochem. (1989) 179, 481–487
11. Montreuil, J., J. Tonnelat, S. Mullet: Préparation et propriétés de la lactosidérophiline (lactotransferrine) du lait de Femme. (1960) Biochim. Biophys. Acta 45, 413–421
12. Sawatzki, G., I.N. Rich: Lactoferrin stimulates colony stimulating factor production *in vitro* and *in vivo*. (1989) Blood Cells 15, 371–385
13. Shau, H., A. Kim, H. Golub: Modulation of natural killer and lymphokine-activated killer cell cytotoxicity by lactoferrin. (1992) J. Leuk. Biol. 51, 343–349
14. Willnow, T.E., J.L. Goldstein, K. Orth, M.S. Brown, J. Herz. Low density lipoprotein receptor-related protein and gp330 bind similar ligands, including plasminogen activator-inhibitor complexes and lactoferrin, an inhibitor of chylomicron remnant clearance. (1992) J. Biol. Chem. 267, 26172–26180
15. Zimecki, M., J. Mazurier, M. Machnicki, Z. Wieczorek, J. Montreuil, G. Spik: Immunostimulatory activity of lactotransferrin and maturation of $CD4^-CD8^-$ murine lymphocytes. (1991) Immunol. Lett. 30, 119–124
16. Zimecki, M., J. Mazurier, G. Spik, J.A. Kapp: Lactotransferrin induces phenotypic and functional changes in immature B cells. (1995) Immunology 86, 122–127
17. Zucali, J.R., H.E. Broxmeyer, D. Levy, C. Morse: Lactoferrin decreases monocyte-induced fibroblast production of myeloid colony stimulating activity by suppressing monocyte release of interleukin 1. (1989) Blood 74, 1531–1536

EFFECTS OF ORALLY ADMINISTERED BOVINE LACTOFERRIN ON THE IMMUNE SYSTEM OF HEALTHY VOLUNTEERS

Koji Yamauchi, Hiroyuki Wakabayashi, Shinichi Hashimoto,
Susumu Teraguchi, Hirotoshi Hayasawa, and Mamoru Tomita

Nutritional Science Laboratory
Morinaga Milk Industry Co., Ltd.
Zama-City, Kanagawa, 228, Japan

A protective effect of bovine lactoferrin (Lf) during lethal bacteraemia has been reported in mice. Also, protective effects of orally administered bovine Lf have been reported in cases of intractable stomatitis in cats and Cryptocaryon irritans infection in red sea bream. In this study, we examined the effects of orally administered bovine Lf on the immune system of healthy volunteers.

Ten healthy male volunteers (age range of 31 to 55 years old) were given bovine Lf (2 g/body/day) for 4 weeks. Blood samples were drawn before, during and after administration of Lf. Phagocytic activity and superoxide production activity of polymorphonuclear leukocytes (PMN) were evaluated from the number of PMN phagocytizing polymer particles and by the dichlorofluorescein (DCFH) oxidation assay, respectively. The expression levels of CD11b, CD16 and CD56 molecules on leukocytes were quantified using flow cytometry.

The phagocytic activity of PMN increased during the period of Lf administration in 3 of the 10 volunteers. In 2 of the 3 volunteers in which the phagocytic activity increased, PMN expressed CD16 at higher levels corresponding to the increase in phagocytic activity observed. In addition, the superoxide production activity of PMN increased in 1 of the 10 volunteers. The CD16+ lymphocytes increased in 3 of the 10 volunteers, whereas the CD11b+ lymphocytes and CD56+ lymphocytes increased in 4 volunteers including the same 3 volunteers who showed an increase in CD16+. These results suggest that the proportion of natural killer (NK) cells among the lymphocytes might have increased in these subjects.

It was demonstrated that the phagocytic activity or superoxide production activity of PMN or the proportions of CD11b+, CD16+ and CD56+ in lymphocytes were influenced by Lf administration in 7 of the 10 volunteers, while the effects of Lf on the immune system differed in individual cases. These results suggest that Lf administration may influence primary activation of the host defense system.

Advances in Lactoferrin Research, edited by Spik *et al.*
Plenum Press, New York, 1998.

1. INTRODUCTION

Lf has been shown to function not only as an iron-binding protein but also as a modulator of various immunological responses and as a protective agent against various infections[1,2]. Recently it has reported that bovine Lf displays host defense effects in vivo. A protective effect of bovine Lf during lethal bacteraemia has been reported in mice treated intravenously with a high dose of Lf 24 hours before challenge with Escherichia coli[3]. Also, protective effects of orally administered bovine Lf have been reported in cases of Cryptocaryon irritans infection in red sea bream[4]. And, orally administered bovine Lf improves intractable stomatitis in feline immunodeficiency virus (FIV)-positive and FIV-negative cats and enhances the phagocytic activity of neutrophils in healthy and ill cats[5]. In this study, as an initial step in exploring the protective effects of bovine Lf in humans, we examined the effects of Lf on the activation of immunologically competent cells in human volunteers who ingested a food supplement containing bovine lactoferrin.

2. MATERIALS AND METHODS

2.1. Administration of Bovine Lf to Human Volunteers: Study Design

The subjects were ten healthy male volunteers. The age range of the healthy volunteers was 31 to 55 years old and the average age was 45 years old. Informed consent was obtained from each of the subjects in advance. We performed this study using a dietary food supplement tablet containing Lf. Each tablet contained 150 mg of Lf derived from cow's milk. Each of the volunteers in this study took 7 tablets in the morning and 7 in the evening for 4 weeks. The total intake from these 14 tablets was approximately 2 g of Lf per day. Blood samples were drawn before administration of Lf and 2 weeks and 4 weeks after the start of Lf administration.

2.2. Reagents

Fluorescent Monodisperse Carboxylated Microspheres (FITC-beads) were obtained from Polyscience, Inc. (Qarrington, PA, USA). 2',7'-dichlorofluorescein diacetate (DCFH-DA) was purchased from Eastman Kodak (Rochester, NY; USA). Phorbol myristate acetate (PMA) was obtained from Sigma Chemicals (St. Louis, MO, USA). Lysing reagent was purchased from Becton Dickinson (San Jose, CA, USA). PE-labeled anti-human CD11b, anti-CD56 monoclonal antibody, and FITC-labeled anti-human CD16 monoclonal antibody were obtained from PharMingen (San Diego, CA, USA).

2.3. The Phagocytic Activity of Polymorphonuclear Leukocytes (PMN)

The phagocytic activity of PMN was examined by use of a method described by Kamakura et al.[6] Briefly, 0.1 ml of collected human peripheral blood was added to 0.1 ml of a suspension of FITC-beads (5×10^7 particles/ml). After incubation at 37°C for 15 minutes, the mixture was incubated at 4°C for 15 minutes with 2 ml of lysing reagent to lyse the erythrocytes. The remaining white blood cells were washed and resuspended in Ca^{2+}, Mg^{2+}-free phosphate-buffered saline (PBS(-)). Flow cytometry analysis was performed with EPICS-PROFILE II, Coulter (Hialeah, FL, USA). Measurement was made after gating on PMN. The PMN were easily distinguished from lymphocytes and monocytes by their characteristic forward/sideward scatter. The phagocytic activity of PMN was evaluated from the ratio of cells capable of phagocytizing FITC-beads.

2.4. Superoxide Production Activity of PMN

Superoxide production activity was examined by use of a method described by Kamakura et al.[6] Briefly, the red blood cells in the human peripheral blood collected were allowed to sediment at room temperature for 40 minutes after mixing with a three-fourths volume of 1% dextran in PBS(-). The cells in the leukocyte-rich supernatant obtained were washed and resuspended in PBS(-). A 0.02 ml aliquot of the suspension was added to 0.13 ml of 5 mM DCFH-DA and incubated at 37°C for 15 minutes. Then, 0.02 ml of PMA (25 mg/ml) and 0.03 ml of EDTA (20 mM) were added and the sample mixture was incubated at 37°C for 15 minutes. DCFH-DA taken into the cells was hydrolyzed to DCFH. DCFH is oxidized to highly fluorescent DCF by superoxide generated by the PMN. The intensity of cellular DCF fluorescence was monitored by flow cytometry.

2.5. Measurement of CD11b, CD16, and CD56

The cells in the leukocyte-rich supernatant obtained from the collected blood were washed and resuspended in PBS(-) containing normal goat serum. A 0.1 ml aliquot of the suspension was added to 0.02 ml of PE-labeled anti-human CD11b, anti-CD56 monoclonal antibody, or FITC-labeled anti-human CD16 monoclonal antibody and the samples were incubated at 37°C for 15 minutes. After washing with PBS(-), measurement of CD11b, CD16 and CD56 expressed on PMN and lymphocytes was performed by flow cytometry.

3. RESULTS

3.1. PMN Functions

The phagocytic activity was estimated from the proportion of PMN phagocytizing particles. The phagocytic activity increased in three of ten subjects during the 4-week period of Lf administration. In one subject with high phagocytic activity before Lf administration, the activity decreased during Lf administration. Obvious changes were not observed in the other 6 subjects. High levels of CD16 were expressed on the PMN in 4 of the 10 volunteers during the period of Lf administration. In 2 of the 3 volunteers in whose cases the phagocytic activity increased, PMN expressed CD16 at higher levels corresponding to the increase in phagocytic activity observed. The superoxide production activity increased during the period of Lf administration in one of ten healthy volunteers (Table 1).

3.2. The Proportions of CD11b+, CD16+, and CD56+ Lymphocytes and Natural Killer Cell Activity

The proportion of CD16+ lymphocytes increased in 3 of the 10 volunteers, whereas the proportion of CD56+ lymphocytes increased in 4 volunteers including the same 3 volunteers who showed an increase in CD16+. The proportion of CD11b+ lymphocytes also increased in the same 4 volunteers. We monitored the natural killer cell activity in one healthy volunteer (Subject H). The activity increased during the 4 week period of Lf ad-

Table 1. The effect of orally administered bLf on the immune system of human volunteers

Subject	Age	PMN			Lymphocytes		
		Phagocytic activity	CD16+ intensity	Superoxide production	CD16+ proportion	CD11b+/56+ proprotion	NK activity
A	31	→	→	→	→	→	
B	33	→	↑	→	→	→	
C	37	↑	↑	→	→	→	
D	40	↓	→	→	↑	↑	
E	48	↑	↑	→	→	↑	
F	48	↑	→	→	↑	↑	
G	51	→	→	→	→	→	
H	55	→	↑	→	↑	↑	↑
I	55	→	→	→	→	→	
J	55	→	→	↑	→	→	

The upright arrow indicates that the parameter concerning PMN or lymphocytes increased during the 4-week period of LF administration.
The sideways arrow indicates that the parameter did not change obviously.

ministration in a manner corresponding to the increase in proportion of CD11b+, CD16+ and CD56+ lymphocytes (Table 1).

4. DISCUSSION

This study indicates that orally administration of bovine Lf influenced either PMN functions or the proportions of CD11b+, CD16+ and CD56+ lymphocytes in 7 of 10 healthy human volunteers. PMN contribute to the non-specific immune defense through their capacity to take up particles (phagocytosis) and to destroy them (intracellular killing : superoxide production activity). The phagocytic activity increased in three of ten subjects during the period of Lf administration. In 2 of the 3 volunteers in whose cases the phagocytic activity increased, the PMN expressed CD16 at higher levels corresponding to the increase in phagocytic activity observed. The CD16 antigen is an IgG Fc receptor expressed on granulocytes, monocytes and natural killer cells (7,8). As one of the reasons for the increase in phagocytic activity, it is thought that the levels of the IgG Fc receptor expressed on PMN may have increased. In one subject with high phagocytic activity before Lf administration, the activity decreased during the 4-week period of Lf administration. The superoxide production activity increased 2 weeks after the start of Lf administration but decreased slightly 4 weeks after the start of administration in seven of ten subjects (data not shown). These findings suggest the possibility that orally administered bovine Lf exerts an immunomodulatory effect on PMN functions in humans.

The CD11b antigen (Mac1) has been identified as the C3bi complement Receptor. CD11b antibodies react with myeloid and natural killer cells in the peripheral blood[7,9]. CD56 is a 140 kDa isoform of a neural cell adhesion molecule and CD56 is expressed on cells with natural killer activity[10]. The proportion of CD16+ lymphocytes increased in 3 of the 10 volunteers, whereas the proportion of CD11b+, CD56+ lymphocytes increased in 4 volunteers including the same 3 volunteers who showed an increase in CD16+. These results suggest that the proportion of natural killer cells among the lymphocytes had increased in these subjects.

The natural killer cell activity in one healthy volunteer increased in the same manner as the proportion of CD11b+, CD16+ and CD56+ lymphocytes. The natural killer cell activity can be estimated from the proportion of CD11b+, CD16+ and CD56+ lymphocytes. Therefore, it seems likely that the natural killer cell activity is increased in subjects who showed an increase in CD11b+, CD16+ and CD56+. The in vitro and in vivo stimulation of NK and LAK cell activity by lactoferrin has been reported[11]. In our study, it was suggested that augmentation of NK cell activity may have occurred in some of the subjects administered Lf.

As a result of ingesting Lf, parameters relating to either PMN or NK cells increased in 7 of the subjects, but not in subjects A,G and I. These results suggest that Lf administration may influence primary activation of the host defense system, while the effect of Lf differs in individual cases. It is possible to see definitely the immunomodulatory effects of Lf in the case of a compromised host such as diabetic patients and hemodialysis patients. Lf receptors have been reported to exist on monocyte/macrophage cells. Specific binding Lf to human neutrophils as well as B-lymphocytes also has been reported[12]. Maneva et al. have shown the existence of receptors for bovine Lf on the PMN plasma membrane[13]. Bovine Lf and its fragments may modulate cell function directly in the blood or indirectly after stimulation of PMN in the intestinal mucosa. Further studies are required concerning the absorption and distribution of Lf and its fragments after ingestion and its immunoregulatory role.

REFERENCES

1. Sanchez, L., Calvo, M., and Brock, J.H. (1992) Biological role of lactoferrin. Arch. Dis. Child 67, 657–661.
2. Yamauchi, K., Tomita, M., Giehl, T.J., and Ellison III, R. T. (1993) Antibacterial activity of lactoferrin and a pepsin-derived lactoferrin peptide fragment. Infect. Immun. 61, 719–728.
3. Zagulski, T., Lipinski, P., Zagulska, A., Broniek, S., and Jarzabek, Z. (1989) Lactoferrin can protect mice against a lethal dose of Escherichia coli in experimental infection in vivo. Br. J. Exp. Path. 70, 697–704.
4. Kakuta, I. and Kurokura, H. (1995) Defensive effect of orally administered bovine lactoferrin against Cryptocaryon irritans infection of red sea bream. Fish Pathology 30, 289–290.
5. Sato, R., Inanami, O., Tanaka, Y., Takase, M., and Naito, Y. (1996) Oral administration of bovine lactoferrin for treatment of intractable stomatitis in feline immunodeficiency virus (FIV)-positive and FIV-negative cats. Am. J. Vet. Res. 57, 1443–1446.
6. Kamakura, M., Kouno, J., and Hamaka, T., (1991) Studies of polymorphonuclear leukocytes by flow cytometry. Modern Medical Laboratory (Japanese) 49, 427–433.
7. Knapp, W., Dorken, B., and Rieber, E.P., et al., eds. (1989) Leucocyte typing IV: White cell differentiation antigens, Oxford 22 University Press, New York .
8. Barclay, N.A., Birkeland, M.L., and Brown, M.H., et al., eds. (1993) The 25. Leukocyte antigen factsbook, CD11 section, Academic Press Inc, San Diego, California, p124.
9. Uciechowski, P. and Schmidt, R.E., (1989) Cluster report: CD11. In: Leucocyte typing IV: White cell differentiation antigens. Knapp, W., Dorken, B., and Gilks, W.R., et al., eds. Oxford University Press, New York, p.543–544.
10. Schlossman, S., Bloumsell, L., and Gilks, W., et al., eds. (1995) Leucocyte typing V: White cell differentiation antigens, Oxford University Press, New York.
11. Shau, H., Kim, A., and Golub, S.H. (1992) Modulation of natural killer and lymphokine-activated killer cell cytotoxicity by lactoferrin, J. Leukoc. Biol. 51, 343–349.
12. Iyer, S. and Lonnerdal, B. (1993) Lactoferrin, lactoferrin receptors and iron metabolism, Eur. J. Clin. Nutr. 47, 232–241.
13. Maneva, A.I., Taleva, B.M., Manev, V.V., and Sirakov, L.M. (1994) Bovine lactoferrin binds to plasma membrane receptors on human polymorphonuclear leucocytes, Med. Sci. Res. 22, 863–866.

LACTOFERRIN AND INTERLEUKIN-6 INTERACTION IN AMNIOTIC INFECTION

Katufumi Otsuki, Aki Yoda, Yoshiro Toma, Yukiko Shimizu, Hiroshi Saito, and Takumi Yanaihara

Department of Obstetrics and Gynecology
Showa University School of Medicine
Tokyo, Japan

1. SUMMARY

Lactoferrin (Lf) has been found in most biological fluids including amniotic fluid and cervical mucoids in pregnant women, and released from neutrophils in response to the inflammation. As Lf possesses antimicrobial properties, it is widely considered to be an important component of the host defence against microbial infections. It is known that premature labor is caused by amniotic infection with the increase of prostaglandin production. High concentration of the inflammatory cytokines: interleukin-1 β (IL-1 β), interleukin-6 (IL-6), tumor necrosis factor-α (TNF-α) in the amniotic fluid has been known. However, changes of Lf in amniotic fluid with infection has not been reported. In the present study, Lf concentrations in amniotic fluid were measured under the intra-uterine infections state and the biological significance of Lf was investigated. The effects of Lf on the IL-6 and IL-6mRNA production in cultured amnion cells were also investigated. The concentrations of Lf and IL-6 in amniotic fluid with CAM were 8.76±0.65 μg/ml and 6.92±4.88 ng/ml (n=28) respectively and both were significantly higher ($p<0.01$) than those without CAM [0.86±0.81 μg/ml and 0.34±0.25 ng/ml (n=31)]. Significant positive correlation ($r=0.91$, $p<0.01$) between Lf and IL-6 levels in amniotic fluid was found. IL-6 production induced by lipopolysaccharide (LPS) (100ng/ml) in cultured amnion cells was significantly inhibited ($p<0.05$) under the physiological concentration of Lf in amnion. Total RNA was extracted from the amniotic cells by guianizine solution. RT-PCR procedure and product analysis were performed from one μg aliquote of total RNA. β-actin was used as an international standard and c-DNA samples were followed by 30 cycles of PCR. RT-PCR product of IL-6 mRNA was detected by Southern hybridization. Expression of IL-6mRNA was inhibited by the addition of Lf. From the results, the possibility that Lf might suppress amniotic IL-6 production under the condition of amniotic infection is suggested. It is also suggested that Lf might act as self defence mechanism from intra-uterine infection.

Advances in Lactoferrin Research, edited by Spik *et al.*
Plenum Press, New York, 1998.

2. INTRODUCTION

Many biological functions of Lf have been reported, including bacteriostatic effect[1–5], iron absorption[6,7], lymphocyte growth promoting effects[8,9] and product regulation of macrophages and neutrophils. The bacteriostatic effect of Lf is one of the significant physiological functions in human milk. High concentration of Lf in colostrum might be one factor that explains why breast-fed infants rarely suffer from intestinal infections. Therefore, it has been suggested that Lf plays a possible role in the self defense against local mucosal infections[10,11]. In addition, Lf has been found in amniotic fluid and suppresses the production of inflammatory cytokines[12]. Although several studies of salivary Lf in adults and in children have been reported, there is no information about Lf in amniotic infection, especially those delivered under conditions of amniotic infection. Amniotic infection causes the degradation of chorioamnion with the migration of neutrophils and intrauterine fetal infection. Therefore, in the present study, Lf concentrations were measured in the amnion of the patients who had been delivered with and without chorioamnionitis[13] (CAM), and possible role of amniotic Lf is discussed.

3. SUBJECTS AND METHODS

3.1. Subjects

Twenty-eight pregnant women (28–36 weeks of gestation), who delivered with amniotic infection were subjected and amniotic fluids were obtained by amniocentesis or at delivery. Thirty one samples of amniotic fluid (32–39 weeks gestation) from normal delivery without amniotic infection were also obtained for the control. Amniotic fluid samples were centrifuged and supernatant was stored at −80°C until measurement. Fetal membranes were also obtained at delivery. The amniotic membrane was obtained at the time of elective caesarean section and amnion cells were cultured. After the cells were confluent, lipopolysaccharide and Lf were added to the medium, and successive 24 hours culture was performed. Culture medium was collected and IL-6 concentration was measured. The effect of Lf on the expression of IL-6 mRNA was also investigated. β-actin was used an internal standard.

3.2. Chemicals

Anti-human Lf IgG from sheep and anti-human Lf peroxidase conjugate from sheep were obtained from the binding site (Birmingham, England). Lf from human milk was purchased from Sigma Co (St Louis, Mo, USA). All other reagents were of analytical grade.

3.3. Measurements

Human Lf in amniotic fluid was measured with enzyme-linked immuno-sorbent assay (ELISA) of the sandwich type modified with the method reported by Camille and Irwin[14]. Microplates (96 wells) were coated with anti-human Lf IgG from sheep. Gelatin was added to the plates to prevent no specific binding. Aliquots of amniotic fluid were diluted 100 to 500 times with phosphate buffered saline and assayed. Inter and intra coefficients of variation were 4.7 and 7.0% respectively. The acceptable assay range lies between 5 and 250 ng/ml. IL-6 in the fluid of amnion and cultured amniotic cells were measured by human IL-6 ELISA kit (Fuji revio, Tokyo, Japan).

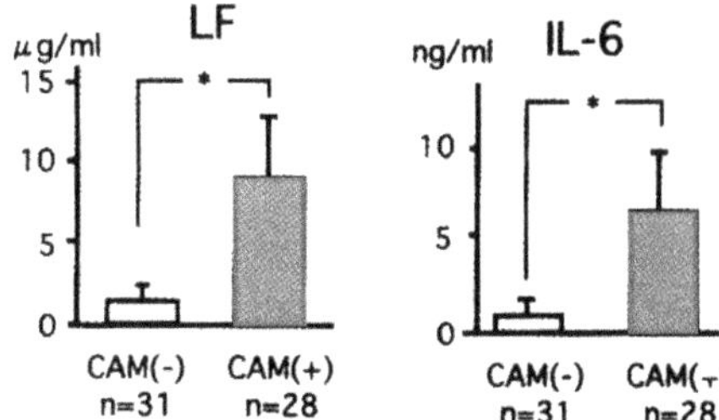

Figure 1. Lf and IL-6 concentrations in amniotic fluid. CAM(−); n= 31; CAM(+); n = 28. Values are expressed as mean ± S.D.; *p<0.01.

3.4. Immunohistochemical Staining of Lf in Fetal Membrane

Immunohistochemical localization of Lf in fetal membrane was investigated by streptavidine-biotin method.

3.5. Expression of IL-6 mRNA

Total RNA was extracted from the amniotic cells by guianizine solution. RT-PCR procedure and product analysis were performed from one µg aliquote of total RNA. β-actin was used as an international standard and c-DNA samples were followed by 30 cycles of PCR RT-PCR product of IL-6 mRNA was detected by Southern hybridization.

3.6. Statistics

The data are presented as mean ± S.D. statistical analysis was performed by means of unpaired Student's *t*-test. In all analyses, $p<0.05$ was considered to indicate statistical significance.

4. RESULTS

The concentrations of Lf and Il-6 in amniotic fluid with CAM were 8.76±0.65 µg/ml and 6.92±0.65 µg/ml (n=28), respectively, and both were significantly higher ($p<0.01$) than those without CAM [0.86±0.81 µg/ml and 0.34±0.25 ng/ml (n=31) (Fig. 1).

Significant positive correlation ($r=0.91$, $p<0.01$) between Lf and IL-6 levels in amniotic fluid was found (Fig. 2).

In the fetal membranes obtained from the patients with CAM, strong positive staining was observed in amniotic membrane and chorionic membrane with leucocytes migration. On the other hand, weak staining was observed in the membrane without CAM. Positive

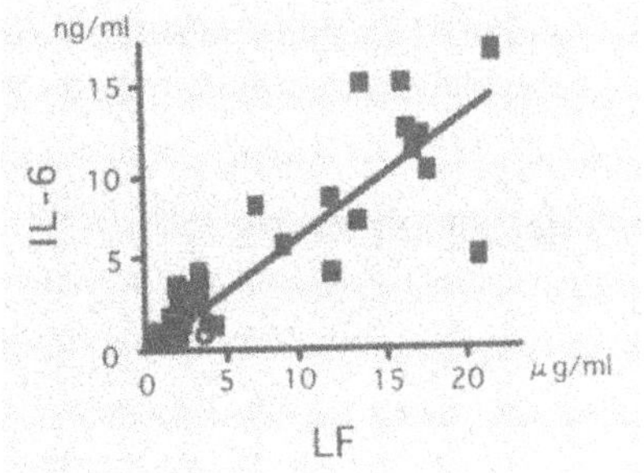

Figure 2. Correlation between Lf and Il-6 concentration in amniotic fluid. Circles are CAM(−); squares are CAM(+), Y=0.72×161.5, r=0.91.

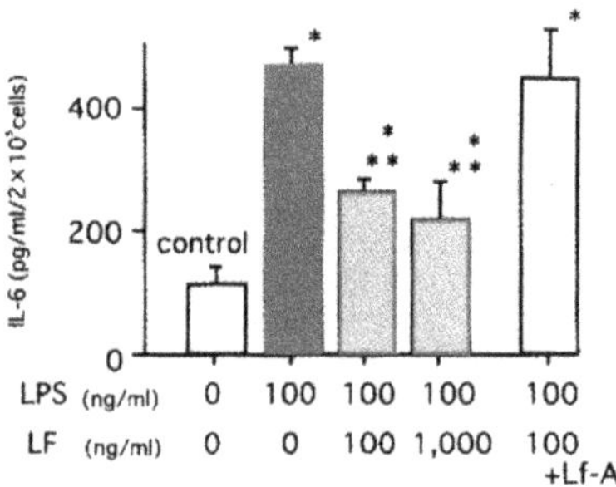

Figure 3. Effect of Lf and LPS induced Il-6 production in amniotic fluid. Values are expressed as mean ± S.D.: *p<0.01 vs control; **p<0.01 vs LPS.

staining of Lf was found in the cytoplasma of epithelial cells of amniotic membranes with CAM. IL-6 production induced by LPS (LPS 100 ng/ml) in cultured amnion cells was significantly inhibited ($p<0.05$) under the physiological concentration of Lf. Inhibitory effect of Lf on IL-6 production was eliminated by the addition of amniotic Lf antibody (Fig. 3). Expression of IL-6 mRNA induced by LPS was inhibited by the addition of Lf.

5. DISCUSSION

Amniotic Lf levels in patients with CAM were firstly demonstrated. Recently the relationship between amniotic infection and preterm labor is well documented. Among several biological functions of Lf proposed, the bacteriostatic effect of Lf is thought to be the most significant one. We have documented that the salivary Lf levels in neonates who delivered under the condition with CAM are significantly higher than those without CAM[15]. Chimura et al.[16] reported the Lf levels in cervical mucous in the cases of CAM. In the present study, the Lf concentrations in amniotic fluids with CAM were measured and significant higher levels of amniotic Lf in the patients with CAM were observed. Significant correlation between Lf and IL-6 concentrations was also obtained. Positive immunohistochemical staining of amniotic membrane with leucocytes migration was found. Therefore, these results strongly suggest that Lf in amniotic fluid may derived partially from amniotic membrane and the infection may be affected in its migration into amniotic fluid. IL-6 production and mRNA expression induced by LPS were significantly inhibited by Lf. Since Lf inhibits the growth of bacteria and fungi *in vitro*, Lf may be a possible contributor to defense against local mucosal infections. Therefore, we presumed that amniotic Lf might act as a local protection factor against infections. At present, however, the physiological functions of amniotic Lf are not fully understood, from the results, the possibility that Lf might suppress amniotic Il-6 production under the condition of amniotic infection is suggested. It is also suggested Lf might act as self defence mechanism from intra uterine infection.

REFERENCES

1. Arnold RR, Brewer A, Gauthier JJ. (1980) Bactericidal activity of human lactoferrin: sensitivity of a variety of microorganisms. Infect Immun, 28, 893–898
2. Staut J, Norell S, Harrington JP (1984) Kinetic effect of human lactoferrin on the growth of *Escherichia coli* 0111. Int J Biochem, 16, 1043–1047.

3. Ellison RT, Giehl TJ, Laforce MF (1988) Damage of the outer membrane of enteric gram-negative bacteria by lactoferrin and transferrin. Infect Immun, 56, 2774–2781.
4. Suzuki T, Yamauchi K, Kawase K, Kiyosawa I, Okonogi S. (1989) Collaborative bacteriostatic activity of bovine lactoferrin with lysozyme against *Escherichia coli* 0111. Agric Biol Chem, 53, 1705–1706.
5. Visca P, Dalmastri C, Verzili D, Antonini G, Chianocone E, Valenti P. (1990) Interaction of lactoferrin with *Escherichia coli* cells and correlation with antibacterial activity. Med Microbiol Immunol, 179, 323–333.
6. Cox T, Mazurier J, Spik G, Montreuil J, Peters TJ (1979) Iron binding proteins and influx of iron across the duodenal brush border. Evidence for specific lactoferrin receptors in the human intestine. Biochim Biophys Acta, 588, 120–128.
7. Lonnerdal B (1990) Iron in human milk and cow's milk - Effects of binding ligands on bioavailability. In: Iron Metabolism in Infants, Lonnerdal B (Ed) CRC Press, Boca Rato, FL, pp 87–107.
8. Broxmeyer H, Gentile P, Cooper S, Lu L, Juliano L, Piacibello W, Meyers LL (1984) Functional activities of acidic isoferritins and lactoferrin in vitro and in vivo. Blood Cells, 10, 397–426.
9. Birgens HS (1991) The interaction of lactoferrin with human monocytes. Dan Med Bull, 38, 244–252.
10. Rudney JD (1989) Relationships between human parotid saliva lysozyme, lactoferrin, salivary peroxidase and secretory immunoglobulin A in a large sample population. Arch Oral Biol, 34, 499–506.
11. Cleveland MG, Bakos MA, Hilton SM, Goldblum RH (1991) Amniotic fluid: the first feeding of mucosal immune factors. Adv Exp Med Biol, 310, 41–49.
12. Machnicki M, Zimecki M, Zagulski T (1993) Lactoferrin regulates the release of tumour necrosis factor alpha and interleukin 6 in vivo. Int J Exp Pathol, 74, 433–439.
13. Blanc WA (1981) Pathology of the placenta, membranes, and umbilical cord in bacterial, fungal and viral infections in man. In: Perinatal Diseases International Academy of Pathology Monograph, Vol 22, Blanc WA Ed, Williams & Wilkins, Baltimore, pp 67–132.
14. Dipaola C, Mandel ID (1980) Lactoferrin concentration in human parotid saliva as measured by enzyme-linked immunosorbent assay (ELISA) J Dent Res, 59, 1463–1465.
15. Katsufumi O, Aki Y, Kazuhiro H, Yukiko S, Hiroshi S, Takumi Y (1997) Salivary lactoferrin in neonates with chorioamnionitis. The Showa Univ J Med Sci, 9–1, unpublished
16. Chimura T, Hirayama T, Takase M (1993) Lactoferrin in cervical mucus of patients with chorioamnionitis. Jpn Antibiot, 46, 318–322.

INHIBITION OF AZOXYMETHANE INITIATED COLON TUMOR AND ABERRANT CRYPT FOCI DEVELOPMENT BY BOVINE LACTOFERRIN ADMINISTRATION IN F344 RATS

Hiroyuki Tsuda,[1] Kazunori Sekine,[1] Joe Nakamura,[1] Yoshihiko Ushida,[1] Tetsuya Kuhara,[1] Nobuo Takasuka,[1] Dae Joong Kim,[1] Makoto Asamoto,[1] Hiroyasu Baba-Toriyama,[1] Malcolm A. Moore,[1] Hoyoku Nishino,[2] and Tadao Kakizoe[3]

[1]Chemotherapy Division
National Cancer Center Research Institute
Tsukiji 5-1-1, Chuo-ku, Tokyo 104, Japan
[2]Department of Biochemistry
Kyoto Prefectural University of Medicine
Kawaramachi-Hirokoji, Kamigyo-ku, Kyoto 602, Japan
[3]National Cancer Center Central Hospital
Tsukiji 5-1-1, Chuo-ku, Tokyo 104, Japan

1. SUMMARY

The influence of bovine lactoferrin (bLf) on colon carcinogenesis was investigated in male F344 rats treated with azoxymethane (AOM). In experiment I, 2% and 0.2% bLf, and *Bifidobacterium longum* (*B. longum*) as a positive control at 3% were given in the diet for 4 weeks, along with two s.c. 15 mg/kg injections of AOM on days 1 and 8. The numbers of aberrant crypt foci (ACF) were decreased by both treatments. Similar results were obtained in experiment II of 13 weeks duration. In experiment III, animals were given three weekly injections of AOM and then received 2 or 0.2% bLf, 2% bLf-hydrolysate, or 0.1% bovine lactoferricin (bLfcin) for 36 weeks. No effects indicative of toxicity were noted, but significant reduction in both the incidence and number of adenocarcinomas of the large intestine was observed with almost all the treatments. Thus, the incidences of colon adenocarcinomas in the groups receiving 2 or 0.2% bLf, 2% bLf-hydrolysate, or 0.1% bLfcin were 15%, 25%, 26.3% and only 10%, respectively, in contrast to the 57.5% control value ($p < 0.01$). ACF values also exhibited reduced development. Investigation of beta-glucuronidase revealed decrease in the cecal contents of animals receiving bLf. In

Advances in Lactoferrin Research, edited by Spik *et al.*
Plenum Press, New York, 1998.

addition, demonstration of enhancement of NK activity by bLf indicated that its inhibitory effects could have been related to elevated immune cytotoxicity.

2. INTRODUCTION

Neoplasia of the colon is a leading cause of cancer death in the United States and other developed countries and numerous epidemiological studies and a plethora of experimental evidence point to a role for dietary factors in the development of the disease[1]. The recent drastic increase in the incidence of colon cancer patients in Japan is also thought to be related to the adoption of a more western style diet over the last three decades[2]. This has generated a great deal of interest in which environmental agents might exert an influence and the possibility of employing chemicals which can in fact prevent neoplasia is now attracting increasing attention[3].

Milk and its products are major dietary constituents in the west but because of the range of included ingredients, the question of what role they might play remains to be determined. However, it has recently been established that the whey fraction of bovine milk can significantly inhibit the appearance of ACF in the colons of rats given dimethylhydrazine (DMH)[4]. This fraction is actually composed of five major proteins, alpha-lactalbumin, beta-lactoglobulin, immunoglobulin, serum albumin and lactoferrin. This last, an 80 kDa siderophilic protein which has two iron binding sites per molecule, is well known to have bacteriostatic properties[5]. In addition to sequestration of iron ion necessary for microbes to grow, it activates NK cells[6,7] and neutrophils[8], induces colony stimulating activity[9], stimulates LAK cells[7] and augments macrophage cytotoxicity[10]. It is present in large amounts in mammalian secretions like tears, saliva and seminal fluid as well as being particularly abundant in colostrum[11]. However, despite extensive studies of its anti-microbial properties there is only little information available regarding the influence of lactoferrin on disease processes. This is unfortunate since Bezault et al. have pointed to a protective influence against growth of solid tumors and development of experimental metastases in mice[12]. They argued that the action of lactoferrin might have been mediated by NK cells, in line with its stimulation of NK and LAK cell activity *in vitro* and *in vivo*[7].

In the present study, in view of the high priority now accorded with finding effective chemopreventors of colon cancer in man, we conducted a number of investigations of the influence of lactoferrin on experimental colon carcinogenesis. For this purpose we chose an established model utilizing AOM administration to F344 rats[13], concentrating attention on effects in both the initiation and post-initiation development phases and using both ACF and histopathologically diagnosed tumors as endpoint marker lesions. As a positive control B. longum was selected since it has been repeatedly shown to inhibit colon carcinogenesis, either in terms of tumors[14] or ACF[15,16]. However, the present report is the first, to our knowledge, to demonstrate clear inhibitory potential for bLf, bLf-hydrolysate and bLfcin against neoplasia in the colon in vivo.

3. MATERIALS AND METHODS

3.1. Chemicals

The carcinogen azoxymethane (AOM) (CAS: 25843-45-2) was obtained from Sigma (St. Louis, MO). Bovine lactoferrin[17], bLf-hydrolysate, a protease treated product of bLf[18] and bovine lactoferricin (bLfcin), a peptide comprised of 25 amino acids purified from

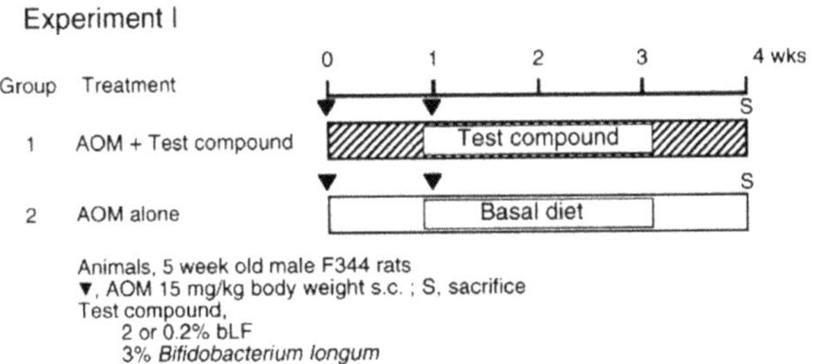

Figure 1. Experimental protocol for experiment I. P, AOM, 15 mg/kg body weight s.c.: S, sacrifice.

bLf-hydrolysate[18], as well as *Bifidobacterium longum* (*B. longum*)[19] were kindly provided by Drs. Tomita and Shimamura of Morinaga Milk Industry Co. Ltd. (Zama City, Japan).

3.2. Animals and Diets

Male F344 rats were purchased from Charles River Japan (Atsugi) at 5 weeks of age and maintained in plastic cages in an air-conditioned room under constant conditions of temperature (22 ± 2°C) and humidity (55 ± 10%). The animals were allowed free access to basal diet (Oriental MF, Oriental Yeast Co., Ltd., Tokyo) and tap water and were used in the experiments after a 1 week acclimation period.

3.3. Experimental Protocol

The following three different experiments were performed. The significance of differences between group means was assessed with the Dunnet's t-test. The data for tumor incidence were analysed using the Fisher's exact probability test. The analyses were all carried out with the JMP software package (Version 3.1, SAS Institute Japan) on a Macintosh.

3.3.1. Experiments I and II. (see Figs. 1 and 2). Experiment I. Four groups of 7 or 8 animals each were given s.c. injections of 15 mg/kg AOM on days 1 and 8 and killed at the end of week 4 for analysis of lesion development, groups 1, 2 and 3, respectively receiving dietary supplementation with 2% and 0.2% bLf, and 3% B. longum throughout the experimental period. Experiment II. A total of 23 animals were divided into three groups receiving

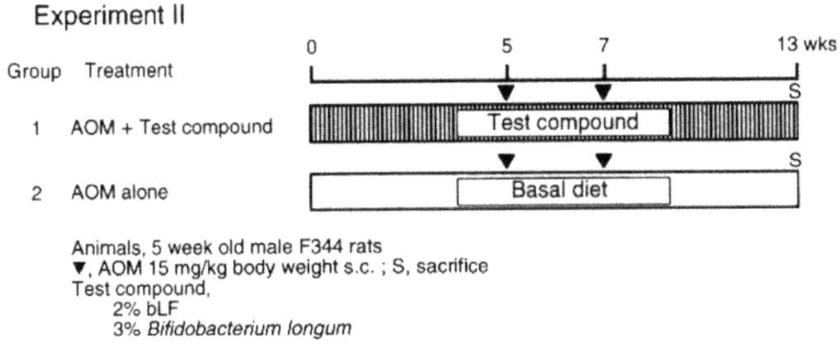

Figure 2. Experimental protocol for experiment II. P, AOM, 15 mg/kg body weight s.c.: S, sacrifice.

dietary supplements with the same materials as in experiment 1 and were given s.c. doses of AOM (15 mg/kg body weight) at weeks 5 and 7. Sacrifice was at the end of week 13.

3.3.2. NK Cell Activity. Effector cells were obtained from freshly isolated spleens at the time of sacrifice in experiment II and used for 4-h 51Cr-release assays as described earlier[20, 21]. NK activity was measured as the percentage specific toxicity with 51Cr-labeled YAC-1 lymphoma target cells at 20:1, 40:1, and 80:1 effector cell to target cell ratios. Percentage cytotoxicity was calculated as follows:

$$\% \text{ cytotoxicity} = \frac{\text{experimental release} - \text{spontaneous release}}{\text{total release} - \text{spontaneous release}} \times 100$$

3.3.3. Experiment III. A total of 215 male rats were divided into 3 groups. Starting at 6 weeks of age, those in groups 1 and 2 were administered AOM (15 mg/kg body weight s.c.) once a week for three weeks. One week after the last injection of AOM, animals in group 1 were fed diets containing 2% bLf [65 rats], 0.2% bLf [20], 2% bLf-hydrolysate [20], and 0.1% bLfcin [20], respectively. Animals in group 2 received the basal diet alone [65]. In group 3, rats were fed the basal [5], 2% bLf [5], 0.2% bLf [5], 2% bLf-hydrolysate [5] or 0.1% bLfcin [5] diets without prior carcinogen exposure. All animals were weighed weekly and checked daily for their condition throughout the experiment. Subgroups of five rats in group 1 receiving the 2% bLf treatment and in group 2 given the basal diet were sacrificed at intervals of 6 weeks from week 6 to week 30, for assessment of ACF development and measurement of the activities of beta-glucuronidase in the cecal contents. Rats demonstrating loss of condition and marked weight reduction (more than 20 g in one week) during weeks 26–39, many due to development of duodenal adenocarcinomas, were sacrificed and autopsied (basal diet, 8 animals; 2% bLf, 2; 0.2% bLf, 1; 2% bLf-hydrolysate, 0; and 0.1% bLfcin, 3). At week 40 all surviving animals were killed under ether anesthesia and carefully autopsied. The major organs, with especial attention to the intestines, were inspected before fixation of tissue samples, including all lesions, in 10% buffered formalin. The tissues were then routinely processed for embedding in paraffin and histopathological examination of hematoxylin and eosin stained sections (see Fig. 3).

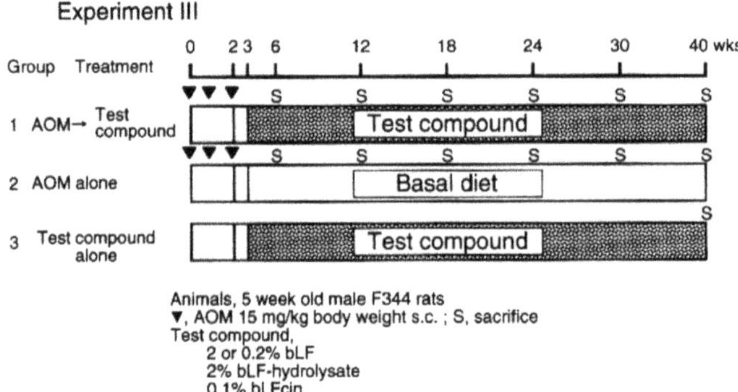

Figure 3. Experimental protocol for experiment III. P, AOM, 15 mg/kg body weight s.c.: S, sacrifice.

3.3.4. Beta-Glucuronidase Activities. Microbial enzymatic activities of cecal contents of the subgroups of animals killed at weeks 6, 12, 18, 24, and 30, were determined immediately after sacrifice by the methods of Rowland et al., as described elsewhere[22].

3.3.5. Analysis of Aberrant Crypt Foci (ACF). For detection of ACF, the colons were removed, inflated with 10% formalin, cut longitudinally from the cecum to anus, folded within filter papers, fixed in buffered 10% formalin and then stained with 0.5% methylene blue in saline according to the procedures described elsewhere[23]. The numbers of ACF/colon and AC/colon were determined by examination under a microscope at a magnification of x40. The criteria used to identify ACF topographically included: (a) increased size; (b) thickened epithelial cell lining; and (c) an increased pericryptal zone relative to normal crypts. Since it has been shown that large size ACF (composed of 4 or more crypts) are more closely related to the occurrence of colon tumors[24], division was made into two categories, 1–3 crypts/focus and 4 or more crypts/focus.

3.3.6. Histopathological Criterion for Invasive Carcinomas. Carcinomas were classified into invasive and non-invasive categories depending on the presence or absence of tumor elements within the submucosa.

4. RESULTS

In Experiments I and II, conducted to determine the effects of bLf and B. *longum* on development of ACF, neither of the treatments was associated with any alteration in body weights or animal condition (data not shown).

4.1. Experiment I

The findings for ACF are summarized in Table 1. In both experiments the two concentrations of bLf demonstrated more pronounced inhibitory effects on lesion numbers than B. longum, although the values for ACF and total numbers of AC were consistently

Table 1. Effects of bLf and *B. longum* on induction of ACF in male F344 rats in experiment I

Group	Test compound	No. of animals	1-3 crypts / focus	4 or more crypts / focus	Total ACF	Total AC
1	2%bLF	8	113.0 ± 16.0[a] (55.5)[b, d]	5.5 ± 5.1 (30.4)[c]	118.5 ± 18.1 (53.5)[d]	230.3 ± 51.0 (51.6)[d]
	0.2%bLF	8	155.3 ± 16.7 (76.3)[c]	8.4 ± 3.4 (46.5)	163.6 ± 17.8 (73.9)[c]	329.1 ±47.6 (73.7)[c]
	3% *B. longum*	7	151.3 ± 21.5 (74.3)[c]	14.3 ± 5.5 (79.4)	165.6 ± 25.4 (74.8)[c]	355.6 ±65.4 (79.6)[c]
2	Basal diet	7	203.5 ± 17.7 (100.0)	18.0 ± 6.3 (100.0)	221.5 ± 21.0 (100.0)	446.5 ±55.8 (100.0)

a, mean±S.D.
b, Data in parenthesis are percentages of the Group 2 (basal diet) values.
c, $P < 0.05$ compared with Group 2
d, $P < 0.01$ compared with Group 2

Table 2. Effects of bLf and *B. longum* on induction of ACF in male F344 rats in experiment II

Group	Test compound	No. of animals	1-3 crypts / focus	4 or more crypts / focus	Total ACF	Total AC
1	2% bLF	7	153.6 ± 27.5[a] (71.0)[b]	17.0 ± 6.6 (55.1)[c]	170.6 ± 27.8 (69.0)[d]	377.3 ± 69.7 (66.7)[d]
	3% *B. longum*	8	187.8 ± 22.1 (86.6)[c]	21.6 ± 7.8 (70.0)	209.4 ± 25.5 (84.7)[c]	454.5 ± 63.0 (80.3)[c]
2	Basal diet	8	216.4 ± 45.1 (100.0)	30.9 ± 8.6 (100.0)	247.3 ± 50.3 (100.0)	565.6 ± 116.6 (100.0)

a, mean ± S.D.
b, Data in parenthesis are percentages of the Group 2 (basal diet) values.
c, P < 0.05 compared with Group 2
d, P < 0.01 compared with Group 2

and significantly decreased with both treatments. In addition to the degree of decrease being greater in the bLf case, large lesions composed of more than 4 crypts/focus were only significant affected with the 2% dose of this compound. No ACF were observed in any Group 3 animals not receiving AOM treatment.

4.2. Experiment II

Data for ACF and for NK cell activity for animals killed at the end of week 13 are given in Table 2 and in Fig. 4, respectively. The results for ACF and AC numbers again demonstrated clearer inhibition by bLf than by B. longum. In both treatment cases, significant increase in the degree of cytolysis as compared to the carcinogen alone control case was observed ($P < 0.01$) but this effect was more pronounced with bLf ($P < 0.01$). No ACF were found without AOM administration.

4.3. Experiment III

Data for body and organ (liver, right and left kidney) weights at the 40 week sacrifice time points are summarized in Table 3. No significant intergroup variation was observed for the groups of animals given AOM. Slight increases were detected in Group 3 without AOM injection compared with Group 1 and Group 2, but no statistically significant influence of bLf, bLf-hydrolysate or bLfcin was apparent.

Data for ACF in the post-initiation kinetic study of bLf influence are illustrated graphically in Fig. 5. At all time points studied the numbers of ACF and total AC were sig-

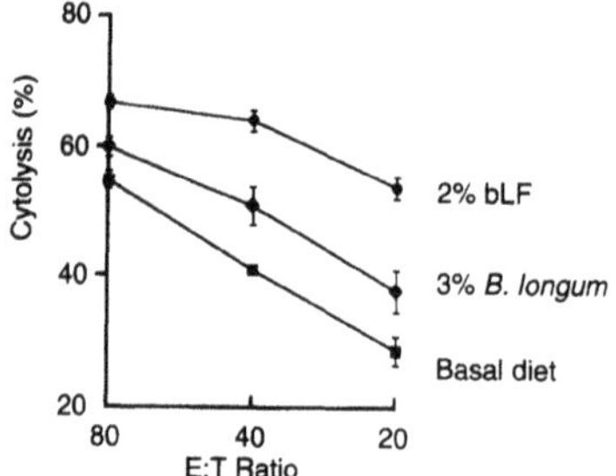

Figure 4. NK cell activities of splenocytes from AOM-injected male F344 rats in experiment II, with or without additional bLf or *B. longum* treatment.

Table 3. Body and organ weights of AOM-treated male F344 rats at week 30 (experiment III)

Group	Test compound	No. of animals	Body (g)	Liver (g)	Kidney (g)	
					Right	Left
1	2% bLF	38	391.7 ± 27.9[a]	9.41 ± 1.14	1.04 ± 0.12	1.06 ± 0.12
	0.2% bLF	19	392.4 ± 31.6	9.32 ± 1.07	1.05 ± 0.10	1.06 ± 0.09
	2% bLF-hydrolysate	20	384.6 ± 24.3	9.31 ± 0.86	1.04 ± 0.09	1.04 ± 0.10
	0.1% bLFcin	17	395.8 ± 18.6	9.47 ± 0.67	1.04 ± 0.06	1.05 ± 0.05
2	Basal diet	32	391.2 ± 33.9	9.82 ± 0.87	1.07 ± 0.08	1.10 ± 0.15
3	2% bLF	5	409.2 ± 27.4	10.45 ± 1.30	1.03 ± 0.08	1.02 ± 0.07
	0.2% bLF	5	399.5 ± 17.4	9.01 ± 0.95	0.95 ± 0.08	1.03 ± 0.09
	2% bLF-hydrolysate	5	402.3 ± 5.6	10.04 ± 0.72	1.06 ± 0.05	1.09 ± 0.04
	0.1% bLFcin	5	413.5 ± 25.2	10.20 ± 0.37	1.11 ± 0.11	1.10 ± 0.05
	Basal diet	5	410.2 ± 32.6	10.15 ± 0.80	1.08 ± 0.06	1.09 ± 0.04

a, mean ± S.D.

nificantly decreased in the 2% bLf treatment group. Division of the lesions into small and large categories revealed similar inhibition effects of bLf on both, with the reduction in numbers of those comprising 4 or more crypts/focus being particularly clear.

Results of the analysis of sequential changes in beta-glucuronidase activity are shown in Fig. 6. Treatment with bLf appeared to cause reduction, this being significant at the 6 and 12 week time points.

Incidences and multiplicities of small and large intestinal tumors are given in Tables 4 and 5, respectively. In the small intestines, no adenomas were found except for a single lesion in a Group 1 rat receiving the 2% bLf-hydrolysate. All the malignant lesions detected were of invasive adenocarcinoma type. A tendency for reduction was observed in Group 1 as compared with Group 2 regarding the incidence and the multiplicity but this was without significance.

With regard to the large intestine, while the development of adenomas was not altered by any of the treatments, both doses of bLf, 2% bLf-hydrolysate, and 0.1% bLfcin significantly reduced the incidence and the multiplicity of carcinomas. In addition, 2% bLf, 2% bLf-hydrolysate, and 0.1% bLfcin caused significant decreases of total tumor incidences (adenomas + carcinomas) and all but 2% bLf-hydrolysate also reduced the multiplicity. Both invasive and non-invasive carcinomas appeared to be equally affected. No

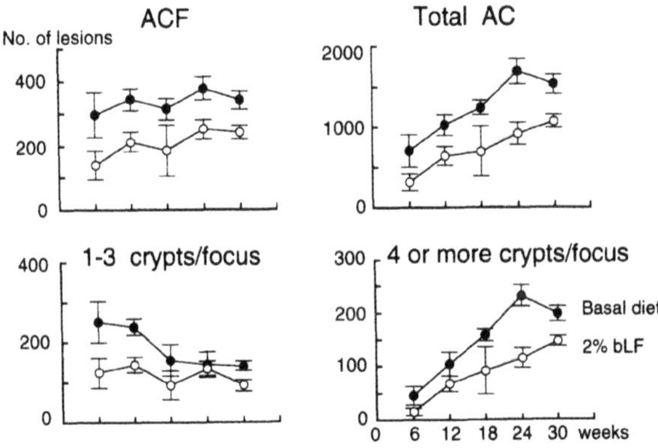

Figure 5. Kinetic study of ACF in AOM-treated male F344 rats given basal diet or 2% bLf in experiment III. Subgroups of five rats were examined at 6 week intervals from week 6 to week 30 for assessment of the development of total ACF, number of lesions with 1–3 crypts/focus and 4 or more crypts/focus, and total AC.

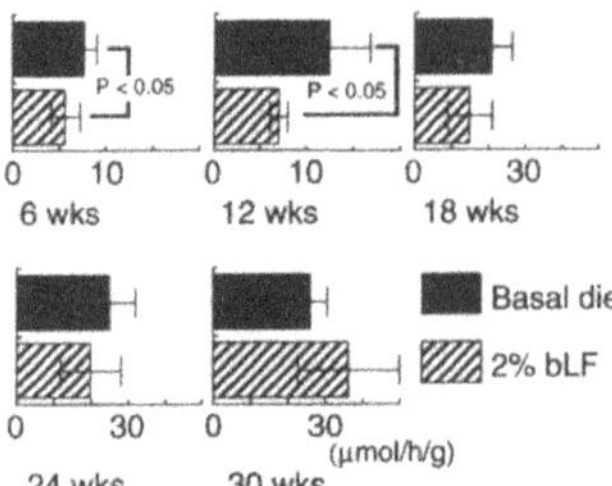

Figure 6. Kinetic study of enzymatic activities in the cecum contents from AOM-treated male F344 rats given basal diet or 2% bLf in experiment III. Animal treatments were as for Figure 5.

Table 4. Effects of bLf, bLf-hydrolysate and bLfcin on the incidences of intestinal tumors in AOM-initiated male F344 rats (experiment III)

Group	Test compound	No. of animals	Adenoma	Carcinoma			Adenoma + Carcinoma
				Non-invasive	Invasive	Total	
Small Intestine							
1	2% bLF	40	0[a] (0.0)[b]	0	5	5 (12.5)	5 (12.5)
	0.2% bLF	20	0 (0.0)	0	4	4 (20.0)	4 (20.0)
	2% bLF-hydrolysate	19	1 (5.3)	0	3	3 (15.8)	4 (21.1)
	0.1% bLFcin	20	0 (0.0)	0	2	2 (10.0)	2 (10.0)
2	Basal diet	40	0 (0.0)	0	9	9 (22.5)	9 (22.5)
Large Intestine							
1	2% bLF	40	17 (42.5)	2	4	6 (15.0)[d]	21 (52.5)[c]
	0.2% bLF	20	7 (35.0)	4	2	5 (25.0)[c]	11 (55.0)
	2% bLF-hydrolysate	19	6 (31.6)	0	5	5 (26.3)[c]	9 (47.4)[c]
	0.1% bLFcin	20	6 (30.0)	0	2	2 (10.0)[d]	8 (40.0)[d]
2	Basal diet	40	14 (35.0)	11	16	23 (57.5)	31 (77.5)

a, No. of animals
b, Values in parenthesis are percentages.
c, $P < 0.05$ compared with Group 2
d, $P < 0.01$ compared with Group 2

Table 5. Effects of bLf, bLf-hydrolysate and bLfcin on the multiplicities of intestinal tumors in AOM-initiated male F344 rats (experiment III)

Group	Test compound	No. of animals	Adenoma	Carcinoma			Adnoma + Carcinoma
				Non-invasive	Invasive	Total	
Small Intestine							
1	2% bLF	40	0	0	0.13 ± 0.33	0.13 ± 0.33	0.13 ± 0.33
	0.2% bLF	20	0	0	0.20 ± 0.41	0.20 ± 0.41	0.20 ± 0.41
	2% bLF-hydrolysate	19	0.05 ± 0.23[a]	0	0.21 ± 0.53	0.21 ± 0.53	0.26 ± 0.56
	0.1% bLFcin	20	0	0	0.10 ± 0.31	0.10 ± 0.31	0.10 ± 0.31
2	Basal diet	40	0	0	0.28 ± 0.55	0.28 ± 0.55	0.28 ± 0.55
Large Intestine							
1	2% bLF	40	0.55 ± 0.75	0.05 ± 0.22[c]	0.13 ± 0.40[c]	0.18 ± 0.45[c]	0.73 ± 0.88[b]
	0.2% bLF	20	0.50 ± 0.76	0.25 ± 0.55	0.10 ± 0.31[b]	0.35 ± 0.67[b]	0.85 ± 0.93
	2% bLF-hydrolysate	19	0.47 ± 0.84	0[b]	0.26 ± 0.45	0.26 ± 0.45[b]	0.74 ± 0.93
	0.1% bLFcin	20	0.40 ± 0.75	0[b]	0.10 ± 0.31[b]	0.10 ± 0.31[c]	0.50 ± 0.76[c]
2	Basal diet	40	0.43 ± 0.75	0.33 ± 0.57	0.50 ± 0.68	0.83 ± 0.90	1.25 ± 0.76

a, mean ± S.D.
b, $P < 0.05$ compared with Group 2
c, $P < 0.01$ compared with Group 2

obvious histopathological differences were evident between tumors in any of the AOM-treated groups. There was no clear dose dependence of the inhibitory potential of bLf.

No evidence of toxicity in the organs of animals treated with any of the test compounds was found and none of the rats in group 3 not administered AOM demonstrated any proliferative lesions in either the small or the large intestines.

5. DISCUSSION

The results of the present study provide clear evidence of an inhibitory potential of bLf and related compounds against development of malignant tumors in the rat colon in the post-initiation phase. This was associated with an equivalent reduction in the numbers of ACF when bLf was given in either the initiation or post-initiation stages.

With regard to lactoferrin effects on neoplasia the literature is extremely limited and there has so far only been one publication, documenting reduction of fibrosarcoma development and metastasis from a transplantable melanoma line in mice[12]. Population studies in man have indicated a protective influence of milk and other dairy foods against various tumors, including colorectal cancer[25,26] although there is a lack of any consensus about this point, presumably reflecting the variety of ingredients which are included in milk preparations. However, experimental studies in rats and mice have generated much support for the idea that some proteinaceous component of milk exerts a beneficial influence. Thus tumor growth was found to be inhibited by dairy products[27]. Inhibition of the initiation stage of colon cancer in rats has been shown with skim milk using ACF as endpoint lesions[15]. Attention has been concentrated on whey protein in mice by the group of Bounous[28], who established purified preparations to be more efficacious against DMH-induced colon carcinogenesis than casein. Papenburg et al.[29] subsequently demonstrated marked retardation in terms of both number and size of DMH-induced tumors as well as enhanced activity of the immune system. Whey protein was also confirmed to exert stronger inhibitory potential than casein in the rat intestine by McIntosh et al.[4] Interestingly, administration of whey protein was further found to reduce tumor growth in patients with metastatic carcinomas[30].

ACF are generally accepted as being relevant end point lesions of colonic cancers for assessment of modification potential in both the rat and other species[31]. Therefore the results of the present study, viewed together with the information available for whey protein in the literature, strongly suggest that bLf holds great promise as a natural chemopreventive agent very suitable for practical use in the human situation. While bLf is present in raw milk at concentrations ranging from 0.02–0.2 mg/ml in the bovine case to over 2 mg/ml in man, it is denatured by the heat treatment applied for pasteurization or UHT processing (personal communication Dr Kawase, Morinaga Milk Industry Co. LTD, Zama City) and therefore it is not contained in commercially available milk beverages. However, it is found at relatively high concentrations in colostrum (approximately 10 mg/ml), as well as in tears, seminal fluid, uterine secretions, and secondary granules in neutrophils, and may therefore be related to the necessity for anti-bacterial and immunoprotective actions. Indeed, lactoferrin is well known to act against bacteria and stimulate the immune response[11,32].

Therefore with regard to the mechanisms underlying the bLf inhibitory influence, attention should clearly be drawn to changes in the immune system. In fact a possible role of augmented NK cell activity in the antitumor action of bLf was earlier pointed out by Bezault et al.[12]. Anti-tumor effects of immunocytes such as NK cells are well established and lactoferrin has been demonstrated to stimulate their activation *in vitro* and *in vivo*[6,7]. This was confirmed in the present experiment (Fig. 4). We earlier showed that this is also the case for *B. longum*[33], reported to be a potent chemopreventor of colon carcinogenesis[14]. In addition, the fact that the effect was only about half that of bLf in the present study (Tables 1 and 2), correlating with the degree of influence on ACF development, might suggest some mechanistic role for the stimulation of NK cell activity. However, the question of whether NK cells can act on preneoplastic lesions remains open to discussion[34]. Increase in glutathione could play a role in whey protein enhancement of the immune response, as emphasized by Bounous et al.[35] and McIntosh[4] et al. Further work in this area is clearly warranted to determine exactly how this promising chemopreventor exerts its protective functions.

One other possibility concerning inhibition of carcinogenesis is through its action on bacteria and related enzymes. Reduction of beta-glucuronidase is observed after milk and Lactobacillus *acidophilus* feeding[36]. This is important because the enzyme is responsible for releasing methylazoxymethanol from the glucuronide-conjugated form. This would be expected to reduce the initiating potential of administered carcinogen and also the generation of DNA damage in tumors which would contribute to progression. It could thus have played a role in experiments I and II where animals were administered bLf at the time of the carcinogen applications. With regard to the inhibition evident in experiment III where treatment was not commenced until after completion of the AOM initiation, beta-glucuronidase de-conjugation of bile acids might be of significance, since these are known to promote colon cancer development in the unconjugated state[16].

In conclusion, the present demonstration of inhibition of development of both ACF and colon adenocarcinomas by bLf, as well as bLf-hydrolysate and bLfcin, is of major interest in terms of chemopreventive potential. The fact that both the non-iron sequestering bLf-hydrolysate and bLfcin forms were also effective may have significance with regard to the mechanisms of action since they do not bind iron, but this has yet to be clarified. The lack of any clear dose dependence for bLf action is also of interest in the present study and this point requires clarification.

ACKNOWLEDGMENTS

This work was supported in part by a Grant-in-Aid for the Second Term Comprehensive 10-Year Strategy for Cancer Control from the Ministry of Health and Welfare, by a Grant-in-Aid for Cancer Research from the Ministry of Health and Welfare in Japan, and a Grant-in-Aid from the Ministry of Education, Science, Sports, and Culture of Japan. Dae Joong Kim, Vladimir Krutovskikh and Malcolm A. Moore were recipients of the Foreign Research Fellowship Program for Invitation of Foreign Researcher from the Foundation for Promotion of Cancer Research. Tomonori Ota is a recipient of a Research Fellowship during the performance of this work from the Foundation for Promotion of Cancer Research.

REFERENCES

1. Giovannucci, E. and Willett, W. C. Dietary factors and risk of colon cancer. Ann. Med., 26: 443–452, 1994.
2. Wynder, E. L., Fujita, Y., Harris, R. E., Hirayama, T. and Hiyama, T. Comparative epidemiology of cancer between the United States and Japan, a second look. Cancer, 67: 746–63, 1991.
3. Wattenberg, L. W. Inhibition of carcinogenesis by minor dietary constituents. Cancer Res. (suppl.), 52: 2085s–2091s, 1992.

4. McIntosh, G. H., Regester, G. O., Le, L. R., Royle, P. J. and Smithers, G. W. Dairy proteins protect against dimethylhydrazine-induced intestinal cancers in rats. J. Nutr., 125: 809–816, 1995.
5. Lonnerdal, B. and Iyer, S. Lactoferrin: molecular structure and biological function review. Ann. Rev. Nutr., 15: 93–110, 1995.
6. Nishiya, K. and Horwitz, D. A. Contrasting effects of lactoferrin on human lymphocyte and monocyte natural killer activity and antibody-dependent cell-mediated cytotoxicity. J. Immunol., 129: 2519–2523, 1982.
7. Shau, H., Kim, A. and Golub, S. H. Modulation of natural killer and lymphokine-activated killer cell cytotoxicity by lactoferrin. J. Leukocyte Biol., 51: 343–349, 1992.
8. Gahr, M., Speer, C. P., Damerau, B. and Sawatzki, G. Influence of lactoferrin on the function of human polymorphonuclear leukocytes and monocytes. J. Leukocyte Biol., 49: 427–433, 1991.
9. Sawatzki, G. and Rich, I. N. Lactoferrin stimulates colony stimulating factor production in vitro and in vivo. Blood Cells, 15: 371–385, 1989.
10. McCormick, J. A., Markey, G. M. and Morris, T. C. Lactoferrin-inducible monocyte cytotoxicity for K562 cells and decay of natural killer lymphocyte cytotoxicity. Clin. Exp. Immunol., 83: 154–156, 1991.
11. Levay, P. F. and Viljoen, M. Lactoferrin: a general review. Haematologica, 80: 252–267, 1995.
12. Bezault, J., Bhimani, R., Wiprovnick, J. and Furmanski, P. Human lactoferrin inhibits growth of solid tumors and development of experimental metastases in mice. Cancer Res., 54: 2310–2312, 1994.
13. Reddy, B. S., Simi, B., Patel, N., Aliaga, C. and Rao, C. V. Effect of amount and types of dietary fat on intestinal bacterial 7 alpha-dehydroxylase and phosphatidylinositol-specific phospholipase C and colonic mucosal diacylglycerol kinase and PKC activities during stages of colon tumor promotion. Cancer Res., 56: 2314–2320, 1996.
14. Reddy, B. S. and Rivenson, A. Inhibitory effect of *Bifidobacterium longum* on colon, mammary, and liver carcinogenesis induced by 2-amino-3-methylimidazo[4,5-f]quinoline, a food mutagen. Cancer Res., 53: 3914–3918, 1993.
15. Abdelali, H., Cassand, P., Soussotte, V., Daubeze, M., Bouley, C. and Narbonne, J. F. Effect of dairy products on initiation of precursor lesions of colon cancer in rats. Nutr. Cancer, 24: 121–132, 1995.
16. Kulkarni, N. and Reddy, B. S. Inhibitory effect of Bifidobacterium longum cultures on the azoxymethane-induced aberrant crypt foci formation and fecal bacterial beta-glucuronidase. Proc. Soc. Exp. Biol. Med., 207: 278–283, 1994.
17. Tomita, M., Bellamy, W., Takase, M., Yamauchi, K., Wakabayashi, H. and Kawase, K. Potent antibacterial peptides generated by pepsin digestion of bovine lactoferrin. J. Dairy Sci., 74: 4137–4142, 1991.
18. Bellamy, W., Takase, M., Wakabayashi, H., Kawase, K. and Tomita, M. Antibacterial spectrum of lactoferricin B, a potent bactericidal peptide derived from the N-terminal region of bovine lactoferrin. J. Applied Bacteriol., 73: 472–479, 1992.
19. Shimamura, S., Abe, F., Ishibashi, N., Miyakawa, H., Yaeshima, T., Araya, T. and Tomita, M. Relationship between oxygen sensitivity and oxygen metabolism of Bifidobacterium species. J. Dairy Sci., 75: 3296–3306, 1992.
20. Winter, B. K., Wu, S., Nelson, A. C. and Pollack, S. B. Renal cell carcinoma and natural killer cells: studies in a novel rat model in vitro and in vivo. Cancer Res., 52: 6279–6286, 1992.
21. Sekine, K., Ohta, J., Onishi, M., Tatsuki, T., Shimokawa, Y., Toida, T., Kawashima, T. and Hashimoto, Y. Analysis of antitumor properties of effector cells stimulated with a cell wall preparation (WPG) of Bifidobacterium infantis. Biol. Pharm. Bull., 18: 148–153, 1995.
22. Rowland, I. R., Mallett, A. K. and Wise, A. A comparison of the activity of five microbial enzymes in cecal content from rats, mice, and hamsters, and response to dietary pectin. Toxicol. Appl. Pharm., 69: 143–148, 1983.
23. Tudek, B., Bird, R. P. and Bruce, W. R. Foci of aberrant crypts in the colons of mice and rats exposed to carcinogens associated with foods. Cancer Res., 49: 1236–1240, 1989.
24. Bird, R. P., Yao, K., Lasko, C. M. and Good, C. K. Inability of low- or high-fat diet to modulate late stages of colon carcinogenesis in Sprague-Dawley rats. Cancer Res., 56: 2896–2899, 1996.
25. Kampman, E., Goldbohm, R. A., Van den Brandt, P. A. and Van't Veer, P. Fermented dairy products, calcium, and colorectal cancer in the Netherlands cohort study. Cancer Res., 54: 3186–3190, 1994.
26. Sorenson, A. W., Slattery, M. L. and Ford, M. H. Calcium and colon cancer: a review. Nutr. Cancer, 11: 135–145, 1988.
27. Tsuru, S., Shinomiya, N., Taniguchi, M., Shimazaki, H., Tanigawa, K. and Nomoto, K. Inhibition of tumor growth by dairy products. J. Clin. Lab. Immunol., 25: 177–183, 1988.
28. Bounous, G., Papenburg, R., Kongshavn, P. A., Gold, P. and Fleiszer, D. Dietary whey protein inhibits the development of dimethylhydrazine induced malignancy. Clin. Invest. Med., 11: 213–217, 1988.

29. Papenburg, R., Bounous, G., Fleiszer, D. and Gold, P. Dietary milk proteins inhibit the development of dimethylhydrazine-induced malignancy. Tumour Biol., 11: 129–136, 1990.
30. Kennedy, R. S., Konok, G. P., Bounous, G., Baruchel, S. and Lee, T. D. The use of a whey protein concentrate in the treatment of patients with metastatic carcinoma: a phase I-II clinical study. Anticancer Res., 15: 2643–2649, 1995.
31. Bird, R. P. Role of aberrant crypt foci in understanding the pathogenesis of colon cancer. Cancer Lett., 93: 55–71, 1995.
32. Brock, J. Lactoferrin: a multifunctional immunoregulatory protein? Immunol. Today, 16: 417–419, 1995.
33. Sekine, K., Kawashima, T. and Hashimoto, Y. Comparison of the TNF-a levels induced by human-derived *Bifidobacterium longum* and rat-derived *Bifidobacterium animalis* in mouse peritoneal cells. Bifidobacteria Microflora, 13: 79–89, 1994.
34. Wei, W. Z., Fulton, A., Winkelhake, J. and Heppner, G. Correlation of natural killer activity with tumorigenesis of a preneoplastic mouse mammary lesion. Cancer Res., 49: 2709–2715, 1989.
35. Bounous, G., Batist, G. and Gold, P. Whey protein in cancer prevention. Cancer Lett., 57: 91–94, 1991.
36. Goldin, B. R. and Gorbach, S. L. The effect of milk and lactobacillus feeding on human intestinal bacterial enzyme activity. A. J. Clin. Nutr., 39: 756–761, 1984.

BOVINE LACTOFERRIN AND LACTOFERRICIN^{TM} INHIBIT TUMOR METASTASIS IN MICE

Yung-Choon Yoo,[1] Shikiko Watanabe,[2] Ryosuke Watanabe,[1] Katsusuke Hata,[1] Kei-ichi Shimazaki,[2] and Ichiro Azuma[1]

[1]Institute of Immunological Science
Hokkaido University
Sapporo 060, Japan
[2]Department of Dairy Science
Faculty of Agriculture
Hokkaido University
Sapporo 060, Japan

1. SUMMARY

The effect of a bovine milk protein, lactoferrin (bLf), and a pepsin-generated peptide of bLf, lactoferricin (Lfcin-B), on inhibition of tumor metastasis produced by highly metastatic murine tumor cells, B16-BL6 melanoma and L5178Y-ML25 lymphoma cells, was examined in experimental and spontaneous metastasis models using syngeneic mice. The subcutaneous (s.c.) administration of bovine apo-lactoferrin (apo-bLf) and Lfcin-B 1 day after tumor inoculation significantly inhibited liver and spleen metastasis of L5178Y-ML25 cells and lung metastasis of B16-BL6 cells, whereas human apo-lactoferrin (apo-hLf) and bovine holo-lactoferrin (holo-Lf) at the dose of 1 mg/mouse did not. Furthermore, both apo-bLf and Lfcin-B, but not apo-hLf and holo-bLf, inhibited the number of tumor-induced blood vessels and suppressed tumor growth on day 8 after tumor inoculations in an in vivo model. However, in a long-term analysis of tumor growth for up to 21 days after tumor inoculation, single administration of apo-bLf significantly suppressed the growth of B16-BL6 cells throughout the examination period, but Lfcin-B showed inhibitory activity only during the early period (8 days). In spontaneous metastasis model, multiple administration of both apo-bLf and Lfcin-B significantly inhibited lung metastasis of B16-BL6 cells, however it was only apo-bLf that exhibited the inhibitory effect of tumor growth at the time of primary tumor amputation (on day 21) after tumor inoculation. The results suggest that apo-bLf and Lfcin-B inhibit tumor metastasis through different mechanisms, and that the inhibitory activity of bLf on tumor metastasis may be related to the property of iron (Fe^{3+})-saturation.

Advances in Lactoferrin Research, edited by Spik *et al.*
Plenum Press, New York, 1998.

2. INTRODUCTION

Lactoferrin (Lf), an iron-binding glycoprotein with a molecular weight of about 80,000, is mainly found in most biological fluids of mammals[1] and released from neutrophil granules during inflammatory responses[2]. A variety of biological functions of human Lf (hLf) and bovine Lf (bLf) have been demonstrated in host defense, especially in immune responses[3] antibacterial activity[4] and transcriptional activation of cells[5]. However, despite the diverse functions attributed to Lf, the complete spectrum and accurate role of its activity in primary in vivo mechanism of host defense have not yet been elucidated.

Bovine lactoferricin (Lfcin-B), a bLf-derived peptide consisting of 25 amino acid residues generated by acid-pepsin hydrolysis[6] was shown to have strong bactericidal activity against a wide range of microorganisms[7] and is considered to be the active domain responsible for antimicrobial activity of bLf. Even though many investigators have shown the biological activities of Lf and Lf-derived peptides, there is a few reports suggesting a direct evidence that these molecules have antitumor activities. Recently, Bezault et al.[8] demonstrated that hLf inhibited experimental metastasis of B16-F10 melanoma cells in mice, and that the antitumor activity of hLf might be mediated by NK cells and was independent of iron saturation. However, biological functions against tumors of bLf, which is 69% identical to hLf in the entire amino acid sequence[9] and possesses similar physiological functions with hLf, is nearly understood.

In the present study, we investigated the effects of bLf and Lfcin-B on tumor metastasis using established tumor models in mice, and partly compared the activity between bLf and hLf. We found that iron-free bLf (apo-bLf) and Lfcin-B, but iron-bound bLfs (holo-bLf and native-bLf), inhibited experimental tumor metastasis (lung and liver colonization), whereas apo-hLf did not exhibit any activity. These results suggest an evidence to our knowledge of an antimetastatic effect attributed to bLf and Lfcin-B. In addition, the results indicate a potential function of this multifunctional milk protein and its peptide in primary host defense against tumor development and progression.

3. MATERIALS AND METHODS

3.1. Reagents

bLf was isolated from bovine milk whey according to the method as described[12]. hLf was purchased from Sigma Chemical Co. Ltd. (St. Louis, MO). Apo-Lf and iron-saturated Lf were prepared by the method as described previously[13]. Purified bovine lactoferricin (Lfcin-B) (FKCRRWQWRMKKLGAPSITCVRRAF), a cationic peptide corresponding to residues 17–41 near the N-terminus of bLf[6] was kindly donated by the Nutritional Science Laboratory, Morinaga Milk Industry Co., Ltd.

3.2. Experimental and Spontaneous Metastasis Model in Mice

Tumor metastasis models using syngeneic mice were caried out as described previously[11,14]. In experimental metastasis model, mice were given i.v. administration of various Lfs (1 mg/mouse) or Lfcin-B (0.5 mg/mouse) 1 day after i.v. inoculation of 4×10^4 L5178Y-ML25 or B16-BL6 cells. The mice were killed 14 days after tumor inoculation, and tumor metastasis was evaluated by counting lung tumor colonies prodiced by B16-BL6 cells or measuring the weight of liver and spleen increased by L5178-ML25 cells. In spontaneous metastasis model, C57BL/6 mice were inoculated s.c.with 5×10^5 B16-BL6 cells into the right footpads and the primary tumors were surgically removed 21 days after tumor inoculation[11]. The tumor-bearing mice were administered i.v. with bLf (1 mg/mouse) of Lfcin-B (0.5 mg/mouse) three times at 3 day-intervals before or after tumor amputation. The mice were killed 35 days after inoculation and lung tumor colonies were counted under a dissecting microscope.

3.3. Inhibition Assay of Tumor Growth in Vivo

Three C57BL/6 mice per group were inoculated s.c. with B16-BL6 (4×10^5/site) melanoma cells on the right and left footpads. The mice were administered s.c. with various Lfs (1 mg/ mouse) or Lfcin-B (0.5 mg/mouse) 1 day after tumor inoculation. The diameter of tumor mass was measured from day 8 after tumor inoculation.

3.4. Assay of Tumor-Induced Angiogenesis

The assay of tumor angiogenesis was carried out as described previously[15]. C57BL/6 mice were inoculated i.d. with B16-BL6 melanoma cells (5×10^5) at two sites on the back and administered s.c. with various Lfs (1 mg/mouse) or Lfcin-B (0.5 mg/mouse) 1 day after tumor inoculation. Eight days after tumor inoculation, mice were killed immediately after i.v. injection (0.2 ml/mouse) of 1% Evan's blue solution, and skins were separated from the underlying tissues. Each of the inoculation sites was located under a dissecting microscope, and angiogenesis was quantitated by counting the number of vessels oriented toward the tumor mass.

3.5. Statistical Analysis

The statistical significance of differences between groups was calculated by applying Student's two-tailed t test.

4. RESULTS

4.1. Effect of bLf and Lfcin-B on Liver and Spleen Metastasis of Lymphoma Cells

The antimetastatic effect of various Lfs, such as apo-hLf, apo-bLf and holo-bLf, and Lfcin-B was assessed in liver and spleen metastasis of L5178Y-ML25 tumor cells. Table 1 shows that s.c. administration of apo-bLf (1 mg/mouse) as well as Lfcin-B (0.5

Table 1. Inhibitory effect of bLf and Lfcin-B on liver and spleen metastasis produced by i.v. inoculation of L5178Y-ML25 lymphoma cells

Treatment	Dose (mg/mouse)	Mean weight (g) ± SD (% inhibition)	
		Liver	Spleen
Untreated (tumor control)		2.86 ± 0.7	0.20 ± 0.04
apo-hLf	1	2.84 ± 0.6	0.22 ± 0.01
apo-bLf	1	1.44 ± 0.3 (49.7)*	0.13 ± 0.03 (35)*
holo-bLf	1	2.65 ± 1.2	0.19 ± 0.04
Lfcin-B	0.5	1.53 ± 0.5 (46.5)*	0.15 ± 0.03 (25)*
(Normal mice)		1.00 ± 0.1	0.08 ± 0.02

*$P<0.05$, compared with the untreated group by Student's two-tailed t test.

mg/mouse) 1 day after tumor inoculation significantly inhibited liver and spleen metastasis produced by L5178Y-ML25 lymphoma cells. However, neither apo-hLf nor holo-bLf at the dose of 1 mg/mouse showed any inhibitory effect on liver and spleen metastasis. In addition, native-bLf (1 mg/mouse) did not inhibited tumor metastasis of L5178Y-ML25 cells (data not shown).

4.2. Inhibition of Spontaneous Lung Metastasis by bLf and Lfcin-B

As described in Table 2, s.c. administration of apo-bLf and Lfcin-B 1 day after tumor inoculation caused therapeutic effect on lung metastasis of tumor cells. In addition, apo-bLf administered i.v. and intraperitoneally (i.p.) revealed the same activity to inhibit lung metastasis of tumor cells. However, similarly with the results of Table 1, apo-hLf and holo-bLf did not induce any inhibitory effect of tumor metastasis. Since it was shown that antitumor activity of hLf was mediated by enhancement of NK activity[19], we examined whether the antimetastatic activity of apo-bLf and Lfcin-B was also related to activation of NK cells. However, we failed to obtain the data that could support the results of Table 1 and 2; the lysis activities (%) to NK-sensitive cells (Yac-1) were as follows: 15 ± 2.2, 36 ± 4.8, 27 ± 2.6, 25 ± 3.2 and 27 ± 1.3 for non-treated control, apo-hLf-, apo-bLf-, holo-bLf- and Lfcin-B-treated group, repectively, on day 2 after treatment in which all groups exhibited maximal activity during the period of 5 days after treatment.

4.3. Inhibitory Effect of bLf and Lfcin-B on Tumor-Induced Angiogenesis

Since angiogenesis is an important event for growth and shedding of metastatic tumors in the primary and metastatic sites, we investigated the effect of bLf and Lfcin-B on inhibition of tumor- induced angiogenesis. Table 3 shows that s.c. administration of apo-bLf and Lfcin-B, but not apo-hLf and holo-bLf, 1 day after tumor inoculation resulted in a significant inhibition of tumor-induced angiogenesis and suppression of tumor growth on day 8 after tumor inoculation. These results indicate that apo-bLf and Lfcin-B inhibit tumor metastasis and their antimetastatic effect is related to the inhibition of tumor-induced angiogenesis.

Table 2. Inhibitory effect of bLf and Lfcin-B on lung metastasis produced by i.v. inoculation of B16-BL6 melanoma cells

Treatment	Dose (mg/mouse)	Route	Number of lung metastasis: Mean ± SD (% inhibition)	Number of lung metastasis: Range
Exp 1				
Untreated (tumor control)			125 ± 25	102–129
apo-hLf	1	s.c.	118 ± 7	112–127
apo-bLf	1	s.c.	60 ± 10 (47.8)*	48–70
holo-bLf	1	s.c.	108 ± 35	72–155
Lfcin-B	0.5	s.c.	64 ± 4 (44.3)*	60–68
Exp 2				
Untreated (tumor control)			93 ± 19	80–123
apo-bLf	1	i.v.	49 ± 6 (47.3)*	42–53
	1	i.p.	50 ± 12 (46.2)*	36–66

*P < 0.001, compared with the untreated group by student's two-tailed t test.

Table 3. Inhibitory effect of bLf and Lfcin-B on tumor growth and tumor-induced angiogenesis

Treatment	Dose (mg/mouse)	Tumor size (mm) (mean ± SD)	No. of blood vessels (mean ± SD)
Untreated (tumor control)		8.5 ± 0.7	13 ± 2.6
apo-hLf	1	8.1 ± 0.6	10 ± 2.8
apo-bLf	1	7.3 ± 1.0*	8 ± 1.5**
holo-bLf	1	8.3 ± 0.5	14 ± 4.6
Lfcin-B	0.5	7.5 ± 0.6*	7 ± 1.4***

*P<0.05; **P<0.01; ***P<0.001, comparing with the untreated group by Student's two-tailed t test.

4.4. Effect of bLf Anf Lfcin-B on Tumor Growth in Vivo

To examine their activity to suppress tumor growth in detail, we carried out a long-term analysis of tumor growth inhibition of various Lfs and Lfcin-B for up to 21 days after tumor inoculation in vivo. As shown in Fig. 1, single administration of apo-bLf, but not holo-bLf and hLf, significantly inhibited tumor growth throughout the examination period. However, the administration of Lfcin-B showed just a temporary inhibition of tumor growth until 8 days after tumor inoculation.

4.5. Inhibition of Spontaneous Lung Metastasis by bLf and Lfcin-B

The administration of apo-bLf (1 mg/mouse) 1, 4 and 7 days after tumor inoculation significantly inhibited primary tumor growth and lung metastasis of B16-BL6 melanoma cells (Table 4). In addition, the treatment of apo-bLf after the excision of primary tumors (on day 22, 25 and 28) resulted in a significant inhibition of lung metastasis, and its antimetastatic effect was effective at the dose of 0.3 mg/mouse. The multiple administration of Lfcin-B also exhibited a significant inhibition of lung metastasis in both administration schedules; but the multiple administration of Lfcin-B before the excision of primary tumors failed to suppress tumor growth, supporting the result of Fig. 1. These results reveal that both of apo-bLf and Lfcin-B are able to inhibit spontaneous lung metastasis through different mechanisms.

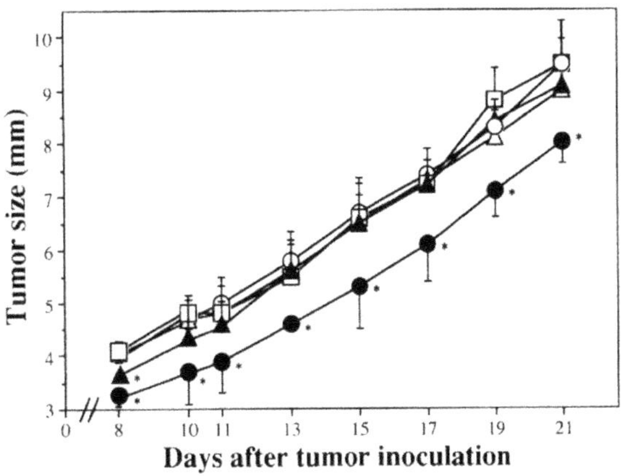

Figure 1. Effect of Lfs and Lfcin-B on the growth of B16-BL6 melanoma cells in vivo. Three C57BL/6 mice per group were inoculated s.c. with B16-BL6 melanoma cells (4×10^5/site) on the right and left footpads. The mice were administered s.c. with Lfs (1 mg/mouse) or Lfcin-B (0.5 mg/mouse) the next day of tumor inoculation. Tumor size was measured for up to 21 days after tumor inoculation. ○, the untreated control; △, apo-hLf; ▲, Lfcin-B; □, holo-bLf; ●, apo-bLf; *P<0.01, compared with the untreated (control) group by student's two-tailed test.

Table 4. Inhibitory effect of apo-bLf and Lfcin-B on lung metastasis produced by B16-BL6 cells in spontaneous metastasis model

			No. of lung metastasis	
Treatment after tumor inoculation (on day)	Dose (mg/mouse)	Primary tumor size (mean ± SD, mm)	Mean ± SD (% inhibition)	Range
Untreated (tumor control)		10.6 ± 0.3	41 ± 22	20–74
apo-bLf	1	9.0 ± 0.8***	13 ± 7 (68.3)*	5–21
1, 4, 7	1	10.2 ± 0.2	9 ± 6 (78.1)**	0–17
22, 25, 28	0.3	10.4 ± 0.8	9 ± 9 (78.1)*	0–22
Lfcin-B				
1, 4, 7	0.5	10.4 ± 0.5	9 ± 4 (78.1)**	5–14
22, 25, 28	0.5	10.4 ± 0.6	16 ± 5 (60.9)*	11–22

*P<0.05; **P<0.01; ***P<0.001, compared with the untreated group by Student's two-tailed t test.

5. DISCUSSION

Among a lot of milk proteins, Lf is considered to be responsible for host primary defense against microbial infections through binding of iron required for microbial growth. In addition, Lfcin-B, a bLf-derived peptide generated by acid-pepsin hydrolysis, was shown to possess a potential capacity to protect microbial infections[7]. Here we showed that apo-bLf and Lfcin-B, but not apo-hLf and holo-bLf, inhibited tumor metastasis and suppressed tumor-induced angiogenesis in mice, and that the antitumor activity of bLf was related to its property of iron saturation.

Since it was shown that immunological mechanisms mediated by NK cells, a well-known antitumor effector, are related to antitumor activity of hLf[8,10], we addressed that the discrepancy of antitumor effect between human and bovine Lf resulted from the differene of their ability to enhance NK activity. Unexpectedly, it was revealed that NK activity was not the main factor determining the difference of antitumor activity between bLf and hLf. On the other hand, hLf was shown to inhibit experimental lung metastasis of B16-F10 cells and the growth of murine fibrosarcoma cells regardless of its iron-saturation[8]. However, in our experiments, bLf showed antitumor activity only in iron-free state.

Considering that iron is required for tumor cell proliferation and iron chelation reduced tumor cell growth[16], it is possible that the inhibitory effects of apo-bLf on tumor growth resulted from iron chelation. However, the reason why apo-hLf, the same iron-free protein with apo-bLf, could not inhibit tumor metastasis remains questionable. Taken together, it may be suggested that each of apo-bLf and apo-hLf possess a distinct character in the physiological properties, such as chemical structure and the strength of iron binding or chelation, and that the discrepancy in antitumor activity between both apo-Lfs come from the character of each protein as well as the difference in tumor systems and cell lines used.

We first demonstrated that Lfcin-B, a cationic peptide with active domain of bLf for its antimicrobial activity[7], inhibited tumor metastasis and angiogenesis. In inhibitory effect on tumor growth, Lfcin-B showed lower activity than apo-bLf ; Lfcin-B inhibited tumor growth just only at the early period (Table 3) whereas apo-bLf maintained the activity during the long-term period (Fig. 1). Many other peptides with similar cationic features with Lfcin-B are shown to exert their lethal effect against microorganisms by disrupting cell membrane functions[17,18]. Thus, the antitumor activity of Lfcin-B is probably related to a similar mechanism with that action.

The present study demonstrated that apo-bLf and Lfcin-B inhibited experimental and spontaneous metastasis of hematogenous and non-hematogenous tumor cells in mice, and that the antitumor activity of bLf was related to its feature of iron-binding. Further studies are now in progress to elucidate more fully the mechanism of the effects of apo-bLf and Lfcin-B on tumorigenesis, and to examine the potential for application of them as a therapeutic agent.

REFERENCES

1. Reiter, B. In " Development in Dairy Chemistry-3, Lactose and Minor Constituents, ed. P.F. Fox, London, pp. 281–336 (1985). Elsevier Applied Sci. Pub.
2. Masson, P. L., Heremans, J. F. and Schonne, E. J. Exp. Med., 130, 643–658 (1969).
3. Hauer. J., Voetsh. W. and Anderer. F.A. Immunol. Lett., 42. 7–12 (1994).
4. Arnold, R. R., Brewer. M. and Cauthier, J. J. Infect. Immun., 28, 893–898 (1980).
5. He, J. and Furmanski, P. Nature, 373, 721–724 (1995).
6. Tomita, M., Bellamy, W., Takase, M., Yamauchi, K. Wakabayashi, H. and Kawase, K. J. Dairy Sci., 74, 4137–4142 (1992).
7. Bellamy, W., Takase, M., Yamauchi, K., Wakabayashi, H., Kawase, K. and Tomita, M. Biochim. Biophys. Acta, 1121, 130–136 (1992).
8. Bezault. J., Bhimani, R., Wiprovnick, J. and Furmanski, P. Cancer Res., 54, 2310–2312 (1994).
9. Pierce, A., Colavizza, D., Benaissa, M., Maes, P., Tartar, A., Montreuil, J. and Spik, G. Eur. J. Biochem., 196, 177–184 (1991).
10. Shau, H., Kim, A. and Golub, H. J. Leukocyte Biol., 51, 343–349 (1992).
11. Yoo, Y. C., Saiki, I., Sato, K. and Azuma, I. Vaccine, 12, 175–180 (1994).
12. Law, B. A. and Reiter, B. J. Dairy Res., 44, 595–599 (1977).
13. Shimazaki, K., Kawano, N. and Yoo, Y. C. Comp. Biochem. Physiol., 98, 417–422 (1991).
14. Yoo, Y. C., Saiki, I., Sato, K. and Azuma, I. Vaccine, 10, 792–797 (1992).
15. Saiki, I., Sato, K., Yoo, Y. C., Murata, J., Yoneda, J., Kiso, M., Hasegawa, M. and Azuma, I. Int. J. Cancer, 51, 641–645 (1992).
16. Weinberg, E. D. Iron depletion: Life Sci., 50, 1289–1297 (1992).
17. Anderson, B. F., Baker, H. M., Norris, G. E., Rice, D. W. and Baker, E. N. J. Mol. Biol., 209, 711–734 (1989).
18. Christenson, B. Fink, J. Merrifield, R. B. and Mauzerall, D. Proc. Natl. Acad. Sci. USA, 85, 5072–5076 (1988)

36

LACTOFERRIN IS SYNTHESIZED BY MOUSE BRAIN TISSUE AND ITS EXPRESSION IS ENHANCED AFTER MPTP TREATMENT

Carine Fillebeen,[1] David Dexter,[2] Valérie Mitchell,[3] Monique Benaissa,[1] Jean-Claude Beauvillain,[3] Geneviève Spik,[1] and Annick Pierce[1]

[1]Laboratoire de Chimie Biologique
UMR n°111 du CNRS
Université des Sciences et Technologies de Lille
59655 Villeneuve d'Ascq cedex, France
[2]Charing Cross and Westminster Medical School
Fulham Palace Road
London W6 8RF, United Kingdom
[3]Unité U422 de l'INSERM
Neuroendocrinologie et Physiopathologie Neuronale
Place de Verdun, 59045 Lille, France

1. SUMMARY

The biological role and origin of human lactoferrin (Lf) within the brain in normal and disease processes are as yet uncharted. In this context the origin and expression of brain Lf in normal and MPTP (1-methyl-4-phenyl-1,2,3,6-tetrahydropyridine)-treated mice were investigated using immunohisto chemistry, PCR amplification and *in situ* hybridization. Lf immunostaining was observed both on sections of mouse lactating mammary gland, which was used as a positive control, and brains from young, adult and aged mice. Lf immunoreactivity was present in the pituitary gland, the hippocampus and the cortex of mouse brains and to a greater extent in older mice. After reverse transcription, Lf transcripts were also found in these brain sections. Lf distribution and expression in the MPTP-induced parkinsonian mouse model were next investigated. A marked depletion of dopamine and its metabolites: dihydroxyphenylacetic acid (DOPAC), homovanillic acid (HVA) and 5-hydroxy indole acetic acid (5-HIAA) occurs in the high dose MPTP-treated mice. The level of Lf expression was found to be greatly increased in the same animals but Lf immunoreactivity detected in the same brain region was not found increased in the affected areas.

Advances in Lactoferrin Research, edited by Spik *et al.*
Plenum Press, New York, 1998.

2. INTRODUCTION

Alterations occur in iron metabolism in various neurodegenerative diseases which lead to excessive iron deposits[1,2] and the formation of highly reactive oxygen species which damage biological molecules and contribute to the cascade of events involved in neuronal death (for review see [3]). Iron is an abundant transition metal in the brain particularly in *the basal ganglia* where its concentration nearly doubles in Parkinson's disease cases[4]. It plays a central role in neuronal development as an essential component of oxidative metabolism and as a co-factor for numerous enzymes. However, iron can be extremely destructive and its intracellular concentration is highly regulated via the IRE-IRE/BP system which plays a crucial role in maintaining brain iron homeostasis. No significant variations in the levels of expression of transferrin, transferrin receptor and ferritin have been found in pathological brain tissue although conflicting results have been also reported[4–6]. The causes and consequences of such excessive iron deposits in the brain are unknown, as is the nature of the responsible iron complex.

Recently, among the iron-binding proteins, Lf[7] which is practically absent from normal human cerebral cortex[8], has received considerable attention since it has been located in brains *post mortem* associated with ageing, and more importantly in enhanced amounts in the specific brain regions adversely affected by the neurodegenerative lesions[8–10], in the mesencephalon in Parkinson's Disease[11] where the loss of Lf-stained cells paralleled the decrease in the total neuronal population, the surviving cells having a higher Lf concentration than normal neurons. Lf was present in the hippocampus and inferior temporal cortex in Alzeihmer's Disease, Pick's Disease and amyloid lateral sclerosis associated with the inflammatory foci[8].

Within the brain Lf is further captured by specific cells via the expression of a specific receptor, the GP-105 protein (LfR) characterized on activated T lymphocytes[12]. The distribution of LfR has been ascertained in specific brain regions in the case of Parkinsonism and LfR is localized on neurons, the cerebral microvasculature and in some cases on glial cells. LfR immunoreactivity on neurons and microvessels was increased in those regions where the loss of dopaminergic neurons is severe and, moreover, in the *substancia nigra* the labelling was more marked for patients with greater nigral dopaminergic loss[13].

Lf and LfR functions within either the normal or pathological brain are unknown, as is whether they are up-regulated in pathogenic conditions. The high concentration of Lf in the *basal ganglia*, and the distribution of Lf in the vicinity of neurological lesions, led us to investigate the origins of this Lf. Lf may cross the blood–brain barrier and/or Lf may be synthesized *in situ*. In this context, we initially investigated the origin and level of expression of Lf found in normal, aged and pathological mouse brain tissues.

3. METHODS

3.1. Immunohistochemistry

Mouse lactating mammary gland, and brain tissues of 18 month, 1 and 8 week old mice, were fixed at room temperature with 4% paraformaldehyde in PBS, frozen in liquid nitrogen and sections (12 μm) were cut. After washes in PBS, the endogenous biotin was eliminated from these sections with a blocking kit (Biosis), prior to immunohistochemistry. Lf immunostaining was observed using rabbit polyclonal anti-Lf antibodies (1/500) diluted in PBS containing 0.25% Triton X100. After washes with PBS, sections were incubated with relevant biotinylated secondary antibodies (1/150) for 1h30 at room temperature, followed by an amplification with the avidin-peroxidase complex using the Vectastain ABC kit (Biosis). The reaction was developed with 0.5 mg/ml diaminobenzidine (Sigma) and 1μl/ml hydrogen peroxide in TBS for 3 minutes.

3.2. *In Situ* Hybridization

In situ hybridization was performed using a ^{35}S-labelled antisense murine Lf riboprobe for 16 h at 55°C. The sense murine Lf riboprobe was used as a negative control. Mouse lactating mammary gland was used as a positive control.

Slides warmed to room temperature, were washed in quenching buffer, digested with proteinase K, rinsed in distilled water, prefixed with 4% paraformaldehyde in PBS and acetylated. The slides were then dehydrated and incubated with the labelled probe (10 ng/μl) for 16 hours. Stringent washes and RNAse treament were performed and the slides were then dehydrated and treated with a photographic emulsion (Kodak NTBII). After 30 days of exposure, slides were developed with D19 (Amersham) and fixed with 30% sodium-thiosulfate.

3.3. RNA Analysis

Extraction of RNA was performed using the guanidium thiocyanate/cesium chloride technique. Murine lactating mammary gland total RNA was used as a positive control and murine liver total RNA as a negative control.

3.4. RT-PCR

5μg of each total RNA preparation were reverse transcribed into first strand cDNA using oligo dT primers and one twentieth of the mixture was amplified by PCR using primer pairs (Eurogentec) designed for specific detection of mouse Lf, Tf and GAPDH (Glyceraldehyde-3-phosphodehydrogenase) DNA target sequence (Table 1). GAPDH is used as internal control, since it is a constitutively expressed house keeping gene. As a positive control, RT-PCR was performed on Tf messenger, since it is highly synthesized by each tissue studied. Liver was used as a negative control since it does not produce Lf. First strand cDNA sequences were amplified by PCR in a DNA thermalcycler programmed for a 94°C denaturation, 55°C annealing and 74°C primer extension step for each cycle using the Tfl Polymerase (Promega). Nested PCR was performed on Lf products using internal primer pairs (Table 1). PCR assays were performed in triplicate. Ten μl of each PCR reaction were loaded on 1% agarose gel stained with ethidium bromide, photographed and scanned.

Table 1. Primer pairs used in the RT- and nested PCR reactions

Protein	Primer 1	Primer 2	PCR product in bp
Lf A	5'-AGTGAGGAGAAGCGCAAGTGTG-3'	5'-AGCCCCAGTGTAGCCTTGGTAT-3'	533
Lf B	5'-TCCAGTCTTGGCAGAGAACCAG-3'	5'-TGGGAGCACACTTGTTCTCACC-3'	328
Tf	5'-ATGTCACTGCCATTCGGAATC-3'	5'-CGCTTCTCCGTTCACAATCTT-3'	199
GAPDH	5'-TCCTGCACCACCAACTGCTTAG-3'	5'-TGGGTGGTCCAGGGTTTCTTAC-3'	575

3.5. MPTP Administration

Male C57/black mice (30 g) received four injections of 2 or 20 mg/kg MPTP-HCl intraperitoneally in 0.2 ml of saline at 2 h intervals. Control mice received saline only. Seven days later animals were sacrificed and hemisected. A half-brain was used to assay monoamines and their metabolites, the other half of the brain was subjected to RNA extraction in order to determine Lf messenger content. Animal care was in accordance with institutional guidelines.

3.6. Determination of Monoamine Levels

Striatal levels of DA, HVA, DOPAC and 5-HIAA were determined by HPLC as previously described[14,15]. Briefly, the striated region of a half-brain was dissected out, weighed and sonicated in 10 vol of 0.4 M perchloric acid containing 1mM EDTA and 0.5% sodium metabisulphite by a Microson tissue disrupter. 1 μg/ml dihydroxybenzyl amine (Sigma) was added to the resulting homogenate (1:9, v/v) as an internal standard. An aliquot of the supernatant was injected onto a Spherisorb ODS-2 reverse-phase column and chromatographic peaks detected by a BAS LC-4B amperometric detector with a thin layer electrochemical cell fitted with a glass carbon working electrode and an Ag/AgCl reference electrode.

4. RESULTS AND DISCUSSION

4.1. Lf Distribution in Mouse Brain Tissue from Young, Adult, and Aged Animals

The investigation of the distribution of Lf was performed on mouse brain tissues from young, 1 week old mice, adult 8 wk old mice and 18 month old mice. Lf immunoreactivity (Figure 1) can be observed in the cerebral cortex (1A) and is always associated with the microvessels with an increased labelling in old mice. In the pituitary gland (1B) Lf immunoreactivity was present and a specific labelling of the hippocampal formation (1C) was visible only in old mice. Therefore, Lf is present in blood vessels of the cortex, the hippocampus and the pituitary gland of old mice. Lf was also present in the young and adult mouse brains, but at a lower level than in old mice. However, Lf immunoreactivity in brain tissue was less strong than in the mammary gland which was used as a positive control (1D) and where a high level of immunostaining of Lf was observed on many epithelial cells.

Both *in situ* hybridization and RT-PCR were carried out to detect the Lf mRNA. For each experiment carried out on the mouse brain tissue, lactating mouse mammary gland tissue was used as a positive control. *In situ* hybridization was performed out using a murine Lf anti-sense riboprobe and the corresponding sense probe was used as a negative control. A high level of Lf mRNA was located in numerous epithelial cells (data not shown). Unfortunately, *in situ* hybridization was not sensitive enough to detect Lf mRNA in brain tissue, although RT-PCR detected Lf mRNA. If 30 cycles of amplification were sufficient to allow detection of Lf messenger in mouse mammary gland, nested PCR had to be performed in the case of mouse brain cDNA and 70 cycles were needed to obtain a visible signal. Mouse liver total RNA was amplified as a negative control (Figure 2A).

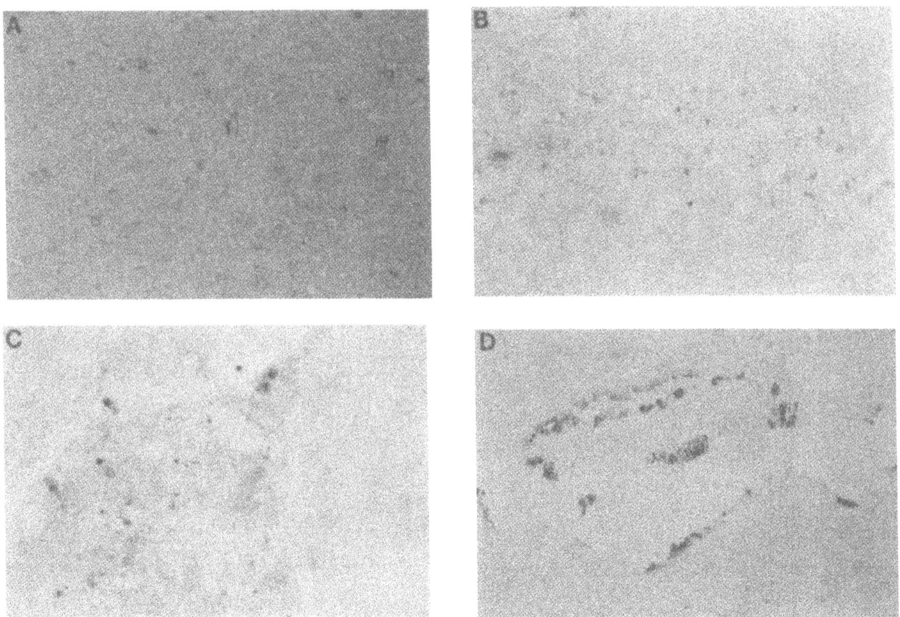

Figure 1. Distribution of Lf immunoreactivity in mouse brain tissue. in blood vessels of the cortex (100x); (B) in the hippocampus of an aged mouse brain section (100x); (C) in the pituitary gland (100x) and (D) in the lactating mammary gland (100x).

In this way Lf mRNA was detected in brains of young, adult and old mice, but was present at a very low level in this tissue compared to the lactating mammary gland. Thus, Lf is synthesized in brain tissue since both Lf protein and mRNA were detected.

4.2. Lf Expression in MPTP-Treated Mice

In order to determine whether Lf synthesis is upregulated in pathological situations, MPTP treatment of mice was used as a model for Parkinsonian pathology[1,16,17]. In both monkeys and mice, MPTP reproduces many of the biochemical, molecular and morphological features of Parkinson's disease. MPTP toxicity depends on its oxidation to the toxic metabolite 1-methyl-4-phenylpyridinium or MPP+ by monoamine oxidase B[18]. MPP+ is accumulated in dopaminergic neurons[19] and inhibits the oxidation of NADH-linked substrates in mitochondria[20] and ultimately inhibits complex I of the electron transport chain, resulting in energy failure[21]. MPP+ also enhances lipid peroxidation, a process dependent on the overproduction of free radicals[22].

Oxidative stress due to MPP+ or in Parkinson's Disease has a dramatic effect on the amount of dopamine and its metabolites. The enzymatic metabolism of dopamine leads to the deaminated metabolites, homo-vanillic acid and 3,4-dihydroxy phenylacetic acid and also to the generation of hydrogen peroxide. The extent of MPTP-induced lesions was calculated by comparing dopamine (DO), dihydroxyphenylacetic acid (DOPAC), homovanillic acid (HVA) and 5-hydroxy indole acetic acid (5-HIAA) concentrations in a sample of striatal tissue of the treated and untreated animals. In Table 2, alterations in the levels of dopamine and its metabolites are evident 7 days after the administration of the higher

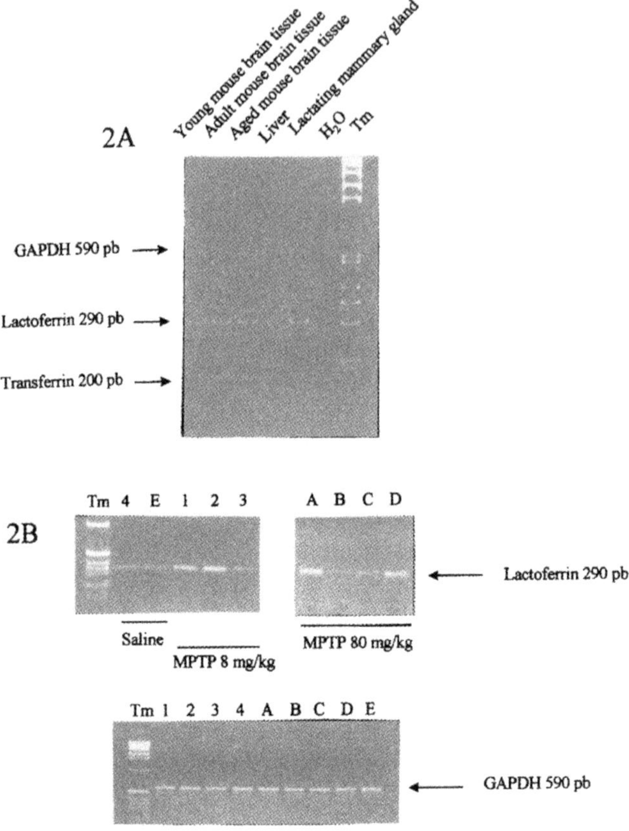

Figure 2. Synthesis of Lf detected by RT- nested PCR (2A): Lf messenger expression by normal, adult and aged mouse brain tissue. Lactating mammary gland RNA is used as positive control, liver RNA as a negative control.

Table 2. Individual mouse responses to MPTP treatment[a]

Treatment	Mouse n°	DO	DOPAC	HVA	5-HIAA
2 mg/kg	1	79.04	4.75	5.34	1.70
MPTP	2	74.33	4.87	3.88	1.35
4 x i.p.	3	71.18	4.38	5.13	1.07
	mean	74.85	4.67	4.75	1.67
	SEM	2.28	0.15	0.43	0.18
	n	3	3	3	3
20 mg/kg	A	0.54	0.12	0.39	0.81
MPTP	B	0.69	0.72	0.62	1.35
4 x i.p.	C	1.13	0.35	0.69	1.28
	D	1.89	0.44	0.30	1.36
	mean	1.06	0.41	0.53	1.20
	SEM	0.30	0.12	0.07	0.13
	n	4	4	4	4
Saline	4	73.87	4.02	4.26	1.48
	E	96.66	4.92	5.32	1.37
	mean	85.27	4.47	4.79	1.42
	SEM	11.39	0.45	0.53	0.05
	n	2	2	2	2

[a]Dopamine (DO), dihydroxy phenylacetic acid (DOPAC), homovanillic acid (HVA), and 5-hydroxy indole acetic acid (5-HIAA) concentrations are expressed in pmol mg wet tissue 7 days after the treatment.

dose, four times 20 mg/kg, of MPTP. A very marked depletion of dopamine occurs with only 1% of the initial amount remaining after the high dose, whereas about 90% remains after the low dose. No significant change occurs in the DOPAC and HVA contents after a low dose of MPTP, but both were significantly reduced after the higher dose, about 20% of each remaining. In addition, there was a loss of 25% of the initial amount of 5-hydroxy-indole acetic acid.

Lf distribution performed on the MPTP-treated mouse brain sections was similar to that previously obtained on the brain sections of the normal and aged animals, no specific increased immunoreactivity was found in the areas affected by the pathological process.

The levels of expression of Lf mRNA were studied in the other half-brains using RT-PCR. As a control RT-PCR was performed on murine GAPDH and the levels of this mRNA were constant in all the treated and untreated brains. The data obtained are shown in Figure 2B and correspond to individual mice. They confirm our previous result since 70 cycles of amplification were necessary to obtain a Lf signal in the controls. The animals treated with the low dose of MPTP gave a Lf signal similar to the controls, in line with the minimal changes seen in levels of dopamine and its metabolites at this dose. 55 cycles of amplification were necessary to obtain a Lf signal in brain from animals treated with the higher dose of MPTP, indicating that Lf transcription is up-regulated in these mice.

However, wide individual variations are visible, reflecting individual susceptibility to the drug. Strikingly, the Lf PCR signal obtained corresponds to the individual levels of reduction in dopamine levels observed (Table 2), supporting the conclusion that the two parameters are closely linked. The results obtained so far in this model are in favour of the synthesis of Lf in response to the oxidative stress caused by MPTP treatment.

5. CONCLUSION

We can now hypothesize that Lf upregulation is an initial response to oxidative stress in brain tissue. Lf could actively participate in the neurodegenerative disorders by capturing the iron in higher concentrations in some specific brain regions and acts in this way as a natural scavenger of reactive oxidative species. Alternatively its synthesis at abnormal levels or its release from necrosing neurons may exacerbate and amplify the lesions leading to a cytotoxic effect resulting in an increase in neuronal death. Further studies will be necessary to elucidate brain Lf function.

ACKNOWLEDGMENTS

This work was supported by the Université des Sciences et Technologies de Lille, the Centre National de la Recherche Scientifique (UMR du CNRS n°111; Prof. A. Verbert), the Ministère de l'Education Nationale and the Conseil Général de la Région Nord-Pas-de-Calais-Picardie (Axe régional: Maladies Neurodégénératives et Vieillissement).

REFERENCES

1. Sofic E., Riederer P., Heinsen H., Beckmann H., Reynolds G. P., Hebenstreit G. & Youdim M. B. H., 1988, J. Neural. Transm, 74, 199–205
2. Connor J.R., Menzies S. L., St Martin S. M. & Mufson E., 1992, J. Neurosci. Res., 31, 75–83

3. Jenner P., 1996, Neurology, 47, 161–170
4. Dexter D., Carayon A., Javoy-Agid F., Agid Y., Wells F., Daniel S., Lees A. J., Jenner P. & Marsden C. D., 1991, Brain, 114, 1953–1975
5. Faucheux B., Hirsch E., Villares J., Selimi F., Mouatt-Pringent A., Javoy-Agid F., Hauw J.J. & Agid Y., 1993, J. Neurochem., 60, 2338–2341
6. Connor J.R., & Benkovic, S.A., 1992, Ann. Neurol., 32, 51–61
7. Montreuil J. & Mullet S., 1960, C. R. Acad. Sci. Paris, 250, 1736–1737
8. Leveugle B., Spik G., Perl D., Bouras C., Fillit H. & Hof P., 1994, Brain Res., 650, 20–31
9. Osmand A.P. & Switzer III R.C., 1991, in Alzheimer's Disease: Basic mechanisms, diagnosis and therapeutic strategies (Iqbal, K., McLachlan, D.R.C., Winblad, B. & Wisniewski, H.M. eds), John Wisley & Sons, pp 219–228
10. Kawamata T., Tooyama I., Yamada T., Walker D. G. & McGeer P. L., 1993, Am. J. Pathol., 142, 1574–1585
11. Leveugle B. , Faucheux B., Bouras C., Nillesse N., Spik G., Hirsch E.C., Agid Y. & Hof P. R., 1996, Acta Neuropathol., 91, 566–572
12. Mazurier J., Legrand D., Hu W. L., Montreuil J. & Spik G., 1989, Eur. J. Biochem., 179, 481–487
13. Faucheux B., Nillesse N., Damier P., Spik G., Mouatt-Pringent A., Pierce A., Leveugle B., Kubis N., Hauw J-J., Agid Y. & Hirsch E., 1995, Proc. Natl. Acad. Sci., 92, 9603–9607
14. Rose S., Nomoto M. & Jenner P., 1989, Biochem. Pharmacol, 38, 3677–3681
15. Ward R., Dexter D., Florence A., Aouad F., Hider R., Jenner P. & Crichton B., 1995, Biochem Pharmacol, 49, 1821–1826
16. Langston J. W., 1996, Neurology, 47, 153–160
17. Jucker M. & Ingram D. K., 1997, Behavioural Brain Research, 85, 1–25
18. Chiba K., Trevor A. & Castagnoli N., 1984, Biophys. Biochim. Res. Comm., 120, 574–578
19. Chiba K., Trevor A. & Castagnoli N., 1985, Biophys. Biochim. Res. Comm., 128, 1228–1232
20. Javitch J. A., D'Amato R. J., Strittmatter S. M. & Snyder S. H., 1985, Proc. Natl. Acad. Sci., 82, 2173–2177
21. Vyas I., Heikkila R. E. & Nicklas W. J., 1986, J. Neurochem., 46, 1501–1507
22. Rojas P. & Rios C., 1993, Pharmacol Toxicol., 72, 364–368

A STUDY OF LACTOFERRIN AND ANTIBODIES AGAINST LACTOFERRIN IN NEUROLOGICAL DISEASES

Silvana Penco,[1] Barbara Villaggio,[2] GianLuigi Mancardi,[3]
Michele Abbruzzese,[3] and Cecilia Garre'[1]

[1]Institute of Biology and Genetics
[2]Department of Internal Medicine
University of Genova
Viale Benedetto XV 6
16132 Genova, Italy
[3]Department of Neurological Sciences
University of Genova
Via De Toni 5
16132, Genova, Italy

1. INTRODUCTION

Lactoferrin (Lf) mediates some of the effects of inflammation and has a role in regulating various components of the immune system[1]. Marked elevation of Lf has been noted in the cerebrospinal fluid of patients with acute cerebrovascular lesions as an index of inflammation[2]. Furthermore, comparative immuno-histochemical analysis has revealed that Lf is present in pathological lesions in a variety of neurodegenerative disorders[3]. Previous studies have reported the presence of antibodies to Lf in the serum of patients with a wide variety of autoimmune disorders such as primary sclerosing cholangitis, ulcerative colitis, systemic lupus erytematosous. The clinical significance of these findings remains to be elucidated[4].

The aim of the present study was to investigate the presence of anti-Lf antibodies in patients with multiple sclerosis (MS). MS is the most common autoimmune disease involving the nervous system; its pathology is characterized by a classic picture of inflammation surrounding venules and extending into the myelin sheath. The cause of the disease is unknown but genetic factors, environment and viral infections are involved[5]. A precise role has been attributed to some cytokines such as interleukin-2 (Il-2), tumor necrosis factor (TNF-α, -β, -γ) and granulocyte-macrophage colony-stimulating activity (GM-CSF). It has been

Advances in Lactoferrin Research, edited by Spik *et al.*
Plenum Press, New York, 1998.

suggested that an increase in GM-CSF expression could be correlated with an increase in the inflammatory process[6].

Our previous study revealed that functional GM-CSF was constitutively present in the cerebrospinal fluid (SF) of all the patients tested. In MS patients variable levels of GM-CSF were found, though the concentration was not significantly different from that seen in all other patients tested. This was quite surprising because it has been suggested that an increase in GM-CSF could be correlated with an increase in the inflammatory process[6]. Since a down-regulation of GM-CSF by Lf in interleukin-1β (Il-1β) stimulated cells has been demonstrated[7], we investigated not only the presence of anti-Lf antibodies, but also the GM-CSF and Lf proteins in the SF of patients with MS versus patients with other non-inflammatory diseases (OND) to look for a possible correlation. The data presented here are from a preliminary study in which a restricted number of patients were analyzed.

2. MATERIALS AND METHODS

2.1. Patients

The study involved 30 patients: 12 MS, and 18 OND. MS patients were unselected in terms of form and stage of activity of the disease, the OND were used as control vs MS. After informed consent has been obtained, SF was collected, filtered with 0.2 mm microfilters, supplemented with 20,000 U/ml of aprotinin and stored at −80°C until used.

2.2. Anti-Lf Assay

Measurement of human anti-Lf in SF was performed by an indirect enzyme immunoassay, in which wells are coated with a preparation of highly purified Lf: during the first incubation, specific autoantibodies in diluted SF will bind to the antigen coating. After washings, a conjugate of enzyme-labeled monoclonal antibody to human IgG binds to surface-bound antibodies in the second incubation. After washings, specific antibodies are traced by incubation with substrate solution. The strips are read as absorbance at 550 nm. This assay detects the IgG class of autoantibodies specific for Lf. Each sample was run in duplicate. Negative and positive controls were always included.

2.3. GM-CSF Protein Assay

Quantitative measurement of human GM-CSF protein in SF was performed by a solid-phase enzyme-amplified sensitivity immunoassay (GM-CSF-EASIA) (Medgenix Diagnostics, Fleurus, Belgium). This assay is based on an oligoclonal system in which several monoclonal antibodies directed against distinct epitopes of GM-CSF are used. The use of several distinct monoclonal antibodies avoids hyperspecificity and endows the assay with high sensitivity, an extended standard range and short incubation times. This assay allows a minimum protein concentration of 3 pg/ml to be detected. Cross-reactions with G-CSF, M-CSF, Il-1α, Il-1β, Il-2, Il-3, Il-4, interferon-α, -β and -γ, TNF-α and TNF-β were insignificant.

2.4. Lf Immunological Detection

SFs were also tested for Lf presence. Immune slot-blot was performed on nitrocellulose membrane using a Schleicher & Schuell microfiltration apparatus. After blocking the membrane with 3% (wt/vol) BSA the antigen was exposed to 1:10 diluted primary rabbit anti-human Lf antibody (Dako Corporation, Carpinteria, CA). A secondary biotinylated goat anti-rabbit antibody was incubated with the membrane according to the manufacturer's instructions (BRL, Bethesda, MD). Streptavidin-β-galactosidase conjugate was added and color development was initiated by adding halogenated indolyl-β-D-galactoside in dimethylformamide. No positive reaction was observed when controls were used to evaluate the presence of non-specific interactions between Lf and various reagents used as controls.

2.5. Evaluation of SF Biological Parameters

Immediately after lumbar puncture each SF sample was tested for: cell count: by phase-contrast microscopy in a Fuchs-Rosenthal chamber. Elevation above the upper limit of the normal range (5 cells/ml) was considered pathological; albumin ratio: [albumin (mg %) in SF/ albumin (mg %) in serum] X 1000. Elevation above the upper normal limit of 7.4 is a sign of damage to the blood–brain-barrier; IgG index: IgG ratio: albumin ratio, where IgG ratio is expressed by [IgG (mg %) in SF: IgG (mg %) in serum] X 1000. Elevation above the upper normal limit of 0.69 is an expression of the intrathecal synthesis of IgG (i.e. local humoral immune activation).

2.6. Statistical Analysis

Concentrations of anti-Lf antibodies and GM-CSF in the SF of the MS and OND groups were statistically evaluated by Student's t test. The correlation between anti-Lf antibodies level and GM-CSF concentration and each of the SF biological parameters was analyzed by means Pearson's correlation coefficient.

3. RESULTS AND DISCUSSION

In this study, GM-CSF protein detected in the SF of MS and OND patients did not show significantly different levels among the samples, in agreement with our previous results. The GM-CSF protein concentration expressed as pg/ml is shown in Table 1.

Lf protein was revealed in all SF tested by immune slot-blot. All samples were Lf positive (data not shown), but a quantitative determination was not performed.

In SF of MS and OND patients anti-LF antibodies were detected by immunoassay. Statistical analysis showed a different distribution of the anti-Lf antibodies between the two classes investigated: a significant increased level was observed only in MS patients ($p = 0.012$) (Table 1).

Previous studies have shown the presence of anti-bovine Lf antibodies in the serum of healthy subjects, whereas anti-human Lf antibodies have been detected only in patients with auto-immune diseases, such as systemic lupus erytematosus and ulcerative colitis[7].

Table 1. Values of GM-CSF and anti-Lf antibodies in SF

	MS (mean ± SD)	OND (mean ± SD)	$p^{\S}$
GM-CSF*	41.29 ± 25.29	36.26 ± 17.53	0.654
Anti-Lf antibodies*	0.0656 ± 0.0067	0.0494 ± 0.0103	0.012

*Measurement units: see Materials and Methods.
§Student's t test.

Table 2. Relationships* between anti-Lf antibodies and GM-CSF or IgG index

	MS		OND	
	r	p	r	p
Anti-Lf antibodies vs GM/CSF	−0.903	0.035	+0.791	0.866
Anti-Lf antibodies vs IgG index	−0.075	0.123	−0.304	0.506

*Pearson's correlation coefficient.

Although the anti-Lf antibodies are probably not involved in the pathogenic mechanism, occurrence of circulating IgG anti-Lf antibodies could be a marker of these diseases[7]. Our results not only confirm the above-mentioned reports, but also broaden our knowledge on autoimmune MS.

When in MS and OND, were separately analyzed, no significant correlation was found between the level of anti-Lf antibodies and cell count, or the value of albumin ratio (data not shown). A significant negative relationship between anti-Lf antibody level and GM-CSF concentration was observed (p = 0.035) in the SM group but not in the OND group (Table 2). It should be noted that, although not statistically significant, only in the MS group was there a trend toward a negative relationship between level of anti-Lf antibodies and IgG index, a parameter expressive of ongoing intrathecal immunological activation.

The observed significant negative relationship between anti-Lf antibody level and GM-CSF concentration is quite surprising and its meaning needs to be further investigated. A wide variety of speculations could be advanced; one possible hypothesis is that Lf, produced in a large amount in response to an inflammatory status, might not only down-regulate GM-CSF expression[7], but also increase the autoimmune response. This also could explain why the GM-CSF concentration in MS patients was not found higher than in OND, as expected.

Our study of course will give more indications analyzing a larger number of patients and selecting MS patients in terms of form and stage of activity of the disease. We think that it is worth continuing this study, since these preliminary results not only reveal, for the first time, the presence of anti-Lf antibodies in cerebrospinal fluid, but also show an increased levels of anti-Lf antibodies in MS, another auto-immune disease. The increased level of anti-human Lf antibodies in SF of MS patients suggests that anti-Lf antibodies could be a useful marker of this disease.

REFERENCES

1. B. Lonnerdal and S. Iyer. Lactoferrin: molecular structure and biological function. Ann: Rev. Nutr 1995, 15: 93–110
2. A. Terent et al. Lactoferrin lysozyme and β-microglobulin in cerebrospinal fluid. Stroke 1981, 12: 40–46
3. B. Leveugle, G. Spik, D.P. Perl, C. Bouras, H.M. Fillit, P.R. Hof. The iron-binding protein lactotransferrin is present in pathologic lesions in a variety of neurodegenerative disorders: a comparative immunohistochemical analysis. Brain Res 1994, 650: 20–31
4. P.F. Levay and M. Vijoen. Lactoferrin: a general review. Haematologica 1995, 80: 252–267
5. L. Steinman. Multiple Sclerosis: a coordinated immunological attack against myelin in the central nervous system. Cell 1996, 85: 299–302
6. J.E. Merrill and G.M. Jonakait. Interaction of the nervous and immune systems in development, normal brain homeostasis, and disease. The FASEB J. 1995, 9: 611–618
7. S. Penco, S. Pastorino, G. Bianchi-Scarra, C. Garre'. Lactoferrin down-modulates the activity of the Granulocyte Macrophage Colony Stimulating Factor promoter in Interleukin-1β-stimulated cells. J Biol Chem 1995, 270: 12263–12268

ANTIBODIES TO LACTOFERRIN

A Possible Link between Cow's Milk Intolerance and Autoimmune Disease

J. H. Brock,[1] A. Lamont,[1] D. J. Boyle,[1] E. R. Holme,[1] C. McSharry,[1]
J. E. G. Bunn,[2] and B. Lönnerdal[3]

[1]Department of Immunology
Western Infirmary
Glasgow G11 6NT, United Kingdom
[2]Department of Child Health
University of Newcastle
Newcastle upon Tyne NE2 4HH, United Kingdom
[3]Department of Nutrition
University of California, Davis
Davis, California 95616

1. INTRODUCTION

Individuals with various forms of autoimmunity often develop autoantibodies to neutrophil components collectively known as ANCA (anti-neutrophil cytoplasmic antibodies). Two forms can be distinguished according to the pattern of localisation of these autoantibodies. In one form (c-ANCA) staining is predominantly cytoplasmic[1], whereas in the other (p-ANCA) staining is perinuclear[2]. In the latter IgG antibodies predominate and one of the neutrophil antigens involved is lactoferrin[3]. Why lactoferrin should act as an autoantigen is not clear.

Lactoferrin is also present in human and bovine milk, the proteins from the two species sharing about 70% sequence homology[4]. It is thus likely that the two proteins share common or similar epitopes, and it is thus possible that exposure and immunological reaction to bovine lactoferrin during ingestion of cows milk or milk products in infancy could prime the immune system to subsequently react against human lactoferrin. To test this we have determined levels of IgG antibodies to human and bovine lactoferrins in infants on different feeding regimes, with or without cow's milk intolerance, and in p-ANCA positive and negative adults. These have been compared with antibodies to β-lactoglobulin, a

Advances in Lactoferrin Research, edited by Spik *et al.*
Plenum Press, New York, 1998.

highly antigenic protein of bovine milk which, unlike lactoferrin, does not have a human homologue[5].

2. MATERIALS AND METHODS

2.1. Patients and Sera

Sera were obtained from:

- 23 infants with cow's milk intolerance/allergy or failure to thrive and 23 normal control children;
- 25 babies from the Gambia, 40–48 weeks old, that had been exclusively breast fed, and had not been exposed to cow's milk or milk products;
- 10 healthy Swedish babies that had been fed infant formula supplemented with bovine lactoferrin (1 mg/ml) between 1 month and 6 months of age, as part of a study to examine the effect of lactoferrin on iron absorption[6]. Sera were obtained at the start of the supplementation study (age 1 month) and at 4 and 6 months of age;
- 24 pANCA+ve adults with autoimmune disorders and 24 healthy ANCA−ve adult controls.

2.2. Detection of Autoantibodies

A sandwich ELISA using plates coated with 1 μg/ml bovine lactoferrin (Fina Research, Feluy, Belgium), human lactoferrin (Sigma, Poole, England) or bovine β-lactoglobulin (Sigma) was used. Results were expressed as units of antibody activity against a standard positive sample with an arbitrary value of 1000 units/ml. The lower limit of detection was 30 units/ml.

2.3. Effect of Antibodies on Iron-Binding by Lactoferrin

IgG was purified from a hyperimmune rabbit antihuman lactoferrin antiserum and a control normal rabbit serum by affinity chromatography on protein A Sepharose. Human apolactoferrin was 70% saturated with iron using ferric nitrilotriacetate[7] and after overnight incubation at 4°C to ensure complete binding was incubated for 30 minutes at 4°C with the appropriate purified IgG. 'Free' iron was then determined by the bleomycin assay, which detects free or loosely-bound iron but does not react with iron bound to lactoferrin[8].

3. RESULTS

As expected antibodies to human lactoferrin were detected in most of the ANCA+ve patients (Fig. 1), whereas the controls were either very low or negative ($\leq$30units/ml). Antibodies to β-lactoglobulin were detected in both groups, but titres of antibodies to bovine lactoferrin were significantly higher ($p<0.01$) in the ANCA+ve patients than in the controls. This is of particular interest, since exposure to bovine lactoferrin and β-lactoglobulin will occur simultaneously via cow's milk, and the presence of ANCA must therefore selectively enhance reaction to bovine lactoferrin.

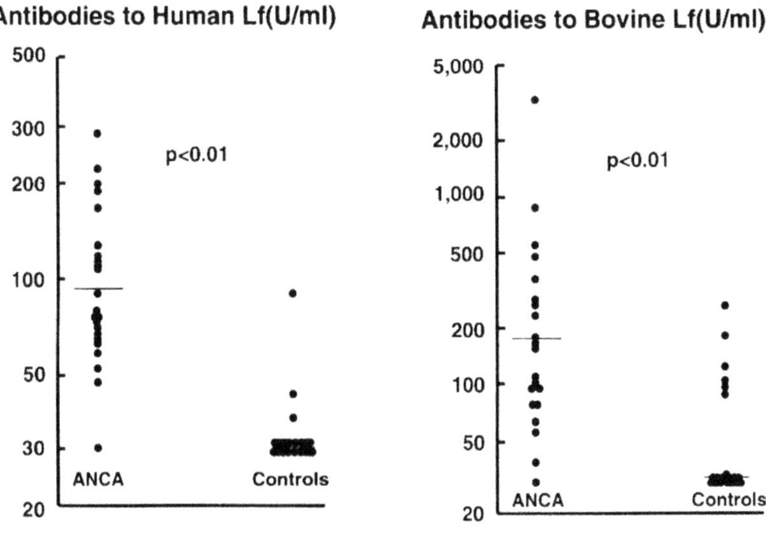

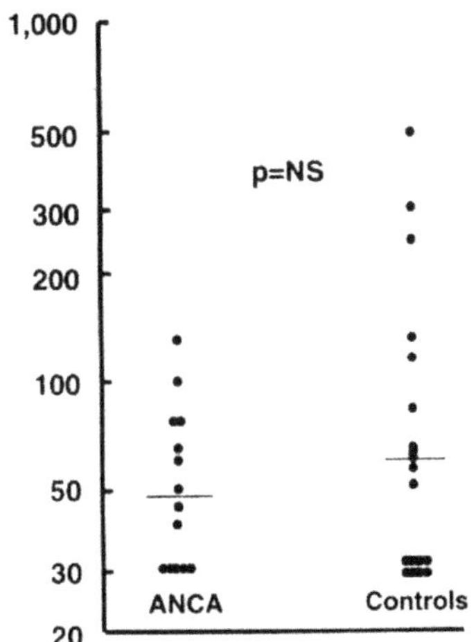

Figure 1. Antibodies to lactoferrin and β-lactoglobulin in ANCA patients and controls.

In infants low titres of antibodies to human lactoferrin were detected in about half of both the controls and the CMI group, with no significant difference between them. However, antibodies to human lactoferrin were detected in most of the breast-fed Gambian infants, and titres were significantly higher ($p<0.05$) than in the other two groups. In contrast antibodies to β-lactoglobulin were highest in the CMI group ($p<0.01$ compared to controls), and were virtually undetectable in the breast-fed infants. The distribution of antibodies to bovine lactoferrin was similar to those to β-lactoglobulin, although several of the breast-fed infants did have low levels of anti-bovine lactoferrin antibodies.

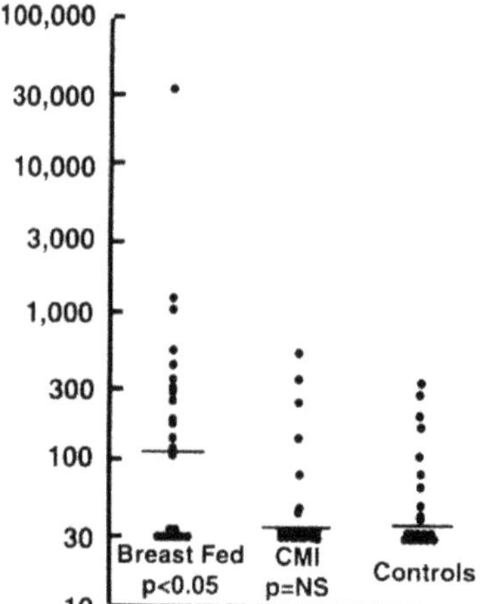

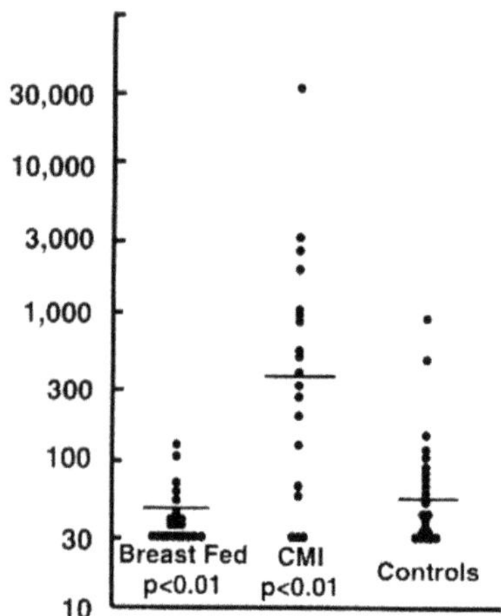

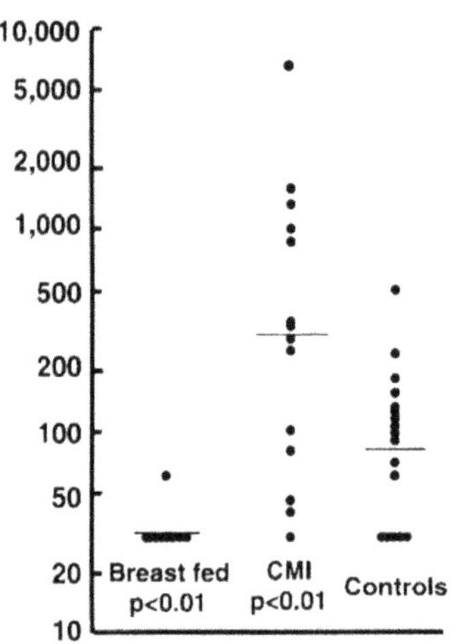

Figure 2. Antibodies to lactoferrin and β-lactoglobulin in infants with cow's milk intolerance (CMI), exclusively breast-fed infants, and controls.

The infants fed formula supplemented with bovine lactoferrin developed high levels of anti-bovine lactoferrin antibodies (Fig. 3). In the control group there was only a slight increase, and antibodies to β-lactoglobulin and human lactoferrin were relatively low and increased only slightly during the trial.

Although there was no correlation within each group between levels of antibodies to human and bovine lactoferrin, a significant correlation was found ($p<0.05$) when the all the groups were considered together.

Purified IgG from a rabbit antibody to human lactoferrin, when incubated with iron-loaded lactoferrin, caused the iron to become reactive in the bleomycin assay (Fig. 4). Normal rabbit IgG had no such effect.

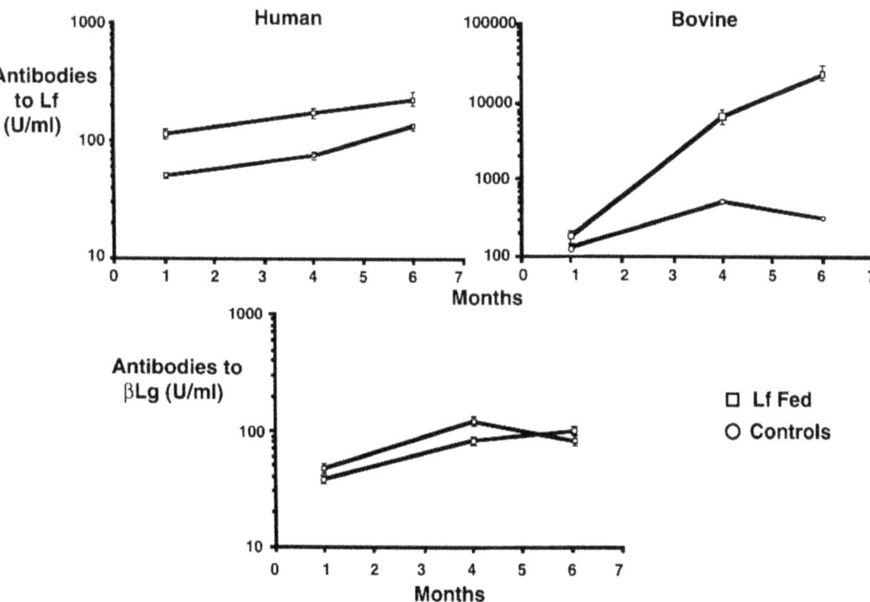

Figure 3. Antibodies to lactoferrin and β-lactoglobulin in infants fed formula supplemented with bovine lactoferrin (1 mg/ml) and controls over a 6-month period.

4. DISCUSSION

This study provides evidence that development of anti-lactoferrin autoantibodies in ANCA patients may be due, at least in part, to prior stimulation by bovine lactoferrin in cow's milk. High titres of anti-bovine lactoferrin antibodies were present in infants with CMI, in whom gut damage may allow antigenically intact lactoferrin or fragments thereof to enter the circulation. Likewise infants fed high levels of bovine lactoferrin developed high titres of antibodies. Even in healthy infants small amounts of gut contents can cross

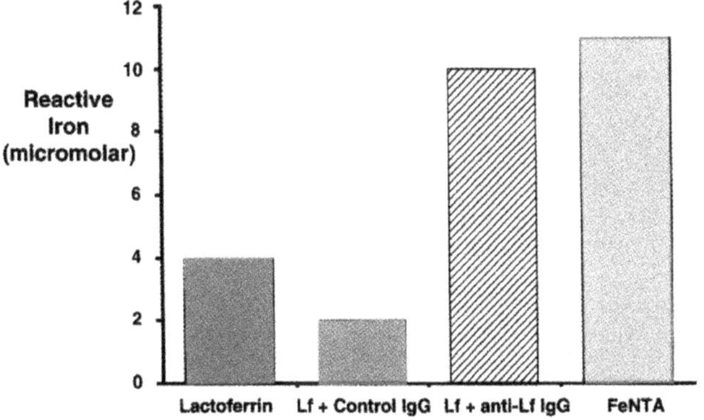

Figure 4. Effect of anti-lactoferrin IgG on iron-binding by lactoferrin.

the intestinal wall, and it may well be that in the lactoferrin-fed infants the lumenal concentration was sufficient to permit an immunogenic dose to enter the circulation.

Further evidence that lactoferrin is highly antigenic comes from the finding that many of the exclusively breast-fed Gambian infants developed autoantibodies to human lactoferrin. This apparent antigenicity of human lactoferrin in breast-fed infants was surprising; some may have suffered gastrointestinal infection, permitting leakage of intact milk proteins into the circulation. Furthermore, antibodies to human lactoferrin were also found in some of the Scottish infants. Thus even human lactoferrin may be immunogenic in infants in certain circumstances.

Most of the patients with ANCA had antibodies to bovine lactoferrin and to β-lactoglobulin, as well as to human lactoferrin. However, a crucial finding was that whereas patients and controls had similar levels of anti-β-lactoglobulin antibodies, antibodies to bovine lactoferrin were significantly higher in the ANCA patients than in the controls. Since cow's milk is the only likely source of exposure to both proteins it is possible that cross-reaction with human lactoferrin, released by degranulating neutrophils during the inflammatory response associated with autoimmune disease, caused an elevation of the titres to bovine lactoferrin.

An overall correlation between the levels of antibodies to human and bovine lactoferrins was found, although this was not significant within individual groups. However, there may well be a closer correlation if T-cell epitopes were examined.

The ability of anti-lactoferrin antibodies to interfere with iron binding by lactoferrin suggests a possible pathogenetic role for these antibodies, in addition to their previously-demonstrated ability to cause neutrophil priming[9]. Even if iron is not completely released, its ability to react in the bleomycin assay indicates that it would be made available to catalyse the production of harmful free radicals from reactive oxygen compounds.

In conclusion, this work suggests that exposure to bovine lactoferrin through ingestion of cow's milk might predispose to the subsequent development of autoantibodies to human lactoferrin. This should be borne in mind when recommending infant feeding practices to mothers from families with a history of autoimmune disease. It is has to be set against the potential benefits of adding bovine lactoferrin to infant formulas.

ACKNOWLEDGMENTS

DJB was supported by a Wellcome Trust studentship.

REFERENCES

1. Van der Woude FJ, Rasmussen N, Lobatto S, Permin H, Van der Giessen M, Rasmussen N, Wiik A, Daha MR, Van Es LA, Van der Hem GH, The TH. Autoantibodies against neutrophils and monocytes: tool for diagnosis and marker for disease activity in Wegener's granulomatosis. Lancet 1985; i: 425–9.
2. Falk RJ, Jennette JC. Anti-neutrophil cytoplasm antibodies with specificity for myeloperoxidase in patients with systemic vasculitis and idiopathic necrotizing crescentic glomerulonephritis. N Engl J Med 1988; 318: 1651–1657.
3. Peen E, Almer S, Bodemar G, Rydén B-O, Sjölin C, Tejle K, Skogh T (1993) Anti-lactoferrin antibodies and other types of 'ANCA' in Crohn's disease, ulcerative colitis, and primary sclerosing cholangitis. Gut 1993; 34: 56–62.
4. Pierce A, Colavizza D, Benaissa M, Maes P, Tartar A, Montreuil J, Spik G (1991) Eur J Biochem, 196, 177.
5. Taylor CJ, Hendrickse RG, McGaw J, Macfarlane SBJ (1988) Detection of cow's milk protein intolerance by an enzyme-linked immunosorbent assay. Acta Paediatr Scand, 77, 49–54.

6. Lönnerdal B, Hernell O (1994) Iron, zinc, copper and selenium status of breast-fed infants and infants fed trace element fortified milk based infant formula Acta Paediatr, 83, 367–373.
7. Bates GW, Schlabach MR (1973) The reaction of ferric salts with transferrin. J Biol Chem, 248, 3228–3232.
8. Evans PJ, Halliwell B (1994) Measurement of iron and copper in biological systems: bleomycin and copper-phenanthroline assays. Meth Enzymol, 233, 82–92
9. Peen E, Sundqvist T, Skogh T (1996) Leucocyte activation by anti-lactoferrin antibodies bound to vascular endothelium. Clin Exp Immunol, 103, 403–407.

PREVALENCE AND CLINICAL SIGNIFICANCE OF ANTI-LACTOFERRIN AUTOANTIBODIES IN INFLAMMATORY BOWEL DISEASES AND PRIMARY SCLEROSING CHOLANGITIS

C. Roozendaal,[1] G. Horst,[1] K. Pogány,[1] A. W. M. van Milligen de Wit,[2]
J. H. Kleibeuker,[1] E. B. Haagsma,[1] P. C. Limburg,[1] and C. G. M. Kallenberg[1]

[1]Department of Internal Medicine
University Hospital
NL-9715 GR Groningen, The Netherlands
[2]Department of Gastroenterology and Hepatology
Academic Medical Centre
Amsterdam, The Netherlands

1. SUMMARY

Anti-neutrophil cytoplasmic antibodies (ANCA) are autoantibodies directed against cytoplasmic constituents of neutrophil granulocytes. Antibodies with specificity for proteinase 3 and myeloperoxidase are seromarkers for systemic vasculitides. ANCA with specificity for lactoferrin were described in patients with several idiopathic inflammatory diseases, such as the inflammatory bowel diseases and rheumatoid arthritis. However, the clinical significance of anti-lactoferrin autoantibodies is still unclear.

In this study, we determined the clinical significance of anti-lactoferrin autoantibodies in sera from large groups of patients with ulcerative colitis (UC), Crohn's disease (CD), and primary sclerosing cholangitis (PSC). Antibodies to human lactoferrin were detected by ELISA and by immunoblotting, using an extract of sonicated neutrophils as antigen source.

Autoantibodies to lactoferrin were found in 29% of patients with UC, 13% of patients with CD, and 22% of patients with PSC. In inflammatory bowel diseases, the presence of anti-lactoferrin antibodies was not related to treatment, disease activity, duration of disease, or disease extent. In PSC, the presence of autoantibodies to lactoferrin did not correlate with duration of disease or the presence of cirrhosis. However, patients with PSC and coexistent UC had significantly more frequently antibodies to lactoferrin than PSC patients without IBD.

Advances in Lactoferrin Research, edited by Spik *et al.*
Plenum Press, New York, 1998.

In conclusion, autoantibodies to lactoferrin are a common feature of inflammatory bowel diseases and PSC. However, the clinical significance of those autoantibodies is limited as they lack sensitivity and specificity for those disorders. Future research should address the pathophysiological role of anti-lactoferrin ANCA and the influence of anti-lactoferrin ANCA binding on the functional properties of the lactoferrin molecule.

2. INTRODUCTION

Anti-neutrophil cytoplasmic antibodies (ANCA) are autoantibodies directed against certain cytoplasmic constituents of neutrophil granulocytes and monocytes. These antibodies, especially those with specificity for proteinase 3 and myeloperoxidase, are sensitive and specific markers for several forms of systemic vasculitis and idiopathic crescentic glomerulonephritis. [1]ANCA are present also in other idiopathic inflammatory diseases. The prevalence of ANCA as detected by indirect immunofluorescence (IIF) in ulcerative colitis (UC) has been estimated as 50–80%, in primary sclerosing cholangitis (PSC) as 65–95%, and in autoimmune hepatitis as 88–96%. In Crohn's disease (CD), the prevalence is somewhat lower (10–20%). Furthermore, ANCA have been found in patients with rheumatoid arthritis (RA, 30–40%), systemic lupus erythematosus (SLE, 20–50%), juvenile chronic arthritis (20–45%), and HIV-infection (10–40%).

Many different target antigens have been reported to be recognized by ANCA in those diseases, including lactoferrin, bactericidal/permeability-increasing protein, cathepsin G, lysozyme, azurocidin, α-enolase, and β-glucuronidase. However, conflicting results have been obtained by different groups, and it is obvious that more antigens will be detected in the future. Defining the antigenic specificities of ANCA will allow the analysis of the diagnostic and, possibly, pathophysiologic significance of a specific autoantibody.

In this study, we investigated the prevalence and significance of specific anti-lactoferrin antibodies and their possible role as a marker for certain disease characteristics in patients with UC, CD, and PSC.

3. PATIENTS AND METHODS

3.1. Patients

We determined the prevalence and clinical significance of ANCA and anti-lactoferrin autoantibodies in sera from three large groups of patients: one group with UC (n=96), one group with CD (n=112), and one group with PSC (n=69). Forty-five (65%) PSC patients had associated inflammatory bowel disease (UC, n=37 (54%), CD, n=8 (12%)). Diagnosis of UC, CD, and/or PSC was based on accepted clinical, biochemical, histological, endoscopic, and for PSC also cholangiographic criteria.

3.2. Methods

3.2.1. Indirect Immunofluorescence. Detection of IgG ANCA by IIF was performed on ethanol fixed granulocytes. Serum samples were diluted 1:20 in phosphate buffered saline and tested at two-fold dilutions up to 1:640. A FITC-conjugated rabbit anti-human IgG antibody (Dakopatts, Copenhagen, Denmark) was used for detection of bound IgG. A titer ≥1:40 was considered positive.

3.2.2. Antigen-Specific ELISA. The presence of IgG anti-lactoferrin antibodies was detected by ELISA. Lactoferrin (Sigma Chemical Co., St. Louis, MO, USA) was coated 5 μg/ml in phosphate buffered saline. Serum samples were applied at a dilution of 1:100 and in twofold dilutions up to 1:800. Alkaline phosphatase conjugated goat anti-human IgG (American Qualex, San Clemente, CA, USA) was used for detection of bound IgG.

3.2.3. Western Blotting and Immunodetection. Anti-lactoferrin antibodies in samples from patients with inflammatory bowel diseases were detected also by immunoblotting. Immunoblotting was performed as described previously,[2] using an extract of sonicated granulocytes as antigen source. To identify the protein band corresponding with granulocyte lactoferrin, we used a polyclonal rabbit anti-human lactoferrin antibody (Dakopatts, Copenhagen, Denmark).

4. RESULTS

4.1. Indirect Immunofluorescence

ANCA were detected by IIF on ethanol fixed granulocytes in 56/96 UC serum samples (58%). Median titer of the ANCA-positive samples was 1:160 (range 1:40–1:640). Twenty-one out of 112 CD samples (19%) were positive for IIF-ANCA (median titer 1:80, range 1:40–1:640). Furthermore, IIF-ANCA were detected in 46/69 (67%) samples from patients with PSC (median titer 1:160, range 1:40–1:640).

4.2. Anti-Lactoferrin ELISA and Immunoblotting

Autoantibodies to lactoferrin were found by ELISA in 22/96 (23%) patients with UC, 12/112 (11%) patients with CD, and 15/69 (22%) patients with PSC.

Samples from patients with inflammatory bowel diseases were tested also on Western blot, using a rough granulocyte extract as antigen source. A polyclonal anti-human lactoferrin antibody reacted strongly with a broad band of approximately 80 kD, identical to the 80 kD band recognized by 25 (26%) UC and 12 (11%) CD samples, suggesting that the samples recognizing the 80 kD band were directed against lactoferrin. Six of these UC samples and 2 CD samples were not positive by ELISA. On the other hand, 3 UC samples and 2 CD samples were positive by ELISA but negative in the immunoblot.

4.3. Clinical Significance

As the association between ANCA and the various vasculitides is based on the presence of antibodies of defined specificity, we studied the relation of anti-lactoferrin autoantibodies with several clinical characteristics of inflammatory bowel diseases and PSC.

4.3.1. Treatment. Since it is possible that the use of steroids influences the prevalence of ANCA by suppressing immunoglobulin production, we first tested whether the prevalence of ANCA and anti-lactoferrin antibodies was different in patients with or without steroid treatment. In UC, as well as in CD and PSC, patients with or without steroid treatment had comparable prevalences of IIF-ANCA and anti-lactoferrin anitbodies.

4.3.2. Disease Activity or Severity. In Wegener's granulomatosis, anti-proteinase 3 antibodies are a marker of disease activity. During relapses of the disease, ANCA titers are usually high, whereas during remission, they are low or absent. Therefore, we tested whether ANCA or anti-lactoferrin antibodies were associated with disease activity of inflammatory bowel disease, as measured by the Truelove-Witts score (UC) or the Van Hees index (CD). Mean disease activity score was not different for the IIF-ANCA- or anti-lactoferrin-positive or negative groups. Patients classified as having a relapse did not have higher prevalences of IIF-ANCA or anti-lactoferrin antibodies than patients being in remission. However, UC patients with a relapse had higher IIF-ANCA titers (patients in remission: median titer 1:80, patients with a relapse: median titer 1:320).

PSC usually follows a progressive course, ultimately leading to biliary cirrhosis and hepatic failure, and has no relapse - remission pattern. Therefore, we considered the presence of cirrhosis as a measure of severity of disease. Hoewever, neither IIF-ANCA, nor anti-lactoferrin antibodies were associated with the presence of cirrhosis.

4.3.3. Duration of Disease. ANCA might be a feature of longstanding dysregulation of the immune system in immune-mediated disease. However, in none of the three patient groups, IIF-ANCA or anti-lactoferrin were associated with duration of disease.

4.3.4. Localization of Inflammatory Bowel Disease. In CD, it has been stated that ANCA more frequently are found in patients with involvement of the colon.[3] In UC, ANCA have been mentioned to be associated with treatment-resistant left-sided colitis.[4] Therefore, we tested whether ANCA or anti-lactoferrin antibodies were associated with localization of disease in UC or CD. No correlation was found between IIF-ANCA or anti-lactoferrin antibodies and colonic involvement or left-sided inflammation.

4.3.5. Coexistence of PSC and Inflammatory Bowel Disease. Since PSC is often associated with inflammatory bowel disease and since ANCA frequently occur in patients with UC and CD, the presence of ANCA in PSC might be linked to concomitant inflammatory bowel disease. In this study, prevalences of IIF-ANCA in PSC did not differ between patients with and without inflammatory bowel disease. However, antibodies to lactoferrin were more frequently detected in PSC patients with UC than in those without inflammatory bowel disease. Only 8% of PSC patients without inflammatory bowel disease had anti-lactoferrin antibodies, whereas 32% of PSC patients with UC had antibodies to lactoferrin. This is comparable to the 29% prevalence of anti-lactoferrin antibodies in UC without PSC. Thus, the presence of antibodies to lactoferrin in PSC appears to be related to the coexistence of UC.

5. DISCUSSION

In this study, we reported the presence of autoantibodies to lactoferrin in serum samples from patients with inflammatory bowel diseases and PSC. Anti-lactoferrin antibodies in PSC were shown to be associated with coexistent UC. Except from this association, we found no clinical differences between inflammatory bowel disease or PSC patients with or without anti-lactoferrin antibodies.

ANCA were detected by standard IIF methods and anti-lactoferrin antibodies were detected by ELISA and immunoblotting. Results were usually, but not always, identical for the different methods. This might depend on differences in antigens used for these tests. The antigen for ELISA was human milk lactoferrin, whereas the lactoferrin in immunoblotting was derived from granulocytes.

Although in this study no associations between anti-lactoferrin antibodies and clinical parameters of inflammatory bowel diseases and PSC were found, such associations have been reported for other inflammatory diseases.

Antibodies directed against lactoferrin have been detected in 5–20% of RA patients.[5–7] A fair correlation exists between serum anti-lactoferrin antibodies and antibodies to lactoferrin present in the synovial fluid.[5,6] Anti-lactoferrin antibodies were associated with vasculitis in RA.[7]

Anti-lactoferrin antibodies also occur in 10–20% of patients with SLE.[8–11] Pozzi *et al.* reported that anti-lactoferrin antibodies were especially associated with lupus nephritis.[9] However, this could not be confirmed by the same group and by others.[10] De Bandt *et al.* found a consistent pattern of high anti-lactoferrin ELISA titers at the time of diagnosis of lupus nephritis, with decreasing titers during treatment and rising titers during flare.[11]

Anti-lactoferrin antibodies were described in 67% of HIV-positive patients, both in asymptomatics and symptomatics. These patients had no vasculitis and no other known anti-lactoferrin associated disease. The titer, as measured by ELISA, was higher in the symptomatic patients. Anti-lactoferrin antibodies did not correlate with PMN count or with lactoferrin levels in the blood.[12]

In addition to the role of anti-lactoferrin as a seromarker for several inflammatory diseases, it has been shown that those antibodies can influence pathophysiological processes. Priming of granulocytes with cytokines (TNF-α, IL-1) allows the ANCA target molecules to be expressed on the cellular surface, where they become accessible for binding by ANCA. ANCA directed against proteinase 3, myeloperoxidase, and lactoferrin are able to bind to their target antigen on the cellular membrane of primed neutrophils. This binding stimulates those neutrophils to produce oxygen radicals and to degranulate.[13] The process of neutrophil activation by ANCA is dependent on the second Fc-receptor on neutrophils (FcγRII).[13] FcγRII has high affinity for human IgG3 molecules. Interestingly, it was demonstrated that anti-lactoferrin antibodies in inflammatory bowel diseases were mainly of the IgG1 and IgG3 subclass.[14]

Peen *et al.* showed that endothelial cells were able to bind lactoferrin and that anti-lactoferrin antibodies bound to lactoferrin on endothelial cells were able to activate non-stimulated granulocytes to a chemoluminescence response.[15]

Another mechanism through which ANCA might have influence on the pathophysiolocigal process is inhibition of functions of the target molecule. The epitopes towards which ANCA are directed on the various target antigens are under investigation. Antibodies to proteinase 3 have been shown to influence binding of proteinase 3 to its natural inhibitor, alpha-1 antitrypsin.[16] Disturbance of this binding could lead to prolonged enzymatic activity of the proteinase 3 molecule, which leads to tissue damage.

One of the functions of lactoferrin is the inhibition of bacterial growth by iron-withholding and by direct binding to lipopolysaccharide (LPS), followed by permeabilization of the bacterial cell wall. Binding of antibodies to the iron- or LPS-binding site of lactoferrin might influence these functions and cause bacterial overgrowth. However, Audrain *et al.* reported that anti-lactoferrin positive serum samples from six patients were not able to inhibit iron binding by lactoferrin,[17] suggesting that the epitopes recognized by anti-lactoferrin antibodies of those patients were not located in the iron-binding site of lactoferrin.

Until now, it is not known how autoantibodies to lactoferrin are induced. Possibly, the antibodies are raised against foreign proteins and cross-react with human lactoferrin. Bovine lactoferrin has high homology with human lactoferrin. In patients with disturbances of the immune system, drinking cow's milk could lead to formation of antibodies to bovine lactoferrin. These anti-bovine lactoferrin antibodies might recognize epitopes that are also present on the human lactoferrin molecule. Brock *et al.* have investigated the correlation between antibodies to bovine and human lactoferrin (this issue).

The concept of molecular mimicry, that is the hypothesis that antibodies to bacterial proteins cross-react with homologous epitopes on human proteins, could apply to antibodies against the mycobacterial heat shock protein hsp65, that shows homology with human lactoferrin.[18] However, Esaguy *et al.* could not find a positive correlation between serum titers for antibodies to hsp65 and antibodies to human lactoferrin,[19] whereas Peen *et al.* showed that immunization of rats with hsp65 did not raise antibodies to human lactoferrin.[20]

In conclusion, autoantibodies to lactoferrin develop in several immune-mediated diseases. However, their diagnostic and prognostic role seems limited. Further research should establish how and why autoantibodies to lactoferrin develop and elucidate their possible pathophysiological effects in an inflammatory environment.

REFERENCES

1. Kallenberg CGM, Brouwer E, Weening JJ, Cohen Tervaert JW. Anti-neutrophil cytoplasmic antibodies: current diagnostic and pathophysiolical potential. Kidney Int 1994;46:1–15.
2. Mulder AHL, Horst G, Haagsma EB, Limburg PC, Kleibeuker JH, Kallenberg CGM. Prevalence and characterization of neutrophil cytoplasmic antibodies in autoimmune liver diseases. Hepatology 1993;17:411–417.
3. Vasiliauskas EA, Plevy SE, Landers CJ, Binder SW, Ferguson DM, Yang H, Rotter JI, Vidrich A, Targan SR. Perinuclear antineutrophil cytoplasmic antibodies in patients with Crohn's disease define a clinical subgroup. Gastroenterology 1996;110:1810–1819.
4. Sandborn WJ, Landers CJ, Tremaine WJ, Targan SR. Association of antineutrophil cytoplasmic antibodies with resistance to treatment of left-sided ulcerative colitis: results of a pilot study. Mayo Clin Proc 1996;71:431–436.
5. Mulder AHL, Horst G, Van Leeuwen MA, Limburg PC, Kallenberg CGM. Anti-neutrophil cytoplasmic antibodies in rheumatoid arthritis: characterization and clinical correlates. Arthritis Rheum 1993;36:1054–1060.
6. Afeltra A, Sebastiani GD, Galeazzi M, Caccavo D, Ferri GM, Marcolongo R, Bonomo L. Antineutrophil cytoplasmic antibodies in synovial fluid and in serum of patients with rheumatoid arthritis and other types of synovitis. J Rheumatol 1996;23:10–15.
7. Coremans IEM, Hagen EC, Daha MR, Van der Woude FJ, Van der Voort EAM, Kleyburg-Van der Keur C, Breedveld FC. Antilactoferrin antibodies in patients with rheumatoid arthritis are associated with vasculitis. Arthritis Rheum 1992;35:1466–1475.
8. Spronk PE, Bootsma H, Horst G, Huitema MG, Limburg PC, Cohen Tervaert JW, Kallenberg CGM. Anti-neutrophil cytoplasmic antibodies in systemic lupus erythematosus. Br J Rheumatol 1996;35:625–631.
9. Pozzi C, Radice, Rota S, Li Vecchi M, Sinico RA. Clinical significance of antilactoferrin antibodies in renal diseases [abstract]. Am J Kidney Dis 1991;18:202.
10. Sinico RA, Pozzi C, Radice A, Tincani A, Li Vecchi M, Rota S, Comotti C, Ferrario F, D'Amico G. Clinical significance of antineutrophil cytoplasmic autoantibodies with specificity for lactoferrin in renal diseases. Am J Kidney Dis 1993;22:253–260.
11. De Bandt M, Meyer O, Haim T, Kahn MF. Antineutrophil cytoplasmic antibodies in rheumatoid arthritis patients. Br J Rheumatol 1996;35:38–43.
12. Defer MC, Follezou JY, Picard O, Damais C. Antilactoferrin autoantibodies associated with HIV infection. C R Acad Sci III 1994;317:675–678.
13. Mulder AHL, Heeringa P, Brouwer E, Limburg PC, Kallenberg CGM. Activation of granulocytes by anti-neutrophil cytoplasmic antibodies (ANCA): a Fc gamma RII-dependent process. Clin Exp Immunol 1994;98:270–278.
14. Peen E, Locht H, Skogh T. IgG subclasses of lactoferrin (LF)-anti-neutrophil cytoplasmic antibodies (ANCA) [abstract]. Sarcoidosis 1996;13:278.
15. Peen E, Sundqvist T, Skogh T. Leucocyte activation by anti-lactoferrin antibodies bound to vascular endothelium. Clin Exp Immunol 1996;103:403–407.

16. Van de Wiel BA, Dolman KM, Van der Meer-Gerritsen CH, Hack CD, Von dem Borne AEK, Goldschmeding R. Interference of Wegener's granulomatosis autoantibodies with neutrophil proteinase 3 activity. Clin Exp Immunol 1992;90:409–414.
17. Audrain MA, Gourbil A, Muller JY, Esnault LM. Anti-lactoferrin autoantibodies: relation between epitopes and iron-binding domain. J Autoimmun 1996;9:569–574.
18. Aguas A, Esaguy N, Sunkel CE, Silva MT. Cross-reactivity and sequence homology between the 65-kilodalton mycobacterial heat shock protein and human lactoferrin, transferrin, and DR beta subsets of major histocompatibility complex class II molecules. Infect Immun 1990;58:1461–1470.
19. Esaguy N, Freitas PM, Aguas AP. Anti-lactoferrin autoantibodies in rheumatoid arthritis. Clin Exp Rheumatol 1993;11:581–582.
20. Peen E, Enestrom S, Skogh T. Distribution of lactoferrin and 60/65 kDa heat shock protein in normal and inflamed human intestine and liver. Gut 1996;38:135–140.

EFFECT OF LACTOFERRIN ON THE PHAGOCYTIC ACTIVITY OF POLYMORPHONUCLEAR LEUCOCYTES ISOLATED FROM BLOOD OF PATIENTS WITH AUTOIMMUNE DISEASES AND *Staphylococcus aureus* ALLERGY

Vladi Manev,[1] Ana Maneva,[2] and Ljuben Sirakov[2]

[1]Department of Experimental Immunology
Military Medical Academy
Sofia, Bulgaria
[2]Department of Biochemistry
Medical Faculty, Medical University
1431 Sofia, Zdrave 2 St., Bulgaria

1. SUMMARY

Phagocytic number (PN) and phagocytic index (PI) of neutrophils isolated from blood of patients with autoimmune diseases, allergy to *Staphylococcus aureus* and from blood of healthy individuals were examined. Our results concerning the influence of lactoferrin (Lf); (6.7 mg/l) on the PI of PMN showed that: 1) Lf enhances reliable PI of PMN at the 30-th minute starting the phagocytic reaction in patients with autoimmune disease in an active stage, in blood donors treated as healthy with the presence of autoantibodies, in patients with autoimmune diseases and proved autoantibodies against tissue, cell antigens and collagen, 2) Lf influences non-significantly PI of PMN in patients with autoimmune collagen diseases in remission, 3) Lf increases PI of PMN with 19% only in 58% from the assessed patients with *Staphylococcus aureus*, and 4) Lf decreases non-significantly PI of PMN in the healthy controls. Our studies on the effect of Lf on the phagocytic activity of PMN suggest that Lf has stronger effect on the PN compared to the PI: 1) Lf enhances with 86% the PN in patients with *Staphylococcus aureus*, 2) Lf increases PN of PMN in all of the assessed patients with autoimmune collagen diseases in active stage (mean with 72%), and 3) Lf increases PN of PMN in 4 from the 5 investigated healthy controls (mean

Advances in Lactoferrin Research, edited by Spik *et al.*
Plenum Press, New York, 1998.

with 22%). Our results show a "corrective" effect of Lf on the phagocytic functions in the investigated groups of patients. The possible mechanisms, by which Lf increases PN and PI of neutrophils, is discussed: 1) they may concern the antioxidative properties of Lf to block the iron ions in their catalytic inactive form or to take part as ferric-Lf in an oxidative-reduction processes on the plasma membrane and controlling transmembrane transport systems, 2) Lf decreases the negative surface charge and thus enhances the adherent ability of the PMN. Probably to this stimulated adherent ability dues the increased ingestion of bacteria in the presence of Lf, and 3) The "changed" membrane of PMN may have higher number receptors for Lf to bind more molecules of exogenous Lf. The increase of Lf binding which enhances the adherence and aggregation of neutrophils, facilitates the phagocytosis.

2. INTRODUCTION

The biological importance of lactoferrin (Lf) in host defence is emphasised by the observed susceptibility of subjects with congenital or acquired lactoferrin deficiency to recurrent infections[10]. Lf, is a multifunctional protein[17,45], *in vivo* it may have pro-[1] and anti-inflammatory activity[16]. Numerous investigations on the Lf participation in the pathogenic mechanisms of autoimmune and allergic diseases are known.

In vivo it was shown that Lf concentration is enhanced in the locuses with inflammatory exudates, for example joints in RA[8] and in patients with allergic diseases[48]. Plasma measurements indicated that asthmatics, even when clinically asymptomatic, have evidence of persistent granulocyte activation[28]. Lf induces P-selectin-dependent granulocyte accumulation and activation, monoclonal antibodies to P-selectin may be therapeutically beneficial in inflammatory disorders[27]. There were found antibodies against Lf in rheumatoid arthritis (RA), systemic lupus erythematosus (SLE), liver autoimmune diseases etc.[33,34,40] It was shown, that antibodies against Lf, connected with the vascular endothelial cells, may activate the phagocytes and to enhance the destruction as a consequence of the phagocytosis in the chronic inflammatory processes[39]. Lf may work as an accessory stimulatory factor of the T-cell autoreactivity associated with mycobacterium-induced arthritis[21,2].

On the other hand it was found an inhibitory effect of Lf on the effector stage of the reactions of delayed hypersensibility[47]. It has been also found increased platelet activating factor (PAF) and Lf release in skin chambers overlying immunoglobulin E-mediated human skin reactions[49]. Lf inhibits manifestations of autoimmune response in mice, and authors suppose, that lactoferrin may be of therapeutically value in treatment of autoimmune disorders[48].

A major part of pro- and antioxidative activity of Lf is connected with the control by Lf of the PMN leucocytes functions.

Lactoferrin is a constituent of specific granules of polymorphonuclear leucocytes (PMN)[32]. Lf might contribute to the bactericidal activity of PMN by two opposing mechanisms. In the apo form it may perform an iron withholding function and prevent growth of phagocytosed bacteria[18]. On the other hand, iron-Lf may induce the production of free radicals which lead to microbial killing within the phagolysosome of phagocytic cells[29]. This ability of Lf is probably confined to acidic conditions, such as those within the phagolysosome. At physiological pH it is more likely that lactoferrin inhibits radical production, by scavenging catalytic "free" iron[4].

Lf participation in the regulation of the neutrophilic functions includes also interaction of the exogenous Lf with membrane receptors on the PMN plasma membrane[11,31].

Membrane-bound Lf, as a positive charged glycoprotein decreases the negative charge of PMN and facilitates their adhesion to another cell surfaces[37]. This was considered as a mechanism, by which Lf stimulates the aggregation of PMN and the adhesion to the endothelial walls. Activating the adherent mechanisms, Lf enhances the clearance of PMN by scavenger macrophages[26]. *In vitro*, experiments showed that Lf stimulates to 91% the ingestion of latex particles and *Trypanosoma cruzi* by murine peritoneal macrophages and human monocytes[29]. It has not been established if Lf has similar effect on human PMN. Such stimulation might have an important meaning in patients with deficiency in the mechanisms of phagocytosis. Such disturbances in the phagocytic mechanisms have been reported in patients with autoimmune diseases[12,3,36,43]. Data about correction in the functions of PMN by exogenous injected Lf was reported in patients with congenital absence of Lf in the specific granules of PMN[10,11]. In order to check if Lf has similar effect in patients with autoimmune and allergic diseases we studied the phagocytic activity of PMN isolated from blood from such patients.

3. MATERIALS AND METHODS

3.1. Patients

Phagocytic index(PI) was measured in the following groups of patients:

- 10 patients with autoimmune collagen diseases in an active stage of the disease with the following diagnoses: 5 with RA, 3 with SLE, and 2 with Mixed Conjunctive Tissue Disease
- 6 patients with autoimmune collagen disease in remission: RA (4), SLE (1), MCTD (1).
- 12 patients with allergy to *Staphylococcus aureus*. Some of them have also an allergy to another antigens: home dust (1), Candida (1), allergy to more than one alimentary antigen (2), allergy to cows milk (1).
- 4 with autoimmune chronic active hepatitis
- 7 healthy individuals: men 18–22 years of age with normal erythrocyte sedimentation velocity (ESV) and normal count of leucocytes.
- 4 blood donors considered as healthy, without clinical manifestation, in which were found circulating autoantibodies to collagen (2), aorta (2) and renal tissue (1).
- 8 patients with autoimmune diseases with proved circulating antibodies to tissue antigens and collagen: 4 patients with RA and antibodies to collagen, 1 patient with MCTD with antibodies to ragged red fibbers, 1 patient with SLE with antinuclear antibodies, 1 patient with endocarditis and antibodies to myocardium, 1 patient with autoimmune chronic active hepatitis and antibodies to liver.

All patients were in the active stage of disease (excluding those with autoimmune collagen diseases in remission) and were not cured by hormonal anti-inflammatory drugs, 15 days before the examination. Patients with autoimmune collagen disease in remission were cured prolonged time with steroid and non-steroid anti-inflammatory drugs.

Phagocytic number (PN) of PMN was measured in:

- 8 patients with autoimmune collagen diseases: 3 with SLE, 3 with RA and 2 with MCTD
- 11 patients with allergy to *Staphylococcus aureus*
- 5 healthy individuals

3.2. Isolation of PMN

PMN were isolated from venous blood by density gradient centrifugation using "Polysep" (DSO "Pharmachim", Bulgaria) as previously described[31]. PMN were resuspended at a final concentration 2.5×10^6 cells/ml in medium 199 containing penicillin and streptomycin with final pH 7.4.

3.3. Bacterial Preparation

Killed culture *Staphylococcus aureus* strain 209 was used as an antigen. Bacteria were washed twice with phosphate buffered saline (PBS) pH 7.4 and suspended at a final concentration 2.5×10^8 cells/ml in PBS pH 7.4. Then bacteria were opsonized by adding 5% de complemented AB-serum in Hanks buffer (pH 7.4) containing 0.1% (v/v) gelatine: 0.1 ml bacterial suspension was mixed with 0.9 ml 5% AB-serum, incubated for 30 min at 37°C, washed and resuspended in Hanks (pH 7.4) / 0.1% gelatine.

3.4. Phagocytic Reaction

0.2 ml bacterial suspension was mixed in polypropylene tubes with 0.2 ml (0.5×10^6) PMN. 87 pmol (0.2 ml = 6.7 μg/ml) Lf was added in the sample tubes, and 0.2 ml of the same molar concentration bovine serum albumin (BSA), was added in the control tubes. Mixtures were stirred 30 min at 37°C. Reaction was stopped by adding 2.5 ml of ice cooled PBS pH 7.4. Adherent to the PMN membrane non-absorbed bacteria were separated by twice centrifugation for 5 min at $160 \times g$, 4°C.

3.5. Phagocytic Activity

Phagocytic activity was measured using the phagocytic index (PI) and phagocytic number (PN) of PMN. The PI-number of phagocytic cells per each 100 PMN cells (%) and the PN-mean number bacteria inside each PMN cell, (200 PMN) were counted in Burkers camera after staining with May-Grunwald-Gimsa for each measurement of PN. PI was measured at the 15-th and 30-th minute starting the phagocytic reaction. PI of patients with *Staphylococcus aureus,* autoimmune diseases and circulating autoantibodies to tissue allergens and collagen was measured at the 60-th minute starting the incubation.

3.6. Measurement of the Circulating Autoantibodies

Some of the patients were tested for circulating autoantibodies to different tissues and cell antigens according to Steffen[41]. Collagen type VII, red ragged fibbers, aorta, myocardium, renal tissue and liver were used as antigens. Antigens from ragged red fibbers, aorta, kidney, myocardium, were isolated from human tissue immediately after death, and nuclear antigens from calf thymus. Tissues were washed many times in ice cooled phosphate buffered saline then homogenised and stored lyophilized.

4. RESULTS

In our study, we found different PI of PMN in the patients and the healthy donors at the 15-th and 30-th minute from the beginning of the phagocytic reaction. These differences were observed in the absence of Lf or the presence of 6.7 μg of Lf .

4.1. PI of PMN in the Absence of Lf

The higher PI of PMN, in the absence of Lf was found in the all groups of patients and in the healthy subjects with proved autoantibodies at the 15-th minute starting the phagocytosis.

In the healthy controls, PI reaches maximal mean values: 67.60±11.27 (x ± SD, n=10) at the 30-th minute from the incubation. This value is higher than the measured PI of the all groups of patients and healthy subjects with autoantibodies. It is 39% higher than those of patients with autoimmune collagen diseases in the active stage ($p < 0.001$); 34% higher than PI of patients with allergy ($p < 0.001$), and 40% ($p < 0.001$) higher, than the value of patients with circulating autoantibodies. Under the same conditions, the PI of patients with autoimmune collagen diseases in remission and with autoimmune chronic active hepatitis is nearest to those of the healthy donors (Table 1).

Measurement of PI at the 30-th minute in patients with autoimmune collagen diseases shows that PMN of patients in remission have higher ability for phagocytosis than PMN of patients in the active stage of autoimmune collagen diseases : 51.41 ± 25.45 and 31.60 ± 11.09 ($p < 0.025$) (Table 1).

4.2. PI of PMN in the Presence of Lf

At the 15-th minute from the incubation, Lf enhances the number of phagocyting PMN (PI) only in the patients with autoimune collagen diseases in the active stage (20.24%) and in those with autoimmune chronic hepatitis (2%) but the effect is not statistically reliable.

In all investigated groups of patients, Lf enhances PI at the 30-th minute from the phagocytic reaction but the effect is statistically significant in patients with autoimune collagen diseases in active stage (42.01%, $p < 0.002$) in the healthy subjects with circulating autoantibodies (96%, $p < 0.001$) and in the patients with autoantibodies (59%, $p < 0.002$) (Tables 1–2). Lf enhances PI about 19%, in patients with allergy to *St. aureus*, but the effect is not statistically reliable, because it was shown only in 58% from the patients, 7

Table 1. Effect of lactoferrin on the phagocytic index (PI) of polymorphonuclear leucocytes from blood of healthy people and different groups of patients

Diagnoses	Without Lf	With Lf	% Difference[b]
Healthy	60.28 ± 14.52	54.12 ±12.94	11 ($p > 0.1$)
Autoimmune collagen diseases (active)	31.60 ± 11.09	44.88 ± 14.13	42 ($p < 0.025$)
% Difference[a]	48 ($p<0.001$)	17	
Autoimmune collagen diseases (remission)	53.41 ± 25.45	57.00 ± 23.93	7 ($p > 0.1$)
% Difference[a]	11	5	
Active chronic autoimmune hepatitis	49.75 ± 11.32	69.50 ± 12.76	40 ($p > 0.1$)
% Difference[a]	17	28 ($p < 0.01$)	
Staphylococcus allergy	44.73 ± 22.53	53.27 ± 24.88	19 ($p > 0.1$)
% Difference[a]	26	2	
Healthy with autoantibodies	19.75 ± 4.79	38.75 ± 6.40	96 ($p < 0.002$)
% Difference[a]	67 ($p < 0.001$)	28 ($p < 0.05$)	
Patients with autoantibodies	30.75 ± 17.00	49.00 ± 12.27	59 ($p < 0.05$)
% Difference[a]	49 ($p < 0.001$)	9	

[a]% Difference between the values obtained for healthy subjects and patients.
[b]% Difference between the values obtained with and without addition of lactoferrin.
Statistical analyses were made using the Student-Fischers t-test.

Table 2. Phagocytic index (PI) of PMN from blood of patients with autoantibodies in the presence or absence of lactoferrin (Lf)

Diagnosis	Antibody	PI without Lf	PI with Lf	% Difference
MCTD	Anti muscle	75	88	17
SLE	Anti nuclear	37	72	85
RA	Anti collagen	26	38	80
RA	Anti collagen	7	15	114
RA	Anti collagen	30	58	98
RA	Anti collagen	30	44	47
Endocarditis (autoimmune)	Anti myocardium	12	30	150
Active chronic autoimmune hepatitis	Anti liver	29	47	93

from 12 investigated. In patients with autoimmune chronic active hepatitis, in the presence of Lf, PI increases about 40%, but the effect is not statistically reliable, because it was observed only in 2 from 4 patients (Table 1).

In the presence of Lf the differences in the PI of the phagocyting PMN observed between patients and healthy subjects decrease at the 30-th min from the incubation (Tables 1–2).

4.3. PN of the PMN in the Absence of Lf

The mean values of PN for PMN isolated from blood from patients and healthy subjects are almost equal (Table 3).

4.4. PN of PMN in the Presence of Lf

In the presence of Lf an increase of PN of PMN isolated from blood of healthy donors occurred. The stimulatory effect appeared in 4 from the 5 investigated cases, but statisti-

Table 3. Effect of lactoferrin on phagocytic number (PN) of PMN

Patients	PN without Lf	PN with Lf	% Effect	P
SLE	12.05 ± 4.44	22.00 ± 10.14	83	<0.001
SLE	9.40 ± 3.68	15.00 ± 7.25	60	<0.001
SLE	8.72 ± 2.91	17.20 ± 5.93	97	<0.001
RA	7.38 ± 3.82	13.26 ± 4.69	80	<0.001
RA	11.11 ± 4.59	18.70 ± 6.38	50	<0.001
RA	7.40 ± 3.76	14.18 ± 6.29	92	<0.001
MCTD	10.85 ± 3.61	17.31 ± 7.34	60	<0.001
MCTD	7.40 ± 2.09	11.07 ± 4.52	50	<0.002
x ± SD:	9.29 ± 3.61	16.09 ± 6.56	72	<0.001
Allergy to *St. aureus* (n = 12)	9.17 ± 3.97	17.06 ± 7.45	86	<0.01
Healthy people (n = 5)	9.42 ± 5.04	11.15 ± 5.49	23	>0.05
	10.68 ± 6.58	10.34 ± 4.46	0	>0.05
	9.11 ± 4.24	12.24 ± 3.92	34	<0.001
	8.46 ± 4.53	11.84 ± 4.66	40	<0.001
	8.33 ± 3.43	9.50 ± 4.00	14	>0.05
x ± SD:	9.20 ± 4.77	11.09 ± 4.50	22	>0.1

cally reliable are the measurements only for two of them. The mean effect of increased PN by Lf in the healthy people is about 22% (9.20 ± 4.77, without Lf, and 11.09 ± 4.50, with Lf; $p > 0.1$) (Table 3).

Lf enhances PN of PMN from blood of patients with autoimmune collagen disease in the active stage from 50 to 97%. Mean values of PN in the presence of Lf (16.09 ± 6.56) are 86 % higher, compared to those in the absence of the protein (9.21 ± 3.61) ($p < 0.001$)-(Table 3).

In patients with allergy to *Staphylococcus aureus*, Lf enhances PN in all of the investigated patients, where the percent of stimulation is 86%. Mean values of PN, measured in the absence of Lf are 9.17 ± 3.97 and in the presence of Lf (17.06 ± 7.45) ($p < 0.01$).

5. DISCUSSION

PMN from patients with autoimmune diseases and *Staphylococcus aureus* allergy were analysed in the present study for the following reasons: 1) In patients with RA and SLE, different defects in the phagocytic mechanisms were found *i)* in particular, deficiency in the functions of Fc-receptors on the cells of the reticuloendothelial system and a decreased ingestion of the immune complexes by these cells, have been described[23,38]. *ii)* a decrease of the phagocytic activity of PMN isolated from blood of patients with autoimmune collagen diseases has been established[3,9,6]. Such patients have predisposition to infections, because of extremely loss of secretory products with antibacterial activity from PMN, which leads to impairing bacterial killing and digestion[9]. 2) It would be of interest to identify the factors which stimulate the phagocytic ability of PMN isolated from blood from such patients 3) It has been established that *in vitro*, Lf stimulates the phagocytic activity of murine macrophages and human blood monocytes to intracellular parasites and latex particles[29]. 4) There is no data indicating that Lf enhances the ingestion of antigens during phagocytosis by PMN from patients or healthy individuals.

Our study on patients with autoimmune collagen diseases in the active stage of disease showed reliable decreased values of PI compared to healthy people. It was established that PMN functions in collagen diseases might be corrected by suitable therapy[6,9]. Our results support the existence of such possibility: PI in patients with RA in remission, cured prolonged time by hormonal and non steroid anti inflammatory drugs, does not differ reliable from those of healthy subjects (Table 1).

The results concerning the influence of Lf on the PI of PMN showed that: 1) Lf significantly increases PI of PMN at the 30-th minute from the phagocytic reaction in patients with autoimmune diseases in active stage, in blood donors, admitted as healthy with a presence of autoantibodies and in patients with autoimmune diseases and proved autoantibodies against tissue or cell antigens and collagen. 2) Lf influences non-significantly PI of PMN in patients with autoimmune collagen diseases in remission. 3) Lf enhances PI of PMN in 58% of the investigated cases with allergy to *Staphylococcus aureus*. 4) Lf decreases non-significantly PI of PMN in the healthy controls (Tables 1–2).

Our studies on the effect of Lf on the phagocytic activity of PMN of patients with allergy to *Staphylococcus aureus* show that the protein enhances more PN (86%) than PI of PMN (19%) (Tables 1, 3). Lf also increases PN of PMN in most of the investigated healthy individuals, although its effect is not statistically significant, but a similar effect on PI was not observed (Tables 1, 3). Lf increases, about 72% PN of PMN in all examined cases with autoimmune collagen diseases in an active stage. These facts, suggest that Lf has a major effect on PN than on PI of PMN (Tables 1–3).

Some studies show that the deficiency in the specific granules of PMN leads to abnormality in their cell functions: chemotaxis, adherence, aggregation and ability to decrease the cell-surface charge in response to FMLP-stimulation[13]. When exogenous Lf was added to the cells, adherence and aggregation became normal[10]. Our results show that Lf "corrects" at a different extent the decreased values of PI and enhances PN in the investigated groups of patients in the active stage of the autoimmune or allergic diseases, in the clinical healthy subjects with circulating autoantibodies and in the patients with circulating autoantibodies (Tables 1, 3).

In our experiments, Lf has higher "corrective" effect on PI in clinical healthy patients with circulating autoantibodies, 96% increased of PI at the 30-th minute starting the incubation compared to the healthy without antibodies (Table 1). PMN isolated from blood of clinically healthy people with circulating autoantibodies in the absence of Lf have a stronger decrease in the PI at the 30-th minute from the circulation. It was observed in the healthy controls, but a similar "exhaustion" although at a lower degree, was observed in the other patients in the active stage of the disease (Table 1). A probable reason for the differences of the phagocytosis during the time might be modification of the cell membrane of PMN. Such modifications might be caused by oxidative destructions of functionally important membrane structures. It is known, that in the autoimmune diseases, PMN continuously release in their micro environment, proteases and oxidative products of the system myeloperoxidase–hydrogen peroxide–halide[27,46] which may disturb their own structures connected with a control of the phagocytosis[25]. Introduction of antioxidants has an effect of stabilisation and enhancement of the phagocytic functions[25]. Free oxygen radicals scavengers inhibit the iron-induced functional defect of phagocytosis[44]. Lf which protects the plasma membrane of PMN from oxidative stress after binding to the cell membrane[20] may be considered as an antioxidant. Lf may block the iron ions in their catalytic inactive form[15] or control the transmembrane transport systems[42]. It may be supposed that *in vivo* Lf stimulates phagocytic functions of PMN by similar mechanisms.

Lf decreases the negative charge of the cell surface and thus enhances the adherent ability of the PMN[5,37]. Probably, the increased ingestion of bacteria in the presence of Lf is due to this stimulated adherent ability. This effect of enhancement of PN, appeared in all of the studied groups of patients or healthy subjects. It is possible that PMN plasma membranes of patients with autoimmune collagen diseases in the active stage of the disease and *St. aureus* allergy are capable to bind higher doses Lf, compared to healthy people, because of the localisation of a higher number Lf receptors of the PMN surface. Similar enhancement of the number of Lf receptors has been observed in another diseases[22]. Present studies suggest that Lf is capable to bind to the nuclear DNA and to activate the transcription[24]. Such investigations on PMN are not known, although it has been observed Lf bound to the nuclear DNA of the PMN[14]. It may be supposed that after binding to the membrane receptors on the PMN surface[11,31], Lf controls their expression on the plasma membrane.

Series of reports mentioned the plasma lactoferrin concentration as an index of granulocyte activation and function in a number of inflammatory conditions[7,8,28]. Antibodies to Lf in RA, SLE, autoimmune liver diseases have been found[33,34]. Neutrophil cytoplasmic antibodies were not detected in non-autoimmune liver diseases[34]. It was proved that these antibodies against Lf bind to the vascular endothelium and activate the leucocytes[39]. Lf is capable to shape immune complexes with these or with another antibodies[19]. PMN have receptors for the Fc-domain of the antibodies and could bind antibodies and antigen-antibody complexes. It is possible that *in vivo,* Lf may bind to the PMN surface by its own membrane receptors or as a compound of an immune complex. Such binding gives a possibility for higher number Lf molecules to occupate the PMN membrane surface and to enhance the effect of Lf on the adherence and aggregation of the PMN, and thus to facilitate the phagocytic reaction. Under the experimental conditions, PMN were washed for removal of antigen-antibody complexes. A possibility exists that some autoantibodies remain bound to the PMN receptors and shape immune complexes with the exogenous Lf during the phagocytic reaction. By this hypothetical mechanism, Lf might stimulate the phagocytic activity of PMN.

REFERENCES

1. Adamik B., Wlaszczyk A., 1996, 50 (1) 33–41.
2. Aguas AP., Esaguy N., Sunkel CE., Silva MT., Infect. Immun., 1990, 58 (5) 1461–1470.
3. Al-Hadithy H., Isenberg DA., Addison IE., Glodstone AH., Snaith ML., Ann. Rheum. Dis., 1981, 43, 33–36.
4. Aruoma OI., Halliwell B., Biochem. J., 1987, 241, 273–278.
5. Asako H, Kurose I., Wolf RE, Granger DN., Microcirculation, 1994, 1 (1) 27–34.
6. Attia WM., Shams AH., Ali MKH., Clark HW., Brown T., Bellanti JA., Ann. Allergy,1982, 48, 279–282.
7. Baynes RD., Bezwoda WR., Adv. Exp.Med. Biol., 1994, 357, 133–141.
8. Bennett RM., Quartey EAC., Holt PJL., Arthritis Rheum., 1973, 16, 186–190.
9. Boghossian SH., Isenberg DA., Wright G. et al., Ann. Rheum. Dis., 1984, 43 (4) 541–550.
10. Boxer LA., Coates TD., Haak RA., Wolach JB., Hoffstein S., Baehner RL.The New Engl. J. Med., 1982, 12, 404–410.
11. Boxer LA., Haak RA., Yang HH., Wolach JB., Whitcomb JA., Butterick CJ., Baehner RL. J. Clin. Invest., 1982, 70, 1049–1057.
12. Brandt L., Hedberg H., Scand. J. Haematol., 1969, 6, 348–353.
13. Breton-Gorius J., Mason D., Buriot D., Vlide J., Griscelli C., AM. J. Pathol., 1980, 99, 413–419.
14. Briggs RC., Glass WF., Montiel MM., Hnilica LS., J. Histochem. Cytochem., 1981, 10, 1128–1136.
15. Britigan BE., Hassett DJ., Rosen GM., Hammill DR., Cohen MS., Biochem. J., 1989, 264, 447–455.
16. Britigan BE., Serody JS., Cohen MS., Adv. Exp. Med. Biol., 1994, 357, 143–156.
17. Brock J., Immunol. Today, 1995, 16 (9), 417–419.
18. Bullen JJ., Armstrong JA., Immunol., 1979, 36, 781–791.
19. Butler JE., Biochim. Biophys. Acta., 1973, 29, 341–346.
20. Cohen MS., Mao J., Rasmussen GT., Serody JS., Britigan BE., J. Infect. Dis., 1992, 166, 1375–1378.
21. Esaguy N., Freire O., van-Embden JD., Aguas AP., Scand. J. Immunol., 1993, 38 (2) 147–152.
22. Faucheux BA., Nillesse N., Damier P., Spik G.,Mouatt-Prigent A., Pierce A., Leveugle B., Kubis N., Hauw JJ., Agid Y., Proc. Natl. Acad. Sci. USA, 1995, 92 (21) 9603–9607.
23. Fields T., Gerardy EN., Ghebrehwet B., Bennett RS., Lawley TJ., Hall RP., Plotz PH., Karsh JR., Frank MM., Hamburger MI., J. Rheumatol., 1983, 10 (4) 550–557.
24. Fleet JC., Nutr. Rev., 1995, 53 (8) 226–227.
25. Hakansson L., Venge P., Immunology, 1982, 47, 687–694.
26. Heifets L., Katsuyuki I., Goren MB., J. Reticuloendothel. Soc., 1980, 28, 391–403.
27. Kurose I., Yamada T., Wolf R., Granger DN., J. Leukoc. Biol., 1994, 55 (6) 771–777.
28. Lerche A., Bisgaard H., Christensen JD., Venge P., Dahl R., Sondergaard J., Allergy, 1988, 43, 139–145.
29. Lima MF., Kierzenbaum F., J. Immunol., 1985, 134 (6) 4176–4183.
30. Maher RJ., Cao D., Boxer LA., Petty HR., J. Cell Physiol., 156 (2) 226–234.
31. Maneva A., Sirakov LM., Manev V., Int. J. Biochem., 1983, 15, 981–984.
32. Masson PL., Heremans JF., Schonne E., J. Exp. Med., 1969, 130, 643–658.
33. Mulder AH., Horst G., van Leeuwen MA., Climburg P., Kallenger CG., Arthritis Rheum., 1993, 36 (8) 1054–1060.
34. Mulder AH., Horst G., Haagsma EB., Limburg PC., Kleibeuker JH., Kallenger CG., Hepatology, 1993, 17 (3) 411–417.
35. Nisimoto Y., Otsuka-Murakami H.,Biochem. Biophys. Acta, 1990, 1040, 260–266.
36. Okuda K., Okamoto R., Noguchi Y., Tadokoro I., Jpn. J. Exp. Med., 1975, 45, 1–17.
37. Oseas R., Yang HH., Baehner RL., Boxer LA., Blood, 1981, 5, 939–946.
38. Parris TM., Kimberly RP., Inman RD., McDouglas JS., Gibofski A., Christian CL., Ann. Int. Med., 1982, 4, 526–532.

39. Peen E., Sundqvist T., Skogh T., Clin. Exp. Immunol., 1996, 103 (3) 403–407.
40. Skogh T., Peen E., Adv. Exp. Med. Biol., 1993, 336, 533–538.
41. Steffen C., Allgem. Experim. Immunol. Immunopathol.. Stuttgard, G. Theime Verlag, 1968.
42. Sun IL., Crane FL., Morre DJ., Low H., Faulk WP., Biochem. Biophys. Res. Commun., 1991, 176 (1) 497–504.
43. Turner RA., Schumacher HR., Myers AR., J. Clin. Invest., 1973, 5, 1632–1635.
44. Van Asbeck BS., Marx JJM., Stryvenberg A., van Kats J., Verhoef J., J. Immunol., 1984, 851–856.
45. Wang D., Pabst KM., Aida Y., Pabat MJ., J. Leucocyte Biol., 1995, 57, 865–874.
46. Weissmann G., J. Lab. Clin. Med., 1982, 100 (3), 322–333.
47. Zimecki M., Machniki M.,Arch. Immunol. Ther. Exp., 1994, 42 (3), 171–177.
48. Zimecki M., Wieczorek Z., Mazurier J., Spik G., Arch. Immunol. Ther. Exp., 1995, 43 (3–4), 207–209.
49. Zweiman B., von Allmen C., Moskovitz A., Atkins M., Taylor M., Korchak H., J. Leukoc. Biol., 1993, 53 (6), 727–731.

LACTOFERRIN

Its Role in Maturation and Function of Cells of the Immune System and Protection against Shock in Mice

M. Zimecki,[1] J. Kapp,[2] M. Machnicki,[1] T. Zagulski,[3] A. Wlaszczyk,[4] A. Kübler,[4] J. Mazurier,[5] and G. Spik[5]

[1]Institute of Immunology and Experimental Therapy
Wroclaw, Poland
[2]Emory University
Winship Cancer Center
Atlanta, Georgia
[3]Institute of Genetics and Animal Breeding
Jastrzêbiec, Poland
[4]Medical Academy
Wroclaw, Poland
[5]University of Sciences and Technology of Lille
Villeneuve d`Ascq, France

About a decade ago we turned our attention to a potential role of lactoferrin (Lf) in maturation and function of cells of the immune sytem. At that time Lf was mainly investigated for its role in protection of animals against pathogens and activation of some cell types such as phagocytes or natural killer cells. Isolated reports indicated its role in modification of mixed lymphocyte reaction and humoral immune response *in vitro* as well as inhibition of cytokines (interleukin 1).

After discovery of receptors for Lf on lymphocytes and brush border intestinal cells it became probable that Lf may affect maturation and function of cells of the immune system. The aim of the studies, which results are presented in this article, was to reveal the role of Lf which could be played by this protein both in immature (newborn) as well as in adult mice. We studied effects of Lf on maturation (differentiation) and function of T and B-cell precursors, modification of effector T cell function and protective action of Lf in mice subjected to lipopolysaccharide (LPS)- or surgery-elicited shock.

Our preliminary studies on the immunotropic activity of human lactoferrin (hLf) showed that the protein stimulated the humoral immune response to sheep red blood cells (SRBC) in mice when administered prior to immunization. In addition, this activity was

Advances in Lactoferrin Research, edited by Spik *et al.*
Plenum Press, New York, 1998.

also exhibited by administration of thymocytes preincubated with hLf. Lastly, highly purified $CD3^- CD4^- CD8^-$ thymocytes (T cell precursors) preincubated with hLf, also demonstrated that stimulatory activity in the humoral immune response *in vitro* and *in vivo*[1]. Acquirement of this property by immunoincompetent T cell precursors was accompanied by gaining of $CD4^+CD8^-$ phenotype characteristic for helper cells. The data showed that lactoferrin shares properties of other factors and cytokines (thymosin, a proline-rich polypeptide from colostrum, interleukin 1) affecting T cell maturation.

In another series of experiments we showed that hLf can also promote differentiation of immature B cells derived from newborn mice or immunodeficient CBA/N mice which lack the mature B cell subset. These cells acquired, upon contact with hLf, new cell markers such as Lyb5, complement receptor and increased density of surface IgD[2]. Similar activity of hLf and bovine lactoferrin (bLf) we found in an immature B cell WEHI 231. Lactoferrins significantly lowered expression of IgM and interleukin 2 receptors on these cells and the phenotypic changes were associated with an increased resistance of WEHI 231 cells to tolerogenic action of anti-IgM polyclonal antibodies[3]. In this model lactoferrin behaved similarly as interleukin 4. With regard to the action of hLf on immature resident B cells we showed that, together with the phenotypic changes, the cells gained the ability to present antigen to antigen specific T cell lines[2] (Fig. 1).

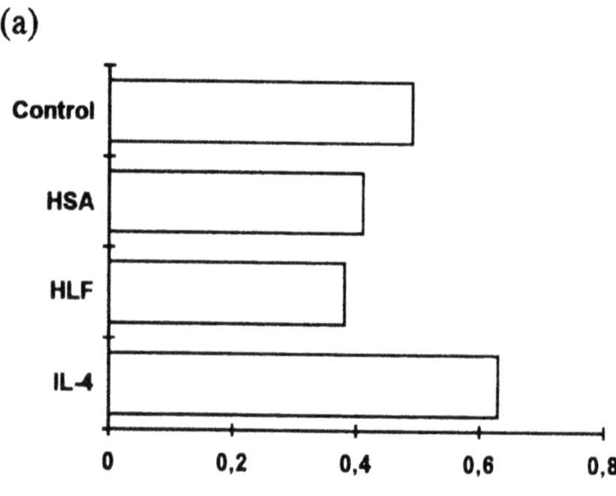

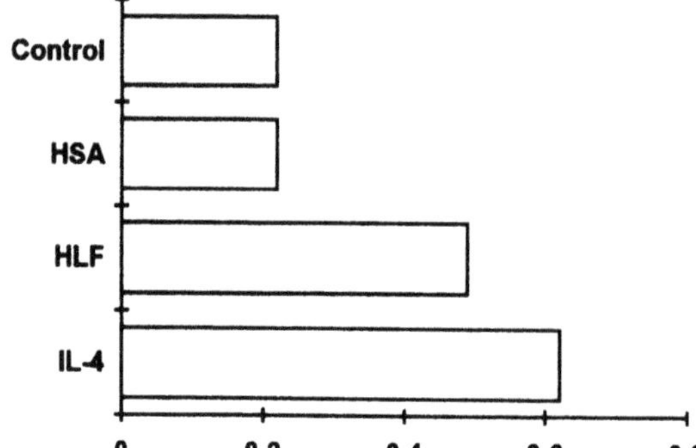

Figure 1. Effects of human lactoferrin and IL-4 on antigen presentation by B cells from newborn CBA mice. B cells from adult CBA (a) and newborn (b) mice were preincubated overnight with HSA (20 g/ml), human lactoferrin (HLF) (20 μg/ml) or IL-4 (200 U/ml), washed twice, and used as APC or Ksins 1 cells. The antigen proliferative response was measured using the XTT reagent, and the optical density (OD) was read at 450 nm. The data are presented as a mean values from triplicate determinations.

Immature B cells were not able to effectively present antigen to T cells. It should be underlined that lactoferrin exhibits similar properties in the model of antigen presentation by immature B cells as B-tropic cytokines (interleukin 1, 4, 6, 7 and 10)[4,5]. Although it is to early to classify lactoferrin into the cytokine family, the analogous actions of lactoferrin and other cytokines in promoting maturation of T and B cells are characteristic features of cytokine redundance.

The actions of Lf were not only restricted to maturation of cells constituting the immune system. We also found that Lf affected activity of antigen-specific T cells in the course of delayed type hypersensitivity (DTH) to SRBC by inhibiting the elicitation of that response[6] (Fig. 2).

Since DTH is mediated by TH1 subset of T cells we compared the action of lactoferrin on the activity of TH1 and TH2 cell lines of the same antigen specificity and found that Lf inhibited both proliferation and cytokine production of TH1 but not TH2 cell lines[8] (Figs. 3 & 4).

Lactoferrin inhibited also other inflammatory reactions, for example elicited by *Mycobacterium bovis*[6].

Lactoferrin was previously shown to protect mice against lethal challenge with *E. coli*[8]. In an attempt to partially elucidate that phenomenon we studied the effect of intravenous administration of bLf before injection of sublethal dose of LPS into mice. The results showed that such a treatment markedly reduced LPS-induced tumor necrosis factor alpha (TNF-α) release and to a lesser degree IL-6 production[9]. Bovine lactoferrin protected also mice from lethal consequences of recombinant TNF-α treatment. That protective effect was mediated by the liver. Since treatment of mice with Lf alone caused a release of interleukin 6 (IL-6) and TNF-α into circulation, the protective effects of Lf could be partially explained by a desennsitization of the immune system. The experiments on the protective effects of Lf were extended to surgery-induced shock in mice. The reason for the initiation of these studies is a high mortality rate in intensive care units caused by postoperative complications (sepsis, organ disfunction). Mice were subjected to minor (thymectomy) or more extensive (splenectomy) operations in general anesthesia and the postoperative shock was measured by the levels of circulating IL-6 and TNF-α. We found that mice

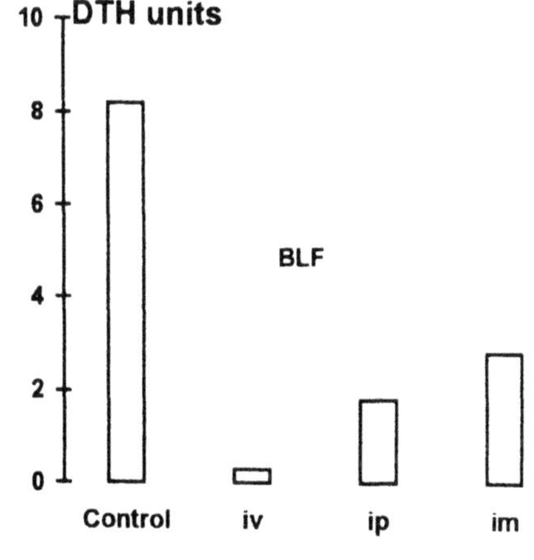

Figure 2. Inhibition of the effector phase of the DTH to SRBC by bLf depending on a route of bLf administration. BALB/c male mice sensitized to SRBC were challenged with the eliciting dose of SRBC into one foot pad. Mice were treated i.v., i.p. or i.m. with 1 mg of bLf 2 h before elicitation of the reaction. The magnitude of foot pad swelling was measured after 24 h. The results are presented as a mean values from 10–12 determinations (mice).

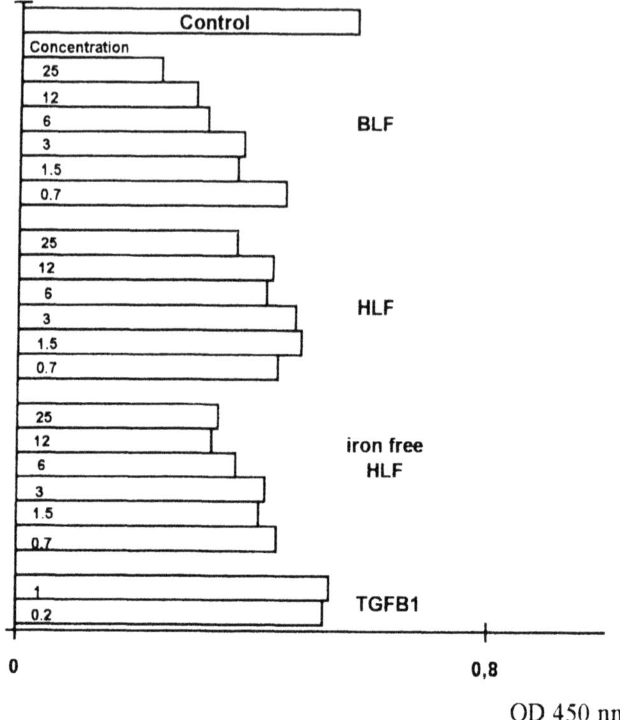

Figure 3. Inhibitory effects of lactoferrin on proliferative response of BP42 cell line (TH1 type). Human (HLF) or bovine (bLf) lactoferrin were added at doses: 25–0.7 μg/ml and TGFβ1 at doses of 1 and 0.2 ng/ml in the beginning of 2-day proliferation assay of BP42 cells. Number of PB42 cells, 2 × 104/well: APC (splenocytes from C57BL mice), 5 × 10^4/well; antigen (pork insulin), 20 μg/ml. The results are presented as a mean OD values from triplicate wells.

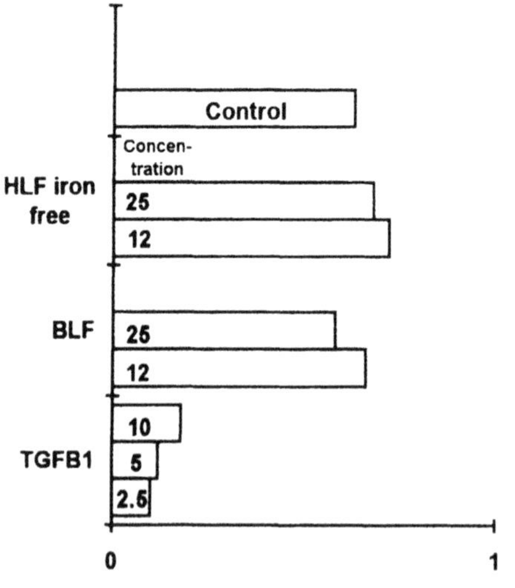

Figure 4. Lactoferrin does not inhibit proliferative response of Ksins 1 cell line (TH2 type). Human (HLF) or bovine (bLf) lactoferrin were added at doses of 25 and 12 μg/ml and TGFβ1 in concentration of 10, 5 and 2.5 ng/ml in the beginning of 2-day assay of Ksins 1 cells, 2 × 10^4/well; APC (splenocytes from CBA mice), 5 × 10^4/well; antigen (sheep insulin), 30 μg/ml. The results are expressed as a mean OD values from triplicate wells.

Table 1. Effect of bovine lactoferrin on serum levels of IL-6 and TNF-α of CBA mice subjected to thymectomy in general anesthesia

Treatment of mice	IL-6 (pg/ml)	TNF-α (pg/ml)
Experiment 1		
0.9% NaCl, 24 h thymectomy, thiopental + thymectomy	102 ± 13.8	156 ± 19.5
bLf (1 mg), 24h before thymectomy, thiopental	1 ± 0.8	116 ± 12.5
bLf (1 mg), 24 h before thymectomy, thiopental + thymectomy	45 ± 3.4	148 ± 23.3
Experiment 2		
0.9% NaCl, 24 h before thymectomy, thiopental + thymectomy	361 ± 55.8	392 ± 19.8
bLf (1 mg), 5 doses every second day, last dose 24 h before thiopental	15 ± 2.7	112 ± 19.7
bLf (1 mg), 5 doses every day, last dose 24 h before thiopental + thymectomy	33 ± 6.5	176 ± 16.4

treated intravenously 24 hours before operation released less IL-6 and TNF-α as measured 4 hours postoperation (Table 1).

Mice were treated with bLf or NaCl per os using a stomach tube.The bLf was given in a volume of 0.5 ml 0.9% NaCl), in one dose or in 5 doses on alternate days. Thymectomy was performed 24 h following the last dose of bLf. The results represent mean values from 5 determinations. The inhibition of IL-6 release by bLf compared with NaCl in experiment 1 was 55% ($p<0.01$) and in experiment 2 was 90% ($p<0.01$). The inhibition of TNF-α level in experiment 2 was 55% ($p<0.001$).

The inhibition of TNF-α release was deeper in the case of thymectomy than after splenectomy. Operation-induced cytokine release could be also reduced by per os treatment of mice. In such cases the doses of Lf could be reduced 5–10 times, but the effectiveness of per os treatment was better when several doses of bLf on alternate days were applied. Weak inhibition of TNF-α release after splenectomy, associated with a more significant IL-6 level reduction, seems to be beneficial taking into account numerous profitable actions of TNF-α in tissue regeneration and wound healing.

Latest data in the human system (to be published) revealed that bLf *in vitro* exhibited very interesting properties in the regulation of constitutive and LPS-induced CD11a adhesion molecule expression on peripheral blood mononuclear cells. Other *in vitro* studies on several categories of patients (sepsis, trauma, surgery) also revealed regulatory activities of lactoferrin with regard to PBMC proliferation and cytokine production (data to be published).

All the findings regarding immunoregulatory activities of lactoferrin, presented above, indicate that lactoferrin meets the requirements of a typical immunoregulator. It seems that lactoferrin may find a broad preventive and therapeutical application as a supplement to newborn's food, a factor correcting abnormalities and deficiences of the immune system and as an anti-inflammatory agent.

REFERENCES

1. Zimecki M., Mazurier J., Machnicki M., Wieczorek Z., Montreuil J. and Spik G. (1991): Immunostimulatory activity of lactotransferrin and maturation of CD4, CD8 thymocytes. Immunol. Lett., 30, 119–124.
2. Zimecki M., Mazurier J., Spik G. and Kapp J.A. (1995): Human lactoferrin induces phenotypic and functional changes in splenic mouse B cells. Immunology, 86, 112–127.
3. Zimecki M., Mazurier J., Spik G. and Kapp J.A. (1995): Lactoferrin diminishes expression of surface IgM and interleukin-2 receptors on WEHI 231 cells and lowers susceptibility of these cells to anti-IgM induced cell death. VIII Meeting of Polish Immunological Society, Wroclaw, abstract, Pol. J. Immunol., 20, 291.

4. Zimecki M., and Kapp J.A. (1994): Presentation of antigen by B cell subsets. III. Effects of interleukins on antigen presenting function and phenotype of immature B cells. Arch. Immunol. Ther. Exp., 42, 355–359.
5. Zimecki M., and Kapp J.A. (1995): Presentation of antigen by B cell subsets. V. Effect of Interleukin 7 (IL-7) and Interleukin 10 (IL-10) on phenotype and antigen presenting function of immature B cells. Arch. Immunol. Ther. Exp., 43, 253–257.
6. Zimecki M. and Machnicki M. (1994): Lactoferrin inhibits the effector phase of the delayed type hypersensitivity to sheep erythrocytes and inflammatory reactions to *M. bovis* (BCG). Arch. Immunol. Ther. Exp., 42, 171–177.
7. Zimecki M., Mazurier J., Spik G. and Kapp J.A. (1996): Lactoferrin inhibits proliferative response and cytokine production by TH1 but not TH2 cells. Arch. Immunol. Ther. Exp., 44, 51–56.
8. Zagulski T., Lipiñski P., Zagulska A., Broniek S. and Jarzabek Z. (1989): Lactoferrin can protect mice against a lethal dose of *Escherichia coli* in experimental infection in vivo. Br. J. Exp. Pathol., 1989, 70, 697–704.
9. Machnicki M., Zimecki M. and Zagulski T. (1993): Lactoferrin regulates the release of tumor necrosis factor alpha and interleukin 6 *in vivo*. Int. J. Exp. Pathol., 74, 433–439.

IMMUNOHISTOCHEMICAL DEMONSTRATION OF LACTOFERRIN IN HUMAN NEOPLASTIC TISSUES

G. Tuccari, G. Giuffrè, C. Crisafulli, and G. Barresi

Department of Human Pathology
University of Messina, Italy

1. INTRODUCTION

By immunohistochemistry, the distribution of lactoferrin (Lf) as well as other iron-binding proteins has been extensively investigated in normal human tissues such as stomach, kidney, lung, pancreas, liver and bone marrow[1]. In addition, Lf has been localized in epithelial elements of major salivary glands[2] as well as in malignant counterparts[3]. Moreover, the present of Lf in prostatic tissue has been documented by radial-immunodiffusion in homogenates of prostatic cancer and benign hypertrophy[4]. In lactating mammary glands as well as in breast carcinomatous tissues, immunohistochemical studies stating the appearance of Lf are conflicting, with a variation in the occurence rate of stained elements[1,5,6]; however, no relationships have been found between Lf and other prognostic factors such as steroid receptors, histologic type and tumour grade.

In the present review, we report immunohistochemical results obtained with polyclonal antibodies against Lf in many different precancerous and neoplastic lesions during a period of fifteen years.

2. MATERIALS AND METHODS

In the period 1984–1997 we have studied bioptic and surgical specimens obtained from neoplastic tissues taken from prostate[7], thyroid[8,9], stomach[10], large bowel[11] and gall-bladder[12].

All specimens, fixed in 10% neutral formalin for 12–24 hours at room temperature, were embedded in paraffin at 55°C and cut into thin sections by routine histological procedure. For the immunohistochemical study, 4–5 μm-thick sections were treated in a moist chamber for 30 min. each time: 1) with 0.1% H_2O_2 in methanol to block the intrinsic peroxidase activity; 2) with normal sheep serum to prevent unspecific adherence of serum

Advances in Lactoferrin Research, edited by Spik *et al.*
Plenum Press, New York, 1998.

proteins; 3) with rabbit anti-human Lf (Dako, Denmark w.d. 1:300); 4) with sheep anti-rabbit immunoglobulin antiserum (Behring Institute; w.d. 1:25); and 5) with rabbit anti-horseradish PAP complexes (Dako; w.d. 1:25). For the demonstration of peroxidase activity the sections were incubated in darkness[13] for 10 min. with fresh 3-3′ diaminobenzidine tetrahydrochloride (100 mg in 200 ml 0.03 per cent hydrogen peroxide in PBS) (Sigma Chemical Co., St. Louis, MO, USA).

The assessment of immunostained sections was always performed on a consensus basis by 2 pathologists (G.T. and G.B.), using a double-headed microscope. To test the specificity of Lf immunostaining, the specific antiserum was replaced by either phosphate-buffered saline, normal rabbit serum or was absorbed with excess of purified human Lf from human liver and spleen (Sigma Chemical Co.): the results obtained were negative.

3. RESULTS

3.1. Prostate

In benign hypertrophy, the glandular epithelium and the secretory product were consistently negative for Lf. A strong staining was always observed in neoplastic cells of differentiated adenocarcinomas and in their endoluminal material; in contrast, a very slite positivity was noted in undifferentiated carcinomas.

3.2. Thyroid

Follicular adenomas showed a constant negativity, whereas follicular and papillary carcinomas exhibited various degrees of positivity for Lf. Incorporated organoid structures observed in anaplastic carcinomas were strong stained; the spindle cell parts of these cancers are always unstained as well medullary carcinomas.

3.3. Stomach

An evident reactivity for Lf was encountered in intestinal type carcinomas, adenomas and incomplete intestinal metaplasia, whereas diffuse type carcinomas, hyperplastic polyps and complete intestinal metaplasia were always stained; mucous neck cells of the antrum and body were also positive for Lf. Iron levels appear to be unrelated to the immunohistochemical presence of Lf.

3.4. Large Bowel

Dysplastic areas of tubular and villous adenomas as well as adenocarcinomas and colloid carcinomas showed a variable cytoplasmic immunoreactivity for Lf, although no staining was noted in some cases; tubular adenomas without dysplasia and colonic inflammatory pseudopolyps were unstined. Metastatic elements present in lymph nodes maintained the Lf immunostaining.

3.5. Gallbladder

In a variable share of adenocarcinomas, a positive immunoreactivity for Lf was encountered in the cytoplasm, while tubulo-villous adenomas, adenomyomas and the normal epithelium of the gallbladder were generally unreactive. In carcinomatous lesions, the staining intensity was variable between different cases or individual tumour cells.

4. DISCUSSION

On the basis of our experience the immunohistochemical evidence of Lf in differentiated adenocarcinomas of the prostate and gallbladder, associated with an absent or very slight staining in undifferentiated, sarcomatous and squamous carcinomas, strongly suggests that Lf may be proposed as a marker of neoplastic differentiation, similarly to that hypothesized by others in adenocarcinomas of the parotid gland[3].

In adenomatous and carcinomatous samples of the stomach and large bowel, a variable percentage of stained neoplastic cells was encountered although unreactive cases were also found. Whether this immunohistochemical inter- and intra-tumoral heterogeneity reflects different cell subpopulations, the stage in the cell cycle and/or the state of activation or instead metabolic abnormalities is presently not fully ascertained. In addition, serum iron levels appear to be unrelated to the immunohistochemical presence of Lf, since hyposideremic anemia has been observed either in Lf-positive or Lf-negative carcinomas; therefore, the suggestion proposed by Loughlin et al.[14] that Lf may represent a mediator of anemia in cancer suffering patients should be considered with caution. Moreover, in chronic disorders with iron deficiency such as coeliac disease, we have demonstrated that Lf mucosal distribution is independent of the subject's iron status[15].

In thyroid pathology, the comparative immunohistochemical analysis between Lf and thyroglobulin showed a similar pattern in follicular, papillary and anaplastic carcinomas; in contrast, in follicular adenomas an evidente discrepancy was noted since Lf was constantly absent, while thyroglobulin was always present. On this way, Lf seems to be an additional useful marker for the differential diagnosis between follicular adenomas, follicular carcinomas and follicular variety of papillary carcinomas. In addition, the cytoplasmic immunostaining for Lf concerning more than one third of cells has been considered an criterion of malignancy in thyroid tumours[16].

The origin of Lf in neoplasias has not yet been completely elucidated. Although we cannot exclude the possibility that this iron-binding protein might be involved in the defense system against tumours, it is probable that the presence of Lf in neoplastic cells of different carcinomas may be related to the production of the protein in the tumour in itself. The intense staining for Lf, as revealed by the immunoperoxidase technique, is an argument for the production by the tumour tissue itself, although this interpretation should be controlled by methods other than morphological analysis. Therefore, we retain that carcinomatous elements may produce Lf in order to have a greater availability of iron for their turnover; however, it is well known that the capability of Lf to bind iron exceeds that of transferrin by a factor of 260[17]. Moreover, iron is an essential nutrient for cells that are dividing rapidly such as tumour cells[18]; in addition, iron is crucial for various metabolic processes in the cell, such as oxidative phosphorylation and RNA and DNA synthesis[19-21]. On the other hand, in order to explain the presence of Lf in neoplastic cells, an alternative hypothesis should be considered since the accumulation of this iron-binding protein may be a consequence of defective or functionally impaired Lf receptors. In fact, specific Lf binding sites have been shown in a human adenocarcinoma cell line HT 29[22]; according to this hypothesis, the cytoplasmic localization of Lf does not reflect its intracellular synthesis by neoplastic elements, but the degree of transmembranous iron transfer.

REFERENCES

1. Mason D.Y. and Taylor C.R. (1978). Distribution of transferrin, ferritin and lactoferrin in human tissues. J. Clin. Pathol. 31, 316–327.
2. Korsrud F.R. and Brandtzaeg P. (1982). Characterization of epithelial elements in human major salivary glands by functional markers: localization of amylase, lactoferrin, lysozyme, secretory component, and secretory immunoglobulins by paired immunofluorescence staining. J. Histochem. Cytochem. 30, 657–666.
3. Caselitz J., Jaup T., and Seifert G. (1981). Lactoferrin and lysozyme in carcinomas of the parotid gland. Virchows Arch. (A) 394, 61–73.
4. Van Camp K, Van Sande M. (1980). Vrije amino-zuren en ninhydrine positieve substanties in menselijk prostaatweefsel. Verh. K. Acad. Geneeskd. Belg. 42, 104.
5. Rossiello R., Carriero M.V. and Giordano G.G. (1984). Distribution of ferritin, transferrin and lactoferrin in breast carcinoma tissue. J. Clin. Pathol. 37, 51–55.
6. Charpin C., Lachard A., Pourreau-Schneider N., Jacquemier J., Lavaut M.N., Andonian C., Martin P.M., Toga M. (1985). Localization of lactoferrin and nonspecific cross-reacting antigen in human breast carcinomas. Cancer 55, 2612–2617.
7. Barresi G. and Tuccari G. (1984). Lactoferrin in benign hypertrophy and carcinoma sof the prostatic gland. Virchows Arch. (A) 403, 59–66.
8. Tuccari G. and Barresi G. (1985). immunohistichemical demonstration of lactoferrin in follicular adenomas and thyroid carcinomas. Virchows Arch. (A) 406, 67–74.
9. Barresi G. and Tuccari G. (1987). Iron-binding proteins in thyroid tumours. An immunocytochemical study. Path. Res. Pract. 182, 344–351.
10. Tuccari G., Barresi G., Arena F. and Inferrera C. (1989). Immunocytochemical detection of lactoferrin in human gastric carcinomas and adenomas. Arch. Pathol. Lab. Med. 113, 912–916.
11. Tuccari G., Rizzo A., Crisafulli C. and Barresi G. (1992). Iron-binding proteins in human coloractal adenomas and carcinomas: an immunocytochemical investigation. Histol. Histopathol. 7, 543–547.
12. Tuccari G., Rossiello R., Barresi G. (1997). Iron binding proteins in gallbladder carcinomas. An immunocytochemical investigation. Histol. Histopathol. 12, (in press).
13. Weir E.E., Pretlow T.G. and Pitts A. (1974). A more sensitive and specific histochemical peroxidase for the localization of cellular antigen by the enzyme-antibody conjugate method. J. Histochem. Cytochem. 22, 1135–1140.
14. Loughlin K.R., Gittes R.F. and Partridge D. (1987). The relationship of lactoferrin to the anemia of renal carcinoma. Cancer 59, 566–571.
15. Tedeschi A., Tuccari G., Magazzù G., Arena F., Ricciardi R., Barresi G. (1987). Immunohistochemical localization of lactoferrin in duodenojejunal mucosa from celiac children. J. Pediatr. Gastroenterol. Nutr. 6, 328–334.
16. Cabaret V., Vilain M.O., Delobelle-Deroide A., Vanseymortier L. (1992). Détection immunohistochmique de la céruloplasmine et de la lactoferrine sur une série de 59 tumeurs thyroidiennes. Ann. Pathol. 12, 347–352.
17. Birgens H.S. (1984). The biological significance of lactoferrin in haematology. Scand. J. Hematol. 33, 225–230.
18. Weinberg E.D. (1984). Iron withholding: a defense against infection and neoplasia. Physiol. Rev. 64, 65–102.
19. Bacon B.R. and Tavill A.S. (1984). Role of the liver in normal iron metabolism. Sem. Liver Dis. 4, 181–192.
20. Bomford A.B. and Munro H. (1985). Transferrin and its receptor: their role in cell function. Hepatology 5, 870–875.
21. Shoji A. and Ozawa E. (1986). Necessity of transferrin for RNA synthesis in chick myotubes. J. Cell Physiol. 127, 349–356.
22. Roiron D., Amouric M., Marvaldi J. and Figarella C. (1989). Lactoferrin-binding sites at the surface of HT29-D4 cells. Eur. J. Biochem. 186, 367–373.

SUMMARY

Bo Lönnerdal

Department of Nutrition
University of California, Davis
Davis, California 95616

Judging from the scientific progress in the area of lactoferrin research that was presented at the Third International Conference, it is evident that this field is "alive and well"—the number of researchers present was higher than at the previous two conferences and the variety of biological activities ascribed to lactoferrin being explored has increased. More tools and sensitive techniques that can be used for this exploration are now available, ranging from powerful analytical instruments, synthetic peptides/epitopes and molecular biology methods to "knock-out" animal models.

An initial problem when arranging scientific meetings on the broad topic area of "lactoferrin" was how to reach active researchers. While computerized literature searches would reveal researchers with an established track record on lactoferrin, on-going, unpublished research is not revealed this way. The problem was compounded, of course, by the fact that research on lactoferrin is found in a very wide range of disciplines, from microbiology, virology, carcinogenesis and immunology to pediatrics, nutrition, food science and structural biochemistry. The previous two international conferences have helped to make a more complete inventory of active research on lactoferrin and its biological functions, which has been documented in the two books that have been published. Researchers with an interest in lactoferrin have met others with the same interest and industries producing lactoferrin, bovine or recombinant human, have found potential customers, all increasing the "contact surface" and helping to attract scientists to the lactoferrin conferences. The breadth of the research presented at this meeting was truly impressive.

It is also evident that research on lactoferrin now is an established "field"; i.e. the turnover of scientists active in lactoferrin research is low. Of the 25 research group represented at the Second International Conference, 21 were present at this meeting. The research presentations can be divided into the following areas, and are summarized as follows:

Advances in Lactoferrin Research, edited by Spik *et al.*
Plenum Press, New York, 1998.

1. LACTOFERRIN STRUCTURE

This was strong area already at the first two conferences. Now, we were given further details about the spatial organization of the molecule, and the tertiary structure of lactoferrin derivatives and recombinant human lactoferrin.

Better techniques were presented which will allow more specific and sensitive analysis. These methods include MALDI/SELDI techniques which can be used to follow the digestive fate of lactoferrin in vivo, persistence of lactoferricin in the gut, formation of immunogenic peptides, etc.

Molecular biology is no aggressively used to probe structural features of the lactoferrin molecule. Several deletion mutants have been produced, site-directed mutagenesis is used to change amino acid residues, and chimeras have been made (human lactoferrin/bovine transferrin) to study the importance of structural motifs for interactions with microorganism.

Some caution was urged regarding the impurity of commercially available lactoferrin preparations (usually not very pure). Further, the N-terminal amino acids have been found to be easily cleaved off by protease activity, resulting in "deletion" variants (Lf^{-N}, Lf^{-2N}, Lf^{-3N}, etc.) being present as "contaminants". These lactoferrin variants may disturb binding studies and also affect the stability of the preparations. The commonly occurring contamination of lactoferrin with lipopolysaccharide (LPS) was re-emphasized, as it may also profoundly affect both binding properties of lactoferrin as well as its biological activity.

2. GENE EXPRESSION: ROLE OF LACTOFERRIN IN DEVELOPMENT

Novel information on the promoter-regulated expression of lactoferrin was presented, particularly on EGF (epidermal growth factor). It was also demonstrated that expression can be enhanced considerably by using genomic sequences of the lactoferrin gene, rather than cDNAs. Expression levels in transgenic mice and cows were now reported to be higher than a gram per liter of milk, without apparent negative effects. This suggests that expression in transgenic animals can be driven to a high level, which may improve the economics of such projects.

A lactoferrin "knock-out" mouse model was presented. Early results demonstrated that the knock-out was lethal at the preimplantation stage of the embryo. The embryo apparently is normal up to the 8-cell stage, but that something occurs between the 8-cell and 16-cell stages. This is when differentiation starts to occur, and further detailed staining studies using labeled lactoferrin showed that lactoferrin binds to the outer cells, possibly to a lactoferrin receptor. It was suggested that lactoferrin may act as a paracrine cytokine at this stage of development. By producing and using recombinant mouse lactoferrin in the culture medium, these investigators were able to "rescue" the embryos and found that lactoferrin may be involved in fetal hematopoiesis (days 14–16) and is also found in glandular epithelial cell. These studies are likely to give detailed insights into the biological significance of lactoferrin in fetal development of the mouse.

Findings on the expression of lactoferrin in the mouse mammary gland were also presented. It appears to be expressed in significant amounts through pregnancy and no pronounced developmental pattern was found during the lactation period. During involution, however, lactoferrin expression was enhanced considerably, which is consistent with earlier findings of very high levels of lactoferrin in "dry secretions" of cows during termination of lactation.

3. LACTOFERRIN–RECEPTOR INTERACTIONS

The importance of the N-terminal arginine-repeat of the lactoferrin for binding of heparin, lysozyme, LPS and DNA was demonstrated by deletion variants, but it was also shown that there is specific binding of lactoferrin to the receptor. This specific binding has a lower dissociation constant (K_d), but it appears that only 20% of the binding sites are of this type.

The concept of "promiscuous" receptors was brought up and illustrated by the LDL-receptor family and the asialoglycopotein receptor. Lactoferrin was shown to be able to block chylomicron uptake and it appears that lactoferrin is taken up by the LRP, being recognized by the arginine-rich domain and an alpha-helix region, most likely identical to the lactoferricin sequence. It seems plausible that there is competition for the receptor during inflammatory events, as several ligands to the receptor are acute phase reactants, possibly shedding light on the involvement of lactoferrin in inflammatory diseases like Alzheimer's and multiple sclerosis (MS).

The binding of lactoferrin to the hepatic asialoglycoprotein receptor appears to occur at or near the carbohydrate-recognition domain and lactoferrin is internalized by a Ca^{2+}-dependent process. This process, however, is galactose-independent and deglycosylated lactoferrin also binds to the receptor. Further studies are needed on this interaction as heterologous system was utilized, i.e. bovine lactoferrin and rat hepatocytes were used.

Further work on bacterial lactoferrin receptors was presented. It was shown that LbpA (lactoferrin binding protein A) is essential for iron uptake, while LbpB is not essential.

4. BIOLOGICAL ACTIVITY OF LACTOFERRIN: ANTIMICROBIAL EFFECTS

Considerable amounts of information on the antimicrobial effects of lactoferrin were presented. This included detailed studies on the mechanisms involved in the binding of lactoferrin to bacteria, in vitro effects of lactoferrin on several pathogens and in vivo studies in mice and human infants. The effect of lactoferrin and lactoferricin on fungi was also demonstrated.

There were several studies presented on the effect of lactoferrin on viruses such as HIV, HCMV, rotavirus and herpes simplex. These results are promising and we were further encouraged by even stronger activity by charge-modified variants (acylation) of lactoferrin. The possibility also exists to conjugate lactoferrin to drugs, which may be a powerful combination to treat viral infections.

5. IMMUNOMODULATING EFFECTS OF LACTOFERRIN

The capacity of lactoferrin to interact with immune function was explored in several ways. It is evident that lactoferrin can affect the production of potent cytokines like TNF-α, IL-1β and IL-6, which may then affect several events, such as cell migration. Interesting information was presented suggesting that lactoferrin can affect protein kinase and its phosphorylation in immune cells, which then may trigger several intracellular events. The consequences of formation of antibodies to lactoferrin, either diet-induced or autoantibodies, were also discussed.

6. EFFECTS OF LACTOFERRIN ON CARCINOGENESIS AND NEURODEGENERATIVE DISEASES

These areas are novel and the results presented are still tentative. Several mechanisms were suggested of this activity, including the involvement of natural killer (NK) cells, reactive oxygen species (ROS) and autoantibodies. It can be expected that more conclusive data will be presented at the next Lactoferrin Conference.

7. THE FOURTH INTERNATIONAL CONFERENCE ON LACTOFERRIN—LOOKING AHEAD

It can be expected that areas that already are strong, like structure of lactoferrin and gene expression, will continue to be so. It also appears likely that there will be further clarification of the "binding/receptor" area in that its/their functional significance will be elucidated. Knock-out experiments are likely to be helpful in such studies. I find it highly likely that there will be a virtual "explosion" in the area of "in vivo" studies on the biological activity of lactoferrin. Recombinant human lactoferrin is now available in significant quantities and animal studies, clinical studies on human subjects and epidemiological studies are likely to give further insights into the multifunctional nature of lactoferrin. It is therefore highly likely that we can expect much new information at the next meeting in Sapporo!

Finally, it is important to stress the words of caution brought up by several investigators (at this meeting and the previous ones):

- *Defined* lactoferrin should be used (LPS-free, untruncated, unmodified).
- *Stringent* criteria should be used for animal studies (positive controls, negative controls, diets, etc.).
- *Stringent* criteria for clinical studies should be used (placebo-controlled, double-blind, sample size calculations, statistical analysis (multiple regression, ANOVA/ ANCOVA)).

Keeping this in mind will ensure that the data obtained will be of high scientific quality and will help to make more definitive conclusions on the effect of lactoferrin on various biological events.

INDEX

GPSR Compliance
The European Union's (EU) General Product Safety Regulation (GPSR) is a set of rules that requires consumer products to be safe and our obligations to ensure this.

If you have any concerns about our products, you can contact us on

ProductSafety@springernature.com

In case Publisher is established outside the EU, the EU authorized representative is:

Springer Nature Customer Service Center GmbH
Europaplatz 3
69115 Heidelberg, Germany

www.ingramcontent.com/pod-product-compliance
Ingram Content Group UK Ltd.
Pitfield, Milton Keynes, MK11 3LW, UK
UKHW020059200726
13856UKWH00002B/294

9781475790696